2nd Virtual Conference on Advances in Electric Drives, Process Control and Automation (VCADPCA 2020)

IOP Conference Series: Materials Science and Engineering
Volume 906

Online
9 June 2020

Editors:

V. Indragandhi
Raja R. Singh
R. Sudha
K.R. Prabhu

ISBN: 978-1-7138-3187-7
ISSN: 1757-8981

Printed from e-media with permission by:

Curran Associates, Inc.
57 Morehouse Lane
Red Hook, NY 12571

CURRAN ASSOCIATES INC. proceedings .com

Some format issues inherent in the e-media version may also appear in this print version.

Institute of Physics
Dirac House, Temple Back
Bristol BS1 6BE UK

Phone: 44 1 17 929 7481
Fax: 44 1 17 920 0979

techtracking@iop.org

Additional copies of this publication are available from:

Curran Associates, Inc.
57 Morehouse Lane
Red Hook, NY 12571 USA
Phone: 845-758-0400
Fax: 845-758-2633
Email: curran@proceedings.com
Web: www.proceedings.com

TABLE OF CONTENTS

IOP Conf. Series: Materials Science and Engineering **906** (2020) 011001 doi:10.1088/1757-899X/906/1/011001

Preface

We would like to present, with great pleasure that the aim of the 2nd Virtual Conference on Advances in Electric Drives, Process Control and Automation (VCADPCA2020) is a platform for researchers, academicians as well as professionals from all over the world to present, discuss and promote advances in knowledge, research and practice in the field of Electric Drives, Process Control and Automation. VCADPCA2020 is offering a fantastic opportunity to attend a global scientific forum from the convenience of your desktop. No travelling, no hotel expenses, no time away from the office. The conference runs fully online, from paper submission, including reviewing, conference discussion and post conference processing. All papers will be referred to double tier approval process, single-blind peer-review and regular check. The online conference is a smart and affordable manner of presenting research results.

Three leading universities VIT, India, Aalborg University Fibigerstræde, Denmark and University West, Sweden are organized this conference successfully on 9th June 2020. Keynote sessions are presented virtually from National and International experts in cutting edge technologies. The selected papers are presented by the authors virtually in the emerging fields like smart grid, power electronics, renewable energy, industrial automation, control and robotics. The panel members evaluated the research ideas and recommended for paper publications in IOP conference proceedings.

Editorial Board

1. Name : Dr. V. Indragandhi
   Institution and address : Vellore Institute of Technology, Vellore, India
   Email address : indragndhi.v@vit.ac.in

2. Name : Dr. Raja Singh R
   Institution and address : Vellore Institute of Technology, Vellore, India
   Email address : rrajasingh@vit.ac.in

3. Name : Dr. Sudha R
   Institution and address : University West, Sweden
   Email address iumha@gmail.com

4. Name : Dr. Prabhu K R
   Institution and address : University West, Sweden
   Email address : Prabhu.k.ramanathan@hv.se

IOP Conf. Series: Materials Science and Engineering **906** (2020) 011002 doi:10.1088/1757-899X/906/1/011002

Peer review declaration

All papers published in this volume of IOP Conference Series: Materials Science and Engineering have been peer reviewed through processes administered by the Editors. Reviews were conducted by expert referees to the professional and scientific standards expected of a proceedings journal published by IOP Publishing.

- **Type of peer review: Other**

An open way of reviewing was used: two reviewers - min requirements - PhD specialists on the topic of the report.

- **Conference submission management system:**

The conference was managed by an international organizing committee, the composition of which is given in the appendix. Conference materials (program, abstracts) were presented in hard copy: program brochure, book of abstracts, and in electronic form (* .pdf format) on the conference website https://itad.com.ua/. Additional information about the conference was presented on the website of the Kharkiv National University of Civil Engineering and Architecture http://kstuca.kharkov.ua/nauka/konferencii-ta-seminari.

- **Number of submissions received:** 129
- **Number of submissions sent for review:** 129
- **Number of submissions accepted:** 123
- **Acceptance Rate (Number of Submissions Accepted / Number of Submissions Received X 100):** 95.3
- **Average number of reviews per paper:** 2
- **Total number of reviewers involved:**67
- **Any additional info on review process:**

The review process was carried out in accordance with the instructions for the reviewer:

REMINDER TO REVIEWER
BASIC REQUIREMENTS FOR REVIEW
Technical criteria
• Scientific novelty: especially scientific rigor, precision and correctness.
• Clarity of expression; exchange of ideas; readability and discussion of concepts.
• Sufficient discussion of the context of the work and appropriate references.
Quality criteria
• Originality: is the work relevant and new?
• Motivation: Does the problem under consideration have valid motivation? All documents must clearly demonstrate scientific interest in the results.
• Repetition: Has a significant portion of the manuscript already been published?
• Duration: is the content of the work of sufficient scientific interest?

IOP Conf. Series: Materials Science and Engineering **906** (2020) 011002 doi:10.1088/1757-899X/906/1/011002

Presentation criteria

• Title: Is it adequate and appropriate for the content of the article?
• Abstract: Does it contain basic information about the article? Is it complete? Is it suitable for inclusion in an abstraction service?
• Diagrams, figures, tables and captions: are they important and understandable?
• Text and math: are they concise and easy to understand? If you recommend shortening, please suggest what to skip.
• Conclusion: Does the article contain a carefully written conclusion summarizing what was learned and why it is interesting and useful?

- **Contact person for queries:** prof. Viktor Sopov E-mail: vpsopov@gmail.com

APPENDIX

ORGANIZING COMMITTEE OF CONFERENCE

Chairman of the organizing committee:

Goncharenko Dmytro - Doctor of Technical Sciences, Professor, Vice-Rector for Scientific and Pedagogical Work of KhNUCEA;

Co-chairs of the organizing committee:

Myronenko Viktor - Dr. Architect, Prof. KhNUCEA, Ukraine;
Sopov Viktor - Doctor of Technical Sciences, Prof. KhNUCEA, Ukraine.

Members of the organizing committee:

Buryak Alexander - Dr. Architect, Prof. KhNUCEA, Ukraine;
Gonchar Olena - Dr. Pedagogical Sciences, Prof. KhDADM, Ukraine;
Huk Valeriy - Doctor of Technical Sciences, Prof. KhNUCEA, Ukraine;
Dubinsky Volodymyr - Dr. Architect, Prof. KhNUCEA, Ukraine;
Izbash Mykhailo - Doctor of Technical Sciences, Professor KhNUCEA, Ukraine;
Kondrashchenko Valeriy - Doctor of Technical Sciences, Professor Russian University of Transport, Russian Federation;
Kostyuk Tetyana - Doctor of Technical Sciences, Prof. KhNUCEA, Ukraine;
Kravets Volodymyr - Dr. Architect, Prof. KhNUCEA, Ukraine;
Miletska Malgozhata - Dr. Architect, Prof. University of Natural Sciences, Lublin, Poland;
Morozova Olena - Dr. Architect, Prof. Belarusian National Technical University, Minsk, Belarus;
Solobay Petro - Dr. Architect, Prof. KhNUCEA, Ukraine;
Usherov-Marshak Alexander - Doctor of Technical Sciences, Prof. KhNUCEA, Ukraine;
Fischer Hans-Bertram - Dr. Eng. Bau House University, Weimar, Germany;
Fomenko Oksana - Dr. Architect, Prof. KhNUCEA, Ukraine;
Chuck Marek - Doctor of Technical Sciences, Prof. University of Warmia and Mazury, Olsztyn, Poland;
Cherkasova Kateryna - Dr. Architect, Prof. KhNUCEA, Ukraine;
Cherkes Bogdan - Dr. Architect, Prof. Lviv Polytechnic National University, Ukraine;
Zhang Ihe - Ph.D., Professor, Director of the Institute of Materials Science and Materials Technology of the Chinese Geological University, Beijing, China;
Shilo Alexander - Doctor of Arts, Prof. KhNUCEA, Ukraine.

IOP Conf. Series: Materials Science and Engineering **906** (2020) 012001 doi:10.1088/1757-899X/906/1/012001

Advancement of Inventive Solar Power Based Frameworks for Rural India

Abhilash Muraleedharan Pillai[1], **R. Rajesh kanna**[2], **R. Raja singh**[2]

[1] Distribution Field Operations, Distribution Power, Dubai Electricity and Water Authority, Dubai.
[2] Advanced Drives Laboratory, Department of Energy and Power Electronics, Vellore Institute of Technology, Vellore, India.

Correspondence: abhilashmpillai@gmail.com

Abstract. The nearness of solar energy at any area in a village or on a farm makes its utilization alluring for such an environment. This paper portrays the advancement and organization of inventive solar power based frameworks for use in rural networks of India. On-going and arranged solar power extends in India are constantly gone for creating plants for network applications. The proposed framework advancement depends on appropriated organization procedure. This methodology utilizes solar power in little units to fulfil a specific requirement for a village, similar to a home in a village, on the farm, or potentially an individual village occupant. Frameworks plans fuse financially savvy, measured, tough, long life, simple to keep up, and advantageous to utilize designs. The procedure proposed for recognizing the assembling configuration is an iterative one. In other words, that at first the model units are conveyed in villages to get input from clients and afterward the plan is changed and reused. The final product of this exertion is the recognizable proof of the last plan and assembling process.

1. Introduction

Solar power emerged as an effectual source of available renewable energy sources and it can be utilized to help maintain your home or company in many ways. Solar energy conversion defines technologies for transforming solar energy into other energy forms, including electricity, fuel and heat. [1]. There are two advancement classes which tackle solar energy, Solar Photovoltaic and Solar Thermal. It shelters light harvesting technologies, including traditional photovoltaic semi-conductor (PV) and emerging photovoltaics [2]-[4]. Solar Photovoltaic (PV) is an invention which captures sunlight by using semiconductors to coordinate electricity [5]. The generation of solar fuel through electrolysis, articulated photosynthesis and other associated forms of photo catalysis aimed at generating Power-rich molecules [6]. Interestingly, Solar Thermal is a technology that uses the warmth of the sun for heating or generating electricity [7]. As the Indian economy appreciates a sustained positive force, rural India remains the country's core, accounting for 67 per cent of the aggregate population and 37 per cent of its GDP [8]. Reformation to solar energy breeding has been propagated to save rural population the time and resources they need to purchase and transit oil and wood, thus rising effectiveness and extending commercial working hours. [9]. Agriculture is the fundamental control of the rural family units and their financial structure [10]. Although India's general economy relies on more than 7 per cent, the fastest among expansive rural economies in the world remains generously behind India [11]. The fundamental obstacle to development in rural profitability and the resulting monetary development is the lack of a fundamental basis, for example, for electricity, clean water, and sanitation [12]. In rural India, nearly 300 million people need access to matrix-related power, advancing the use of age-old energy wells, such as lamp fuel, diesel, wood-let go chulhas, and so on, resulting in enormous government endowments, as well as important welfare and environmental hazards. Solar power offers the opportunity to bridge this gigantic foundation hole and enhance the 30-per-cent social, political, environmental and well-being markers of India's population [13]. While solar power has been around for some time, genuinely high costs required it to be driven by generous capital or government sponsorship, thereby restricting its expansion [14].

Anyway, with a drop in capital expenditure of about 70 per penny over the last four years, solar energy has now advanced to become financially standard, along these lines pulling in private capital and business people. This really makes solar energy the much-expected solution for the millions who live in darkness [15]. In excess of one hundred million people in our planet are detached from the cutting-edge developments since they are deprived of electricity to energise their houses, networks, and organizations. In the event that we amply provide dependable electricity, this will throttle the lifestyle and monetary advancement and change lives [16]. Electricity makes open doors for people to lift themselves out of destitution. Genuine worldwide progress cannot be achieved devoid of electricity, especially in Non-urban areas which constitutes eighty percent of energy backtracked economy [17].
Support for Rural Development was initiated with seventy-five million dollars sponsored in 2015 by the Rockefeller Foundation to tackle the "last mile" power supply deficit by uniting community profit organizations (ESCOs), invention specialists, nearby organizations, national and neighborhood governments, and additionally the private division to construct practical associations around decentralized sustainable power source (DRE) arrangements [18]. For example, DRE arrangements smaller than planned lattices are quick to build and create an adequate ability to energize whole villages, associating efforts also. They were biomass and solar installations, and provide clean power support [19]. Sustainable Energy for Economic Development (SEED) gives specialized, approach, and money related guidance on energy segment improvement to governments and different accomplices [20]. SEED likewise works as an inseparable unit with the private division and legislative experts to guarantee that decentralized sustainable power source arrangements are effectively executed [21]. In Rwanda, SEED helped the administration outline a coordinated national charge intend to convey energy to seventy percent of the republic in 24 months, start from the take-off of the arrangement, and distinguish more than one hundred million USD in investment funds through streamlining of the present energy framework [22].

2. Conceptual framework

Solar power systems support to increase peak demand generating capacity, thus saving the utility during times of high demand from turning on costly and polluting supplemental systems. The many solar power generating systems which are fitted in the service area of a given utility, the lesser amount of capacity that the utility has to create, thereby saving everyone from financing costly new sources of power generation. Supplying green energy from solar electrical system is an excellent way to minimize greenhouse gasses and other problems arising from fossil-fuel electricity. India gets almost 7 electrical power units/ m^2 of solar energy for 10 to 11 months a year [8]. This sunlight richness is enough to install solar plants of at least 19-20 Megawatts for each km^2 of land where received [12]. The energy from the sun is collected using silicon materials generated cells and then converted into electricity. Initial cost is the main factor in making solar cells. It is seen as a renewable energy source since the technology used to transform the power of the sun into electricity does not cause smoke. Acquiring the energy from sun does not damage the environment. Regrettably, by night the sun becomes inaccessible, and in some days, mists and rains and other common conditions keep the powerful beams of the sun from reaching earth. That means it is not always available.
Figure 1 demonstrates the procedure of change of solar energy into the electric power. Solar panels which are also named as modules, accommodate silicon-derived cells that metamorphose falling sun beams into electric current rather than potentiate heat. Solar PV panels are put on the roof in most installations. Generally, a model facing south will provide the most excellent power in all solar panel systems, but there will be ample output from other orientations also. Tree branches and other elements which produce shadows during the sunshine will result substantial disruption in electricity generation. The ratio value of shades and efficacy cannot be overemphasised. If only one of its 36 cells is shaded in a solar panel, more than half would reduce power generation. All roofs cannot provide proper inclination or orientation to benefit from peak sunshine absorption. So, we have systems with panels which pivot according to the path of the sun throughout the whole day. PV systems without any tracking

mechanism are positioned at latitude-path orientation of the site to imbibe the maximum solar shine annually.

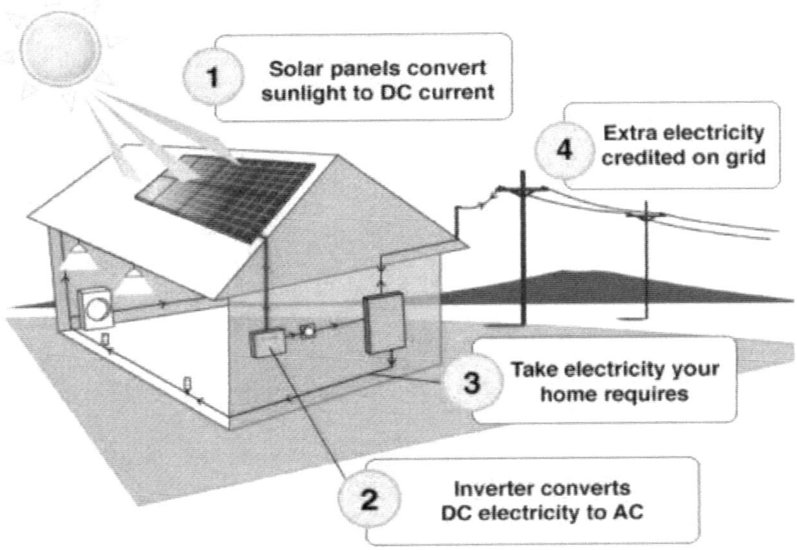

Figure 1. Conversion technique of solar energy into the electric power

Semiconductors are in those cells. Larger solar cell types are used to power streetlight and even larger displays are used to power satellites across the world. In solar warm course, sunshine can be absorbed over into heat energy with the assistance of sunlight gatherers and collectors known as warming gadgets. Bad condition solar warm accessories are used for local and mechanical applications such as solar water radiators, air radiators, solar cookers, and solar dryers.

2.1. Solar photovoltaic

Photo implies illumination and voltaic means electric. Usually, these cells are built from silicon, a module that normally displaces electrons when subjected to sunlight. Evaluation of discharged electrons from silicone cells relies on the power of light. The module is protected with a metallic matrix which allows the electrons to move in such a way as to generate electricity. This flow is directed into a wire that is associated with a Direct current apparatus or a battery. Generally, a cell generates approximately 1.5 watts. Cell units are combined to form a module or Panel that is suitable for generating 3.1 to 109.9 watts of power. Panel boards can be connected in arranged method or paralleled to create a solar cluster, which can create any measure of electricity as area and shine will permit. The modules are generally designed to supply 12 volts of electricity. Solar PV modules are measured on a sunny morning at Solar twelve by their peak Watt yield.

2.1.1. PV tracking system

A PV tracking system is a specialized tool designed to turn the PV modules in such a way that they meet the sun constantly to enhance the irradiation obtained by the PV array (Figure 2). Three well-differentiated components form a solar tracking system those are the mechanism, the driving motors, and the tracking controller.

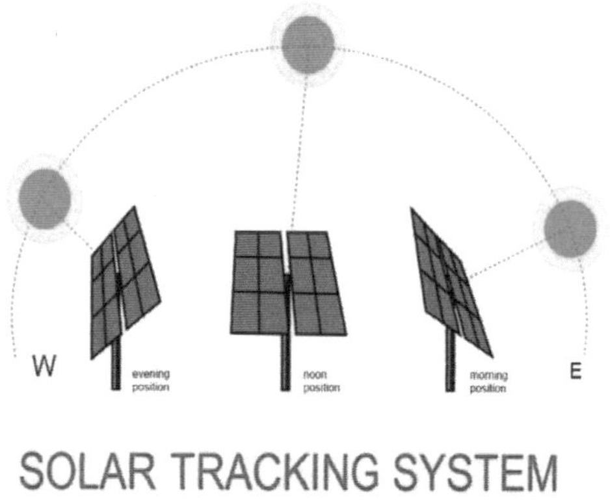

Figure 2. Solar Tracking Mechanism

Figure 3. Battery Backup

2.1.2. PV back-up systems

These are essential because solar modules only generate electricity according to the sunlight strength. alternately most popular strategies for the utilizing excess of solar electrical systems are interfacing the device with the utility companies network or lodging away surplus energy in power banks (battery banks – Figure 3) for evening time or rainy days.

2.2. Inverter

The inverter is usually installed in a reachable place as near to PV system as possible (Figure 4). In a home installation, this equipment is always connected on the main or subpanels of electricity. Since inverters develop a moderate humming, it should be in mind while choosing the placing site. The inverter transforms solar panel DC power into 120/230 V AC and utilised instantaneously by connecting directly to a CB in the power grid panel. The inverter is always connected with electricity production meter and electricity net meter because the electrical loads currently operating will first use supply generated by solar installation. The sense of balance power generated by the solar system pass by electrical panel and out onto the electric grid.

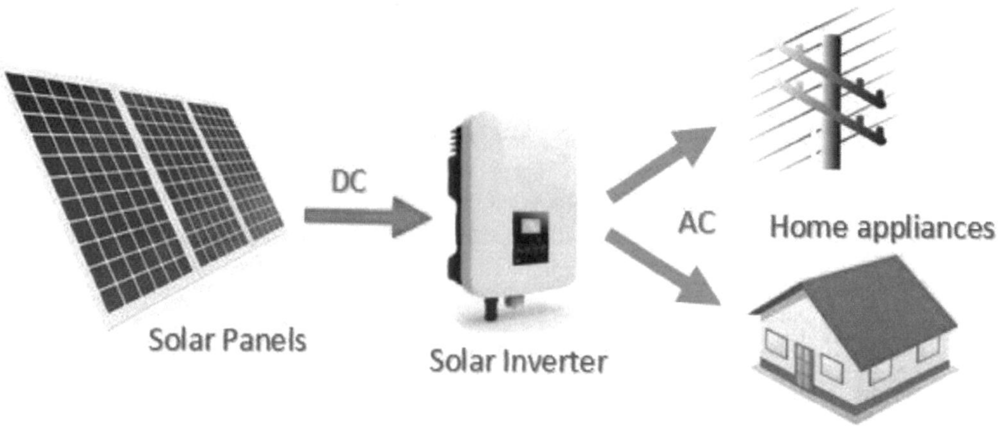

Figure 4. Inverter

2.3. Net metering

The DC supply from the solar panel is remoulded into 120/230 volt AC power and disbursed into the building's utility electric system that is also tied to the national power utility line. The system generated power is "net metered" which means that when the solar array produces electricity, it shrinks the load

IOP Conf. Series: Materials Science and Engineering **906** (2020) 012001 doi:10.1088/1757-899X/906/1/012001

demand from the grid thus utility bill can be reduced, shown in Figure 5. Grid-tied PV systems spontaneously shut off when utility power is disrupted, safeguarding utility line workers from power being back fed during line maintenance. These kind of solar power installations are distinguished as "On-Grid" or "battery-less" and produce the power about 98 percent of today's installed solar energy systems.

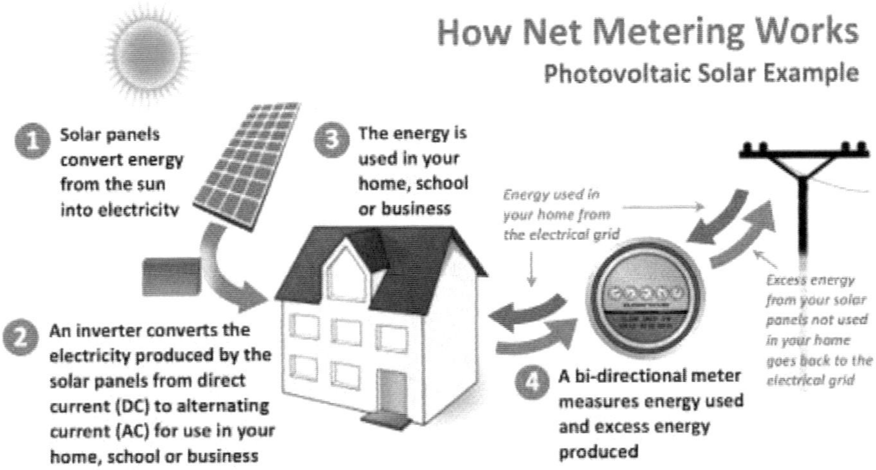

Figure 5. Net Metering

2.4. Potential technology needs

The demands for rural solar power have not been adequately explored. Solar power applications can integrate direct light use, coordinate heat use, and convert solar energy into electric power. This paper is proposing the accompanying conceivable needs considering our perceptions and involvement in rural networks of India.

2.4.1. Solar Power Well

Solar Power Well (SoPoW) concept is like water in the village network. The solar energy is collected and disposed of for this situation, and occupants can charge gadgets or energy storage systems for a fee.

Figure 6. Installation of a Solar Power well in Totora, Liberia.

The creation of the SoPoW would be feasible through a village-friendly course of action so it exists to be made accessible for each watt to the residents at a lower cost. The Installation of a Solar Power well by citizen cooperative and subsequently to be managed by the user's cooperative itself in Totora, Liberia, Figure 6.

2.4.2. Solar Powered Centralized Food Preservation Facility

Agricultural products degrade in ordinary climatic conditions. Minimal temperatures and shakiness cause the accessible life of these objects to expand. Stockpiles with various outlines can be used in rural areas and managed with solar energy invention in their climate. Such stocks can be handled by village cooperatives and subsequently leased to farmers for use, shown in Figure 7. Maintaining sustainability would exceptionally satisfy the growing desire of nation.

Figure 7. Solar Cold Preservation hub

2.4.3. Solar-lit Streets/Corners/Farm ways

Solar lights are frequently used in various locations to provide protection and the light needed to empower a person to achieve undertakings when there is no sunlight. However, they were not found in rural areas.

2.4.4. Solar Powered Rural Homes/Huts

The developed technologies to be used in home include lights, telephones, refrigerators, food cookers, warmers, television and other electronic gadgets, air coolers, water well, and fans.

3. Methodological Analysis

The present world energy utilization is much lower than the measure of energy gotten by the Earth by sun. Looking to the present energy situation, there is an expansive hole between the request and the supply, shortage of irresistible energy assets and with the utilization of non-renewable assets the worry for condition is likewise expanding. The Electric power supply crosswise over India requires quality and amount with a wide insufficiency in supply, an inadequate record for power failure, elevated amounts of transmission and distribution (T&D) misfortunes and a general requirement for broadened and augmented foundation. A lot of power created is squandered because of transmission and dispersion

IOP Conf. Series: Materials Science and Engineering **906** (2020) 012001 doi:10.1088/1757-899X/906/1/012001

wasteful aspects and power robbery. In any case, such issues can't be a justifiable reason motivation to keep more than of India's populace from gaining admittance to power.

This instantaneous consumption of solar energy continues while present day developments made the rural life remarkable. Cases of innovations being used in rural India include cell phones, home appliances, execute cultivation, creature, and harvest transport vehicles. The environmental impact of their use has also emerged as a noteworthy concern. Then again, Inventions for the instantaneous and substantial use of solar energy have found effectively to make this type of energy valuable and pleasant for rural India. The savvy and making creative employments of solar energy will provide practical, reliable, and condition cordial solutions for rural India. In this study, as the name suggests, exploratory research approach is used only to address the research enquiries and does not aim to provide conclusive and definitive remedies to existing issues. Usually this methodology scrutiny is undertaken to analyse a topic that has not yet been distinctly recognised.

Probe work is not meant to provide definitive evidence, but it does enable everyone to understand the issue better. The researcher should be able to change their path when performing exploratory research because of exposure to new data and new visions. "Exploratory research is the initial research, which is the foundation for more final research. It is also used to help in determining research design, methodology of sampling and method of gathering data. Exploratory research aims to address different issues on which negligible or nil precursory research has been carried out. Shapeless interviews with exploratory studies are the most common fundamental form of data collection. Sampling methods are classified into two main types: probability and non-probability. In probability sampling, every inhabitant member has a well-acknowledged, cardinal prospect of contributing in the research. Randomisation is at the heart of the technique of probability sampling. Here probability sampling tactic will be used.

4. Analysis and Recommendation

India is one of the largest producers and consumer of electricity in the world. As of 31 March 2020, the power stations have total equipped size of 370106 MW as per National Electric Grid of India. Out of total installed capacity, 35.8 percent of energy is generated by renewable plants, which also include gargantuan hydroelectric plants. India has an excess capacity for generating electricity but lacks sufficient distribution network. To counter this, India's government initiated a 'Power for All' plan in 2016. India has been able to electrify 99.9 percent of household with only 18700 households remaining without power through its SOUBHAGYA sub-scheme. The initiative was completed by December 2018, delivering the imperative foundation to corroborate non-disruptive electric service to all houses, factories, and mercantile enterprises. Power production capacity in India from nuclear assets is 1.68% of global nuclear power generation capacity which brands it to 15th largest nuclear power producer. India plans to deliver more than 8.99% of nuclear power for demands by 2032 and will attain 25% by 2050. The installed sizes above 1000 KW capacity power generation linked with factory owned plants was 58GW as of 31/03/2019, in financial year 2018-19, a total of 175 Terawatt Hours was produced. Sets of 75GW of diesel power generation are also established in the country, keeping out sets of sizes above 1000kW and below 100 kVA. In addition, there are a substantial number of diesel generator which is having the capacity below 100 kVA to supply for emergency requirements in all regions during power outages. The particulars of renewable power installation capacity in India, excluding large hydropower as of May 2020 is illustrated in table 1.

Table 1. Installed grid interactive renewable power capacity

Source	Total Installed Capacity (MW)	2022 Target (MW)
Wind Power	37693.75	60000
Solar Power	34627.82	100000
Biomass Power	9875.31	10000
Waste to Power	147.64	10000
Small Hydro Power	4683.16	5000
Total	87,027.68	175000

The Government of India has planned to install 40 GW Rooftop Solar Projects (RTP) by 2022, including establishing on the terraces of residence. With effect from October 2019, the 175000 MW provisional or interim target, 83000 MW is in action, 29000 MW under installation, 30 GW is under auction, and remainder 43 GW is under scheme provisions. 175 GW provisional earmark is attained from 100 GW of solar power plant, 60 GW of wind power plant, 15 GW of biomass and small hydro combined. By 2019, 35% of overall energy creation resulted from renewable energy, 13% of power comes from large hydro power plants, 10% of power from wind power plants, 8% of power from Biomass power and bagasse cogeneration. The government plans to allocate 20 GW of solar plants by 2022 but this was already realised before 48 months from proposed conclusion in January 2018, via both solar plants and as roof-top solar panels as well. Currently India has planned to achieve 100000 MW of power from solar, 60000 MW wind power, 10000M W of power from biomass and 5000 MW of power from small hydro plants by 2022.The renewable power installed capacity as on 2013 is described in figure 8.

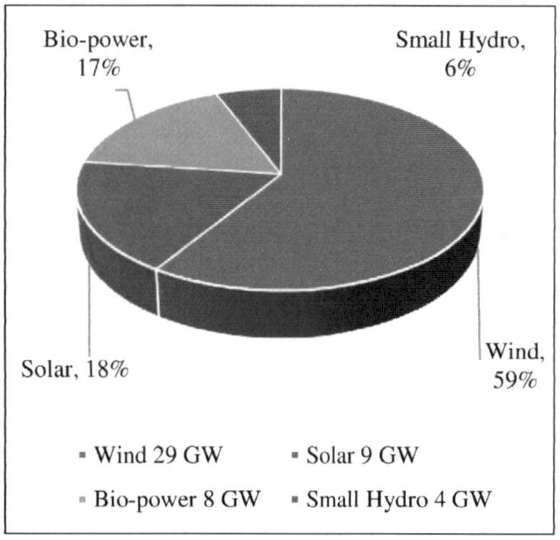

Figure 8. Renewable Power installed capacity

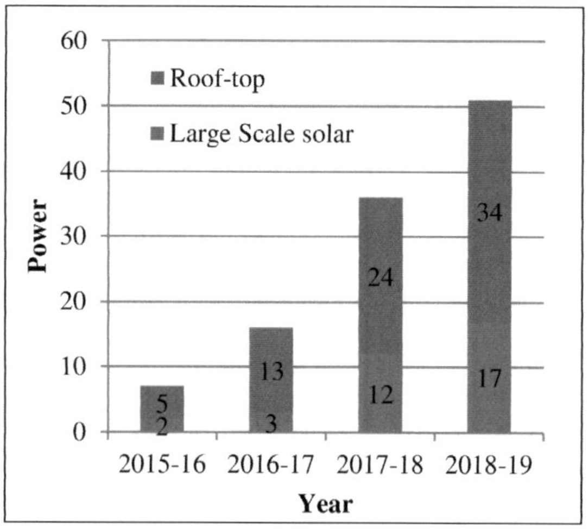

Figure 9. Solar scale up plans of GOI 2022

The aim of the National Solar Mission is to reshape India to a universal superior in solar power, by blueprinting state of affairs for its transmission over the nation. This was propounded to be accomplished in three phases. 2010 to 2013 encompassed first period. The next period expanded till 2017 and the third anticipated in 2017–22. Goals have been set near 1.3999 GW during the first stage, up to 14.999 GW

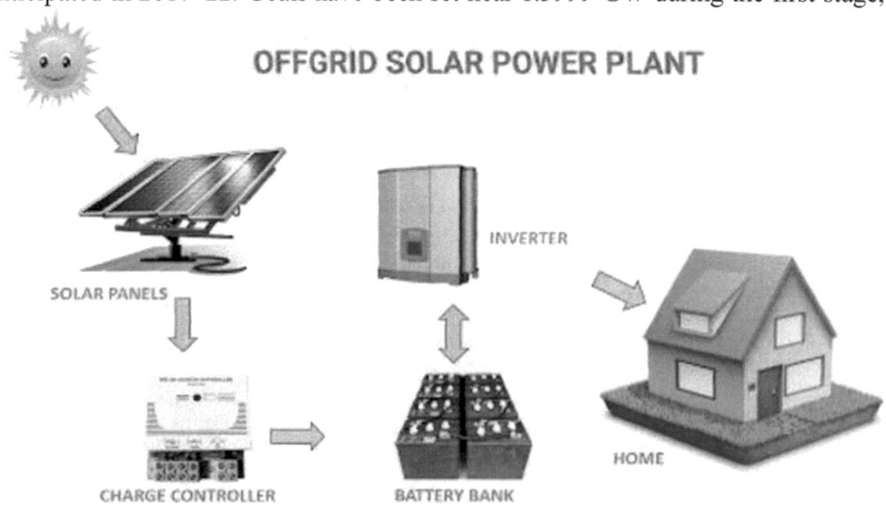

Figure 10. Solar Power Station - Off Grid

by the second time period and 21.999 GW should be executed by the year 2022. Figure 9 shows the reports of the broadening in target accomplishment by solar mission during its operation.

Off grid systems operate independently of the grid but have batteries that can store up the energy from solar produced by the device shown in Figure 10. The device is normally composed of solar panels, battery, charging control, grid division, inverter and device balances. The panels store an adequate amount of daytime sunlight and use the excess power produced at night. Such systems are self-supporting and can deliver electricity for critical loads in regions that do not have a power grid. These systems however need specialized apparatus to work and can be expensive to install. Solar off grid systems can help self-governing, long-term and sustainable production of electricity in rural and rural areas. A number of places in India meet frequent power cuts due to malfunctions in power transmission that can interrupt the functions of enterprises and public institutions. Solar off grid systems will offer a cost-effective and standard long-term standby solution to resolve the difficulties that occur throughout frequent power cuts.

Many rural projects which reduce energy use and result in investment funds for the farmer or independent enterprise are qualified as energy efficiency projects. These include projects such as upgrading lighting or protection or the purchase or replacement of equipment with more efficient units. Qualified renewable energy projects include projects generating electricity from renewable energy sources like wind, solar, biomass, geothermal and hydroelectric. The projects can generate any type of energy including, heat, electricity, or fuel. The administration is setting up a massive program to expand the countryside of India with renewable energy projects, using reserves allocated for rural development and panchayat raj services, as this means to reduce the country's mind-boggling dependency on imported fuel derivatives. The service of new and renewable energy is taking a shot at the arrangement to use the compass and spending plan of the two rural-related services to make these green projects for panchayats and self-improvement gatherings, as indicated by a top-notch government official, who insisted anonymity. The legislature is likewise investigating setting up work connected renewable energy projects in rural zones, as reported by the authority.

5. Conclusion

The vast majority of the recipients of solar home-lighting framework are extremely content with the usefulness of the framework. Technologies which we can design are solar powered automobiles, agricultural machinery, solar air conditioners and temperature control systems, mobile chargers, solar powered cooking accessories and school electrification by solar power and solar power for web bistros. For these structures we may use an iterative approach to understand the assembling outline. In other words, first the prototype units are supplied in villages in order to get input from customers and then we alter the outline accordingly for recognizable evidence of the final design and assembly process. Although this study aims to help the quest to scale up Solar Powered Infra Structure reception by featuring the holes and by depicting the route onward, there is a need to check and accept a part of the detections in other agro-biological and economic settings keeping in mind the end goal to pick up a more extravagant comprehension of the substances over a country as huge and different as India.

References

1. Crabtree, G. W.; Lewis, N. S. (2007). "Solar Energy Conversion". Physics Today 60, 3, 37. doi:10.1063/1.2718755.
2. Light-Induced Redox Reactions in Nanocrystalline Systems, Anders Hagfeldt and Michael Graetzel, Chem. Rev., 95, 1, 49-68 (1995).
3. Materials interface engineering for solution-processed photovoltaics, Michael Graetzel, René A. J. Janssen, David B. Mitzi, Edward H. Sargent, Nature (insight review) 488, 304–312 (2012) doi:10.1038/nature11476.

4. Ponseca Jr., Carlito S.; Chábera, Pavel; Uhlig, Jens; Persson, Petter; Sundström, Villy (August 2017). "Ultrafast Electron Dynamics in Solar Energy Conversion". Chemical Reviews 117 : 10940–11024. doi:10.1021/acs.chemrev.6b00807.
5. Belessiotis & Papanicolaou, V.G & E. (2012). "History of Solar Energy". Comprehensive Renewable Energy. 3: 85–102.
6. Mukherji A (2007) The energy-irrigation nexus and its impact on ground water markets in Eastern Indo- Gangetic basin: Evidence from West Bengal, India. Energy Policy 35: 6413-6430.
7. Mukherji A (2007) Implications of alternative institutional arrangements in groundwater sharing: Evidence from West Bengal. Economic and Political Weekly 42: 2543-2551.
8. MoAFW GoI (2015) All India Report on Agriculture Census 2010-2011.
9. Agrawal S, Jain A (2015) Solar Pumps for Sustainable Irrigation: A Budget Neutral Opportunity. New Delhi: Council on Energy, Environment and Water.
10. Agrawal S, Jain A (2016) Sustainability of Solar-based Irrigation in India:
11. Key Determinants, Challenges and Solutions. New Delhi: Council on Energy, Environment and Water.
12. Agrawal S, Jain A (2018) Financing Solar for Irrigation: Risks, Challenges, and Solutions. New Delhi: Council on Energy, Environment and Water.
13. MNRE GoI (2017) Annual Report, MNRE, 2016-2017.
14. Purohit P, Michaelowa A (2005) CDM potential of SPV pumps in India. Paper 4. Hamburg, Germany: Institute of International Economics.
15. Grant Thornton (2015) Transforming Agriculture through Mechanisation.
16. mnre.gov.in.
17. MNRE GoI (2014) Installation of 10,000 nos. of solar PV water pumping systems for irrigation purpose implemented through NABARD throughout the country.
18. MNRE GoI (2014) Supplementary Guidelines for Implementation of Solar Pumping Programme for Irrigation and Drinking Water Off-Grid and Decentralised Solar Applications Scheme.
19. Guiteras R (2007) The Impact of Climate Change on Indian Agriculture.
20. Ballabh V (1987) Decline of a novel experiment: A case study of group irrigation tube wells in Deoria District. Anand: Institute of Rural Management, unpublished report. GEDA. n.d. Breaking new grounds in clean energy. Gujarat Energy Development Agency.
21. Kandhari R (2011) Mission eclipsed. Down to Earth, March 1-15, 2011.
22. Kishore A (2012) Solar pump project in Rajasthan: A First Information Report. Anand: IWMI-Tata Program, unpublished memo.

Control of battery banks for wind energy conversion system

Aboli A Kulkarni

Department of Electrical Engineering, Vellore Institute of Technology, Vellore, 632014, India
abolikulkarni2000@gmail.com

Abstract. In this paper, the essential methodology to control the battery banks for a renewable power system is explained. The power converter used for the above is a step-down voltage regulator also known as Buck converter. It can be used to garner power from a small wind turbine. The paper also explains the charging process of the battery to track the point where maximum power output is obtained in a wide range of speed variations and proper circuit parameter design. Smart Management of battery charging and control is crucial since the lack thereof, would affect the battery's performance. This method helps avoid the overcharging and overvoltage and also helps build systems that are economic and efficient. This process helps improve the charging efficiency of the battery, also its performance in terms of degradation rate and lifetime. Simple circuit and high reliability are the key factors in the usage of Buck DC/DC converter. A PMSG Stand-alone Wind Turbine System is also modelled and simulated using MATLAB.

1. Introduction

Renewable energy also known as sustainable form of energy, is derived from natural sources or continually renewed processes like sunshine, wind, water flow, combustion and geothermal heat flow. Examples include solar energy, tidal energy, wind energy etc. These alternative sources are being developed and researched on to reduce the environmental impact and carbon footprint that is caused by the exhaustible sources. While renewable energy is often thought of as a new technology, harnessed energy from sustainable sources is utilized for heating, transportation, lighting, and more. One of the most feasible forms of energy, both economically and technically is the wind energy.

For the countryside and remote regions, a small-size stand-alone wind energy conversion system with a battery bank as the energy stockpiling component is essential as it can be easily included in the grid. For these systems, DC micro alternators and permanent magnet (PM) generators are widely used to generate electrical energy. In this paper, a PMSG Stand-alone Wind Turbine System is modelled and the output power variations over a range of turbine speeds are observed. A permanent magnet synchronous generator is used for the energy conversion because of its compatibility with low speeds. It is directly-driven i.e. it can directly be connected to the gearbox which makes the entire architecture considerably less expensive. The electrical power output of the energy system is a function of the the speed of the wind and can be controlled by variations in the pitch angle of the blade (β). The graph characteristics of the output power vs rotor voltage is non-linear in nature. Thus, it is a difficult task to obtain maximum power output from the turbine for all values of the wind speed.

In this paper, the Maximum Power Point Tracking (MPPT) algorithm devised for the same is discussed. The energy sink in this system is a battery with almost steady voltage. The battery absorbs power until the charging current exceeds its threshold value. The value of voltage across the battery remains almost consistent, but the current through it can be varied, therefore it can be considered as a load with a variable resistance. As a storage option, battery banks as load is the most popular choice due to an edge over other options. It proves to be a comparatively inexpensive choice with better efficiency, energy storage period and energy density. Another crucial issue of the stand-alone wind power system is the lifetime of the battery bank. Taking into account the monetary restraints, the lead–acid battery remains the widely used energy stockpiling device for the stand-alone wind energy conversion system. However, the degradation of the lead–acid battery in turn affects the system's reliability.[1] The usage of pulsating currents to charge the battery can increase the lifetime of a battery as well as improve its charging efficiency.

2. Wind energy and wind turbine system

Wind energy, or wind power, is generated with the help of a wind turbine, a device that utilizes the power in the wind to generate electricity. A wind turbine connected to a generator via a shaft is responsible for the conversion of wind energy to electrical energy. Small, individual wind turbine systems can produce around 100 kilowatts of power. The energy produced is enough to sustain small domestic needs. They are also used in places like water pumping stations. In a wind energy conversion system, the difference in pressure of the air between the two sides of the blades of the wind mill cause their rotation. They are connected to the generator which takes in the rotating movements of the shaft and then converts the mechanical energy (kinetic energy of the wind) to electrical energy that helps charge the battery banks to supply power to the load. The output power of a wind turbine system is given by the expression:

$$Pm = \frac{1}{2}\rho\pi Cp\,(\lambda, \beta)\, R^2 {V_w}^3 \tag{1}$$

In the above equation, ρ is the air density, $C_p\ (\lambda,\ \beta)$ is the power conversion coefficient, R is the radius of the blade of the wind turbine, V_w is the speed of the wind and P_m is the power output of the stand-alone wind energy system.

From equation (1), it can be concluded that the power output of the wind turbine depends on the power conversion coefficient C_p for a particular wind speed. C_p is a function of the tip speed ratio (λ) which is defined as the speed of rotation at the tip of the blade by the speed of the wind for a pitch angle (β). Therefore, to obtain maximum power from the wind energy conversion system, the wind turbine should be rotated at different speeds.

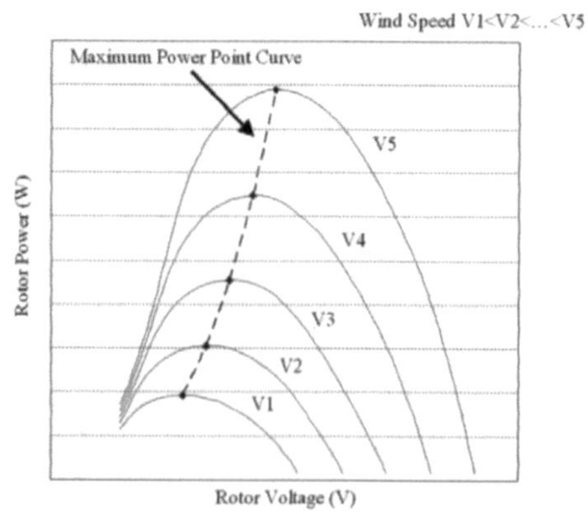

Figure 1. Maximum Power Point Curve of a wind turbine system

The Permanent Magnet Synchronous Generator attached to the blades is responsible for the energy conversion. The output voltage responsible for charging the battery is proportional to the rotor speed. Initially, the PMSG of the wind energy conversion system starts at no-load with the maximum output voltage. With the increase in the output current there is a subsequent increase in the output power but the output voltage drops. Once the output power attains a specific value, the increase in output current will cause a significant output voltage drop leading to a steep output power drop and the PM wind turbine will stall eventually. The dotted line in the above figure represents the MPP curve. It can mathematically be represented as a polynomial equation of order 2 and is a function of the turbine speed for constant *Cp*.

3. MPPT battery charging process

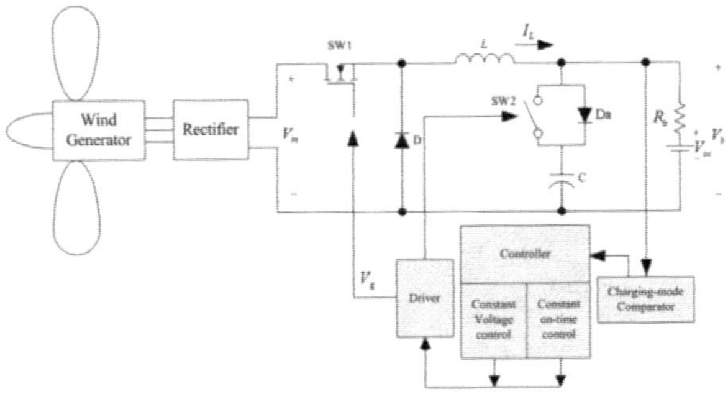

Figure 2. Proposed circuit diagram for MPPT Battery Charger

The battery can be charged in two modes: Pulsating Current Mode (PCM) and Constant Voltage Mode (CVM). The battery operates in the Pulsating Current mode (PCM) when the voltage across the battery (Vb) is below the threshold voltage (Vbth), and the Switch 2 (SW2) is open. Whereas, if the battery voltage crosses its threshold value (Vb>Vbth), the battery charger switches over to the Constant Voltage mode (CVM) with SW2 closed. For the generation of the pulsating current required by the battery bank during the PCM, the charger operates with constant on-time control in the discontinuous conduction mode (DCM). [2] Further, the increase in wind speed in turn increases the output voltage of the wind turbine causing the charger to enter into continuous conduction mode (CCM). The increased current will result in decreased rotor speed as well as the input voltage.

Once the battery is charged and the charger has entered CVM, the output capacitor provides constant voltage across the battery. Further, to prevent the overcharging of the battery, the constant on-time control of the charger is replaced by the voltage-mode control. It helps to regulate the output voltage. During charging period d_1, switch SW1 is closed and there is a linear increase in the inductor current. Whereas, during discharging period d2, switch SW1 is open, I_L decreases and drops to zero. Period d3, the rest duty is meant for stabilization of the chemical actions of the battery allowing it to be ready for the next charging current. This improves the battery charging efficiency. The equations for the charge and discharge cycles when the switches SW1 and SW2 are closed in Figure 2 are given by Equation (2) and Equation (3) respectively.

$$V_{in} - \frac{L\Delta if}{d1} - V_b = 0 \qquad (2)$$

IOP Conf. Series: Materials Science and Engineering **906** (2020) 012002 doi:10.1088/1757-899X/906/1/012002

$$V_b - \frac{L\Delta i f}{d2} = 0 \tag{3}$$

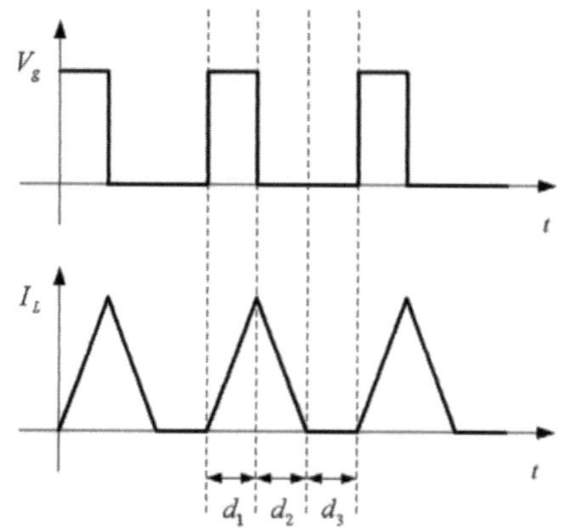

Figure 3. Gate Signal and Inductor current (I_L) waveforms for DCM

From Equation (2) and Equation (3) we can get the amplitude of the charging current. It is given by Equation (4) from which we in turn obtain Equation (5).

$$\Delta i = \frac{(V_{\text{in}} - V_b)d_1}{fL} = \frac{V_b d_2}{fL} \tag{4}$$

$$d_2 = \frac{(V_{\text{in}} - V_b)d_1}{V_b}. \tag{5}$$

Average current flowing through the output inductor is given by Equation (6).

$$I_{\text{avg}} = \frac{1}{2}\Delta i(d_1 + d_2) = \frac{d_1^2}{2fLV_b}(V_{\text{in}}^2 - V_b V_{\text{in}}) \tag{6}$$

Average power charges into the battery during one charge/discharge cycle is given by Equation (7).

$$P_o = I_{\text{avg}} V_b = \frac{d_1^2}{2fL}(V_{\text{in}}^2 - V_b V_{\text{in}}). \tag{7}$$

The rest duty d3 is zero for operation of the battery between continuous conduction mode and discontinuous conduction mode. Therefore, d1+d2=1 (charging period + discharging period =1). On substituting Equation (7) in Equation (5), V_{in} can be derived as,

$$V_{\text{in}} = \frac{V_b}{d_1}. \tag{8}$$

Hence, the previous claim that the average power into the wind turbine's battery is a polynomial equation of order 2 and is a function of V_{in}, is substantiated. The equation is in accordance with the MPP curve of the Wind Turbine System.

4. Maximum power point tracking algorithm

The Maximum Power Point Tracking Algorithm for the stand-alone wind turbine system in this paper is constructed on the basis of speed of the wind. Here, the wind speed and the speed of the rotor of the wind turbine are measured to get the optimal value of λ (tip speed ratio) which in turn helps us achieve the maximum power output of the wind turbine as given in Equation (1).

The MPPT Algorithm proposed in this paper divides one charging, discharging cycle into several intervals. The algorithm helps iterate through the intervals using loop control variable 'k', calculating the output power by multiplying the battery voltage with the current flowing through it. This helps find the maximum output power of a stand-alone wind turbine system for a particular wind speed.

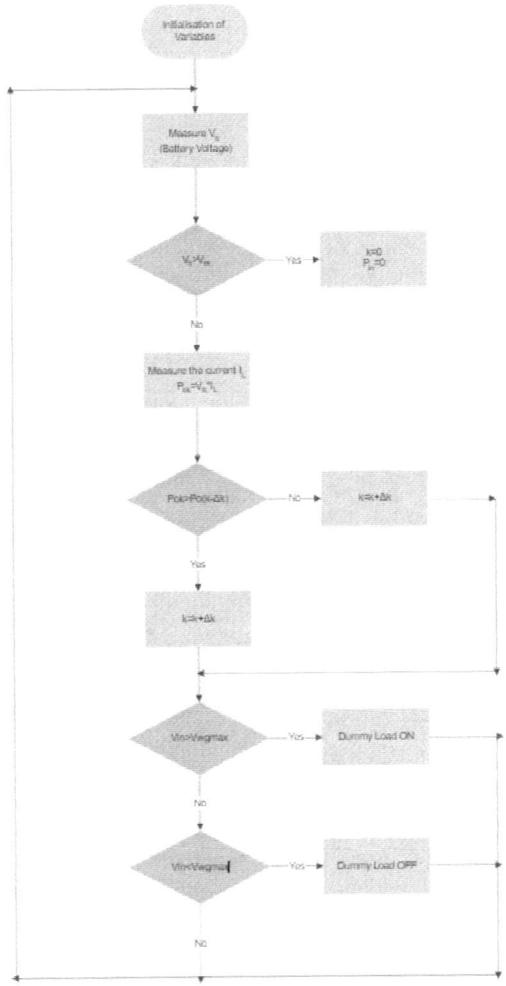

Figure 4. Flowchart for the MPPT Algorithm

5. Experimental Results

The following steps are supposed to be followed to design a buck-type MPPT battery charger for a stand-alone wind turbine system:

- Analyse the wind turbine specifications like the rated voltage, rated power, rotor speed, MPP curve etc.
- Select the appropriate value of the rated battery voltage based on the wind turbine specifications.
- Using equation (8), the rated battery and wind turbine voltage, obtain duty cycle d1.
- Further design the circuit parameters f and L listed in equation (7).

Consider a stand-alone wind turbine system with rated power of 480-W, a PM generator with a rotor speed of 300r/min and 34 poles. A diode rectifier is used to rectify the voltage. A buck-converter is also used, that is a step-down DC/DC converter. It helps reduce the rectified voltage to charge the battery. The circuit diagram for the above is given in Figure 5.

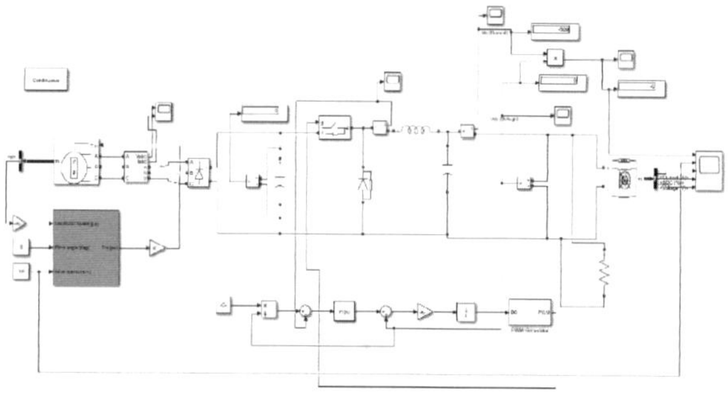

Figure 5. MATLAB Simulation diagram for a PMSG Stand-alone Wind Energy Conversion System

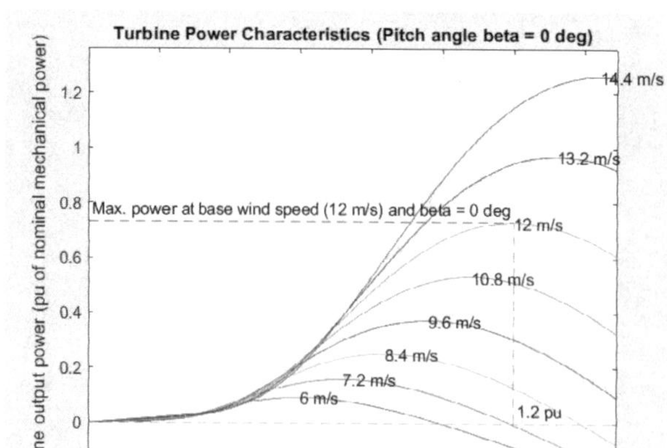

Figure 6. Output Power vs Turbine Speed of the PMSG Stand-alone Wind Turbine System

The above graph is of the PMSG stand-alone wind turbine system for the pitch angle (β) of the blades being 0 degrees. For a particular wind speed, the maximum power output is obtained from the generator for a certain turbine speed or rotor voltage. But after the power attains its maximum value represented by the peak in the graph, its decline is quite appreciable. The output voltage further reduces causing the power to subsequently tend to zero and the PM wind turbine will stall eventually.

The battery in the circuit diagram is considered to be a lead-acid battery with a nominal voltage of 48V. The variations in battery parameters are observed after simulation of the circuit for changes in the wind speed from 16 m/s to 12 m/s.

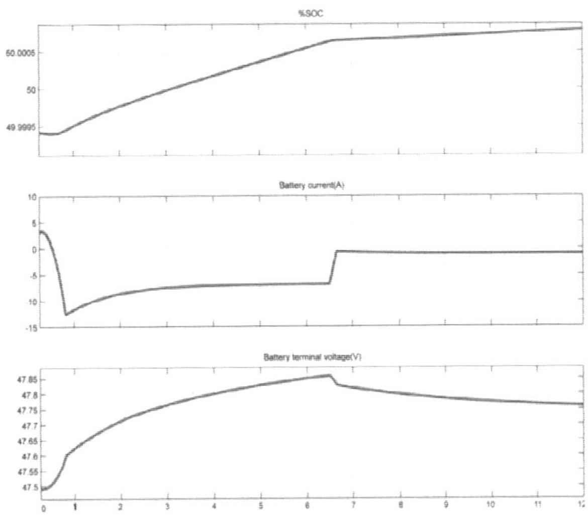

Figure 7. Graphs for Battery Parameters

The graphs above show the changes in the battery parameters like the % SOC (State of Charge), battery current and battery voltage. It is observed that initially the wind turbine voltage helps charge the battery. Once the battery is charged (Vin> Vth) the battery starts supplying current to the load.

6. Conclusion

In this paper, the process of charging a battery for a stand-alone wind turbine system is discussed in detail and the formulae to obtain the various parameters of a Buck DC/DC converter used to charge the battery are derived. The output power equation is also obtained and it is a polynomial of order 2. This is in accordance to the MPP Curve of the system. The control strategy for charging the battery is also discussed. The battery is charged using pulsating current by operating the buck-type charger in the discontinuous mode with constant on-time control to achieve better efficiency. The MPPT Algorithm is also discussed and represented using a flowchart for its better understanding. It is essential to achieve maximum output power at different wind speeds. Furthermore, a PMSG Stand-alone Wind Turbine System is simulated using SIMULINK of the MATLAB Software. The output power of the generator is observed for a variation of wind and turbine speeds and the changes in battery parameters are also analyzed.

References

[1] Lo, Kuo-Yuan & Chen, Yaow-Ming & Chang, Yung-Ruei 2011 MPPT Battery Charger for Stand-Alone Wind Power System *IEEE Transactions on Power Electronics - IEEE TRANS POWER ELECT* **26** 1631-1638

[2] http://www.iraj.in/journal/journal_file/journal_pdf/1-80-141016942691-96.pdf

[3] Grabic S, Celanovic N, and Katic V A 2008 Permanent magnet synchronous generator cascade for wind turbine application *IEEE Trans. Power. Electron.* **23** 3 1136–1142
[4] Soetedjo A, Lomi A and Widodo Puji Mulayanto 2011 Modeling of wind energy system with MPPT control *Proceedings of the 2011 International Conference on Electrical Engineering and Informatics*, Bandung 1-6
[5] Bagen, & Billinton, Roy. 2005 Evaluation of Different Operating Strategies in Small Stand-Alone Power Systems. Energy Conversion *IEEE Transactions* on. **20** 654 – 660
[6] Habibzadeh M, Hassanalieragh M, Ishikawa A, Soyata T and Sharma G 2017 Hybrid Solar-Wind Energy Harvesting for Embedded Applications: Supercapacitor-Based System Architectures and Design Tradeoffs. *IEEE Circuits and Systems Magazine* **17** 29-63
[7] Eltamaly, Ali 2007 Modeling of Wind Turbine Driving Permanent Magnet Generator with Maximum Power Point Tracking System *Journal of King Saud University, Engineering Science* **19** 223-237
[8] Koutroulis E and Kalaitzakis 2006 K Design of a maximum power tracking system for wind-energy-conversion applications *IEEE Transactions on Industrial Electronics* **53** 2 486-494
[9] Effendy ME, Hidayat Khusnul, Mardiyah Nuralif 2018 Circuit Simulation for Wind Power Maximum Power Point Tracking with Four Switch Buck Boost Converter *Proceeding of the Electrical Engineering Computer Science and Informatics*
[10] Patel R, Arora K and Katiyar S 2016 Control Analysis of PMSG Based Wind Energy Conversion System Using Buck-Boost Converter *2016 Second International Conference on Computational Intelligence & Communication Technology (CICT) Ghaziabad* 395-402
[11] Suyampulingam and Subbiah V 2018 PMSG based wind turbine system connected to DC micro-grid with MPPT control *International Journal of Pure and Applied Mathematics* **118** 99-105
[12] Mittal, Sandhu R, Jain K, Dinesh 2009 Controlled operation of variable speed driven permanent magnet synchronous generator for wind energy conversion systems *WSEAS Transactions on Systems* **8**
[13] Bharathi Sankar A and Seyezhai R 2013 Matlab Simulation of Power Electronic Converter for PMSG Based Wind Energy Conversion System *IJIREEICE* **1**(8) 348-353
[14] Husain, Mohammed 2014 Modeling and Study of a Standalone PMSG Wind Generation System Using MATLAB/SIMULINK *Universal Journal of Electrical and Electronic Engineering* **2** 270-277
[15] Kim, Hong-Woo Kim, Sung-Soo Ko, Hee-Sang 2010 Modeling and control of PMSG-based variable-speed wind turbine *Electric Power Systems Research* **80** 46-52
[16] Rolan A, Luna A, Vazquez G, Aguilar D and Azevedo G 2009 Modeling of a variable speed wind turbine with a Permanent Magnet Synchronous Generator 2009 *IEEE International Symposium on Industrial Electronics, Seoul* 734-739

IOP Conf. Series: Materials Science and Engineering **906** (2020) 012003 doi:10.1088/1757-899X/906/1/012003

Self-commissioning of induction motor drives-A critical review

Arati Chougala[1],Keerthi Nemali[2], Sonal Jadhav[3],Dhruval Vora[4],R Raja Singh[5], R Rajesh Kanna[6].

[1, 2,3, 4] Student at Vellore Institute of Technology, Vellore, India.

[5] Associate professor at Vellore Institute of Technology, Vellore, India.

[6] Research scholar at Vellore Institute of Technology, Vellore, India.

Abstract:

Now days efficient performance of machinery is must need especially in industry. Variable speed drives are used to drive the motor efficiently. Control strategy of variable speed drive plays an important role in it. So, it should be selected carefully according to the application. For controlling of motor encoders are used generally, but due to some disadvantages estimation techniques are used in recent time. Also estimation techniques having too types online and offline. For accurate estimation online parameter estimation technique is used, but they are little bit complex. That is also called self-commissioning of the motor. Here in this paper two techniques for self-commissioning of induction motor from different manufactures are compared. For Danfoss frequency injection method and for Rockwell voltage current lag angle method is used. Review on both these methods is presented in this paper.

1. Introduction

In recent decades, much progress has been created in connecting the drives to the industries. The industrial drives, which are a combination of source, power modulator, control unit, motor and load, are strongly correlated with the economic development of any Nations. Every industry needs the technology of motion control to regulate the speed of the machine according to their application as a key part of the next generation of factory automation. Conventionally, commutator motors are utilized in industries, and then brushless motors incorporated with the drive system, which is operated with phase-controlled rectifiers, DC choppers [1]. In recent decades, induction motors are widely adopted in industrial applications, since it has high robustness and consistency with low cost [2]. Variable Frequency Drives (VFD) is specifically designed for permanent magnet motors but it also can be used for all type of AC speed control operations especially induction motors. When an induction motor operates with VFD, it runs at high power factor that avoids overheating of coils due to increase in current. VFDs start induction motor quickly with less inrush current of 100% to 150% of full load current. The VFDs are mainly designed in two types, they are voltage source inverter based VFD and current source inverter based VFD [3]. New VFD technologies mainly focus on improving the controllability and enhancing the efficiency of the system through advancing the adaptive control parameter estimation and energy optimization [4]. Thereby, 11 to 18 per cent of the energy usage of VFD could be reduced [5]. Precise prediction and analysis of machine's parameter are mandatory in controlling the VFD to counterpoise the load output. Hence, it is inevitable to conduct the test on the drive motor under mechanically decoupled state, conventionally known as offline parameter determination method [7]. On the other hand, mechanical decoupling of the load from drive motor is not feasible on every industrial drive systems.

Online parameter estimation techniques are Recursive Least square technique, Model reference adaptive system, Signal Injection technique, Sliding mode Observer, Heuristic or Meta-heuristic algorithms [8]

During acquire the parameters of the motor in offline, the effects of the ascent in skin effect, flux saturation, temperature, will certainly cause variations in the parameters. The heat produced by the motor creates the temperature change in machine and thus distressing the stator and rotor resistance. Besides, there are variations in rotor resistance due to variations of skin effect caused by changes due to rotor frequency and temperature. Finally, the magnetizing inductance is affected due to the saturation of magnetizing flux. Therefore, parameters of a motor are diverged from the nominal values during operation [8].To certify the performance of the IM systems, it is necessary to provide real-time parameters using online parameter identification. Only certain parameters among all the essential motor parameter have an important impact on overall control performance in operation. Thus, online identification usually aims at one or several parameters [8, 9]. The invention of online identification encompasses a method for automated measurement of Ohmic Resistance of Stator, the Main Inductance and Leakage Inductances of the IM by applying a direct Signal which is predetermined by test signal Superimposed with an alternating Signal to a stator winding of the IM [10].

Stage Recursive Least Squares (RLS) is also one of the commonly used algorithms for online parameter Estimation. For applying this technique, some of the necessary details are stator current, voltages and their derivatives. On the process of measuring, the voltages modified by switching of DC inputs which may introduce harmonics and noises at the input. So, the filters included with the necessary compensating device because adding filters may delay the output. The basic Induction Motor model is written in regressive form for applying RLS algorithm. The data obtained from the system gives initially estimated parameter. From that, the covariance matrix is formed. It is updated for each step based on the error value. The errors lumped in the single scalar term, which show larger error value than the estimated parameter, which may not give accurate results. To overcome this inaccuracy The Forgetting factor is added which decompose the past parameter so that algorithm can prioritize the recent data rather than old one. The error percentage of the estimated parameter is less than 5% [9]. The major drawback of RLS algorithms is that the motor parameters vary slowly concerning time, that is, their time derivatives are approximately zero [11]. Two-Stage Recursive Least Squares Algorithm (TSRLS) in this the system is decoupled into two subsystems and then to find parameters of each subsystem using the RLS estimation. Compared to the conventional RLS this algorithm is effectively save computational cost [12].

To attain high control performance of IM drives, some of the classic control strategies are used like field-oriented induction motor, Torque controller or Vector controllers (VC), Hysteresis current controller etc. Here VC is a widely used control technique in induction machine drives systems to accomplish high performance in induction machine applications. Direct Field Orientation controller (DFOC) and Indirect Field Orientation controller (IFOC) are the two various types of VC. DFOC estimates parameters thru terminal measurements. This controller is sensitive to Rotor resistance and Inductance. One of the problems is low speed operation and rotor flux cannot be measured directly. Low speed problem is not inherited by IFOC, it utilize the rotor flux positioning data's for the operation. However, the problem is open circuit time constant, which may often vary to flux and temperature changes. Incorrect time constant leads to wrong calculation of slip frequency and rotor flux positioning [6].

So finding the accurate parameters of IM drives is essential to control the speed of Induction machine. Two methods i.e. Offline Parameter estimation and Online Parameter Estimation can evaluate the Parameters of motors. In Offline Parameter estimation, No load and Blocked rotor test has to be done. Otherwise, parameters are calculated by applying Ac voltage to the motor windings [7].

2. System modelling

In this study a 3 phase balanced symmetrical winding for squirrel cage induction motor with a sinusoidal flux distribution (magneto motive force) and dynamic modelling circuit was performed.

2.1. *Modelling of Induction motor*

Fig. 1 represents a schematic diagram of the modelling of IM with alternating signal injection method [16], to estimate the motor parameters. The squirrel cage IM dynamics is modelled using stationary reference frame dq0.Various quantities are used in their normalized form in this paper and that normalized quantities of the rotor are referred to the stator circuit.An IM parameters are listed in Table 1. Stator and rotor voltages equations [14] in terms of abc transformation

$$v_{abcs} = r_s i_{abcs} + p\lambda_{abcs} \quad (1)$$

IOP Conf. Series: Materials Science and Engineering **906** (2020) 012003 doi:10.1088/1757-899X/906/1/012003

$$v_{abcs} = r_s i_{abcs} + p\lambda_{abcs} \tag{2}$$

The measurement of voltage is very important in estimating the online parameter for which the sensor is generally used to measure the voltage for estimation of other different parameters. [17].the estimation process starts when induction motor reaches to steady state condition, the detailed measurements of steady-state parameters are collected primarily. For accurate control of drives, it is necessary to know exact flux variation in dqo axis. The fundamental reason is that computations are much easier by transforming the 3-φV instantaneous values of voltages and currents within the synchronously rotating reference frame dq0 make calculations much simpler and the second thing is it provides the individual control for both the active current component (on d-axis) and reactive current component(on q-axis). The dqo frame can be found from the Parks transformation. Similarly, in the machine, the torque and flux can be independently controlled. By this way, the coupling effect can be minimized to a certain extent. Additionally, in the dq0-frame, mutual inductance is constant. Thus, it allows to achieve the desired output as the inductance dependent quantities are constant. The online parameter estimation technique requires these quantities to estimate the parameter values. Employing the different estimation techniques resistance, inductance, flux linkage and the angular frequency of the induction motor can be estimated.

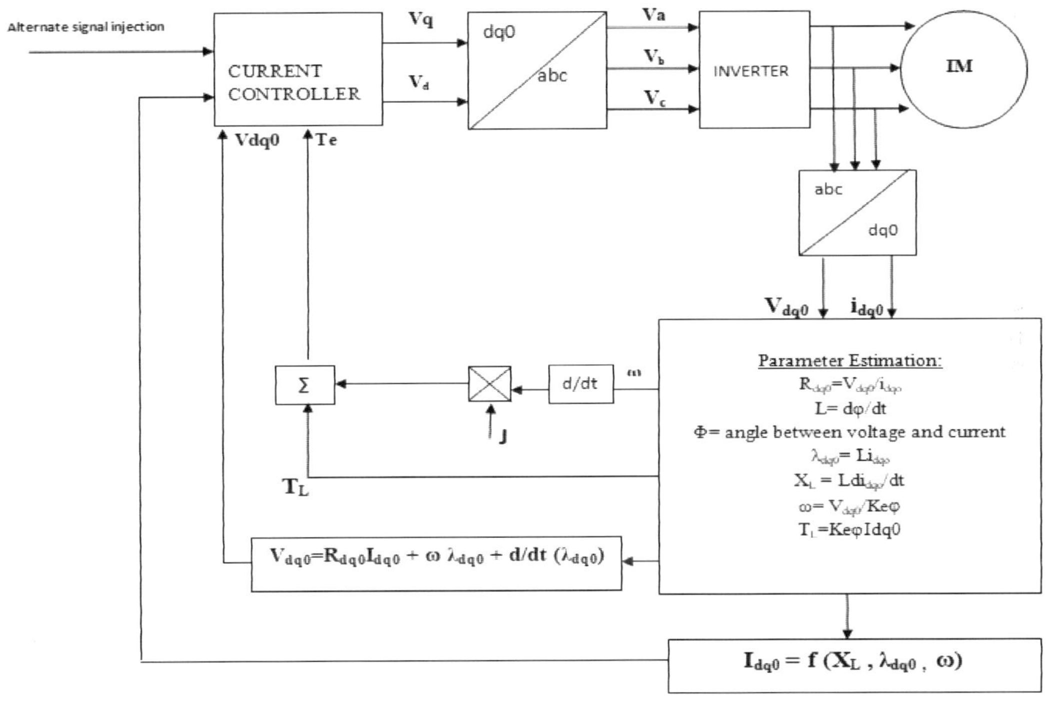

Figure 1.Modelling diagram of the IM with alternating signal injection approach.

Applying these estimated parameters values to induction motor modelling gives Vdqo, Idqo and Te value which is given as a reference to the controller along with D-axis reference currents are injected into the drive system one by one, making sure that results obtained are gathered on each incident when a steady state is reached. This should be observed that the injected d-axis current will produce additional torque reluctance which causes torque ripples for IM. To overcome this situation, the alternating signal injection strategy is performed. The torque ripple caused due to alternating signal injection can be minimized by this compensation strategy. The controller produces controlling signals that are given to the dq0 to abc transformation block and applied to the converter which drives the machine as per the requirement. The following equations are used to construct the model and estimate the parameter of IM. The mechanical part of the motor is modelled through electromagnetic torque as:

$$T_e = J\left(\frac{2}{P}\right)Pw_r + T_L \tag{3}$$

Where J is the rotor inertia.T_L is the load torque which is positive for a torque load on the shaft of the induction machine.

Power system as well as other machine parameters are mostly given in ohms and even fraction of base impedance per unit, it is possible to express the voltage and equation of flux connection in terms of reactance instead of inductance.

$$v_{qd0s} = r_s i_{qd0s} - w\lambda_{dqs} + p\lambda_{qd0s} \tag{4}$$

$$v'_{qd0r} = r'_r i'_{qd0s} + (w - w_r)\lambda'_{dqs} + p\lambda'_{qd0s} \tag{5}$$

$$v_{qs} = r_s i_{qs} + \frac{w}{w_b}\psi_{ds} + \frac{P}{w_b}\psi_{qs} \tag{6}$$

$$v_{ds} = r_s i_{ds} - \frac{w}{w_b}\psi_{qs} + \frac{P}{w_b}\psi_{ds} \tag{7}$$

$$v_{0s} = r_s i_{0s} + \frac{P}{w_b}\psi_{0s} \tag{8}$$

$$v'_{qr} = r'_r i'_{qr} + \frac{(w - w_r)}{w_b}\psi'_{dr} + \frac{P}{w_b}\psi'_{qr} \tag{9}$$

$$v'_{dr} = r'_r i'_{dr} + \frac{(w - w_r)}{w_b}\psi'_{qr} + \frac{P}{w_b}\psi'_{dr} \tag{10}$$

$$v'_{0r} = r'_r i_{0r} + \frac{P}{w_b}\psi'_{0r} \tag{11}$$

Where w_b is base angular velocity used to calculate the inductive reactance.

$$\psi_{qs} = X_{is} i_{qs} - X_M(i_{qs} + i'_{qr}) \tag{12}$$

$$\psi_{ds} = X_{is} i_{ds} + X_M(i_{ds} + i'_{dr}) \tag{13}$$

$$\psi_{0s} = X_{is} i_{0s} \tag{14}$$

$$\psi'_{qr} = X'_{ir} i'_{qr} + X_M(i_{qs} + i'_{qr}) \tag{15}$$

$$\psi'_{dr} = X'_{ir} i'_{dr} + X_M(i_{ds} + i'_{dr}) \tag{16}$$

$$\psi'_{0r} = X'_{ir} i'_{0r} \tag{17}$$

Where,

Table 1. List of Abbreviations

Parameters	Terms
$J =$	Inertia of rotor(Kgm^2)
$T_L =$	Load Torque
$P_r =$	Poles
$\lambda_{abcs}, \lambda_{abcr} =$	Stator and Rotor leakage inductance
$\omega_r =$	Rotor angular speed
$\omega_b =$	Nominal Rotor angular speed
$v_d, v_q =$	d-axis and q-axis stator voltages
$i_d, i_q =$	d-axis and q-axis stator current
$\psi_{ds}, \psi_{qs} =$	d-axis and q-axis stator flux linkages
$\psi_{ds}, \psi_{qs} =$	d-axis and q-axis rotor flux linkages
$r_s, r_r =$	Stator Resistance and rotor resistance (per phase)
$X_{ls}, X_{lr} =$	Stator reactance and rotor reluctance (per phase)
$X_M =$	Magnetizing Reactance

2.2. *Danfoss parameter estimation model*

The basic idea of the Danfoss parameter estimation is applying a testing signal to a phase winding of the IM, measure magnitude and the phase of the resulting current Isa (test signal), then identify the parameters of each subsystem utilizing frequency injection [13]. This parameter estimation using frequency injection save cost and time comparing with the conventional algorithm like RLS, TSRLS.
Consider rotor resistance (Ω) transferred to stator side, dynamic inductance, different parameter equation is given below

$$R_r' = \frac{|U_m'|}{|I_r'|} \tag{18}$$

Where,

R_r'= Stator Side Rotor Resistance.

U_m'= Main inductance voltage drop calculated using L_m.

I_r'= Current through rotor winding.

$$R_r = R_r' . \frac{L_{DM} + L_{lr}}{L_{DM}'} \quad (19)$$

Where,

$$L_{DM} + L_{lr} = L_r \quad (20)$$

L_r =Slip of the asynchronous machine

L_{DM}, L_{DM}'= Dynamic main inductances

L_{lr} = Rotor leakage inductance.

The value of R_r in (19) is independent of current through winding of a stator, but referred value of rotor resistance R_r' is an f (L_{DM}), it depends upon the value of DC-offset. The referred dynamic main inductance L_{DM}', canbe derived with magnetic flux and current. The following equation leads to find the value of dynamic main inductance and rotor resistance,

$$L_{DM}' = \frac{d\varphi'}{dI_m'} \quad (21)$$

Where, Ø = magnetic flux referred to main inductance,
I'_m= Magnetizing current.

$$Ø = L * I \quad (22)$$

$$L'_{DM} = \frac{d(L'_m I'_m)}{dI_m} \quad (23)$$

The required solution referred dynamic main inductance is:

$$L'_{DM} = \frac{dL'_m}{dI_m} I'_m + L'_m \quad (24)$$
$$L'_{DM} = L_{DM}{}^2/(L_{DM} + L_{ir})$$

Equation 25 can be developed and from this L_{DM} is calculated from, L'_{DM}

$$L_{DM} = L'_{DM} + (L_{DM}{}^2 + 4L'_{DM}L_{ir})^{1/2} \quad (25)$$

In this method, a test signal U_{sa} it consists of direct Voltage, superimposed with triangular Voltage is applied to the stator, and the resulting Isa (stator current) is measured. The L_s' (static transient inductance) calculated using induction motor technology, as given below:

$$L'_s = L_m + L_{is} - \frac{{L_m}^2}{L_m + L_{ir}} \tag{26}$$

Moreover, $L_{DS}{}'$(Dynamic transient inductances), according to Communal understanding the field of an IM, as below:

$$L'_{DS} = L_{DM} + L_{is} - \frac{L_{DM}}{L_{DM} + L_{ir}} \tag{27}$$

The R_{T1} value calculated at T1 it re-calculated by R_{T2} at a different temperature T2 using equation (28).

$$R_{T_2} = R_{T_1} \cdot \frac{T_2 + K_T}{T_1 + K_T} \tag{28}$$

Where, K_T constant (for copper material, K_T=235),

T_1, T_2 are measured operating temperatures.

The steps involved in the Danfoss frequency injection method to estimate parameters of an Induction Motor (IM) are given below.

Step 1: To calculate first injection frequency, the test signal (U_{sa}), whose value depend on previously stored values in the memory. Testing voltage (U_{sa}) predetermined direct voltage applied on the phase winding of stator.

Step 2: To measure stator current (I_{sa}) and Ohmic Stator resistance Rs, alternating triangular voltage superimposed with direct voltage of stator.

Step 3: Find the time constant L_s'/ (Rs+Rr') and differential stator current dI_{sa}/dt using Sampled values.

Step 4: Main inductance (Lm) is function of magnetizing current (Im), $L_m=\phi_m/I_m$. Calculate L_s', using

$$U_{sa} = R_s I_{sa} + L'_s(dI_{sa}/dt)$$

Step 5: Determine the correct phase voltage again by U_{ref} (reference voltage) given to the controller. Testing signal (U_{sa}) is triangular signal but can also apply square/sine wave applied till the resulting stator current (I_{sa}) and flux (ϕ) stabilized.

Step 6: calculate a R_r (Ohmic rotor resistance) using stator Resistance (Ω), L'_s (Leakage inductance), L'_{DM} (Main inductance R'_r (Rotor resistance transferred to stator side).

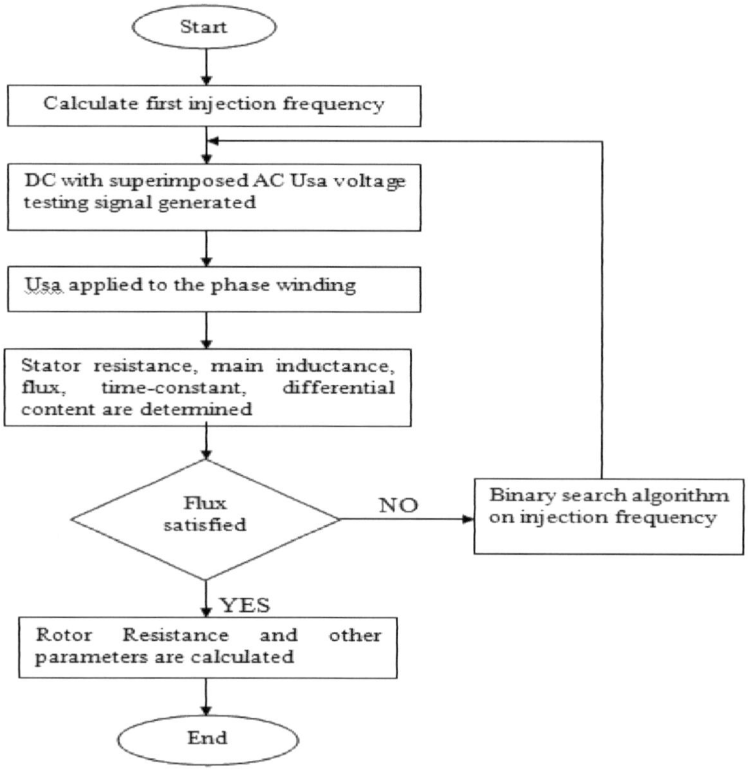

Figure 2.Flowchart of computation of the electric parameter estimation method using the Danfoss frequency injection method.

2.3. Rockwell parameter estimation model

The angle between both high frequency voltage injection and the subsequent high frequency current is used to calculate the instantaneous magnitude of the stator resistance. High frequency voltage reference frame is mapped by three-phase voltage injected. In which the voltage with the higher frequency (V_{dhv}, V_{qhv}) must be coordinated with the direct axis and the three phase currents feedback also analysed by a current reference frame in which the higher - frequency currents (i_{dhv}, i_{qhv}) coordinatedto d or q axis , any one from the both when the estimation is perfect, the and q axis feedback related to higher frequency component ih.

Voltage reference frame i_h decomposed into i_{dhv}andi_{qhv}using Θ_h (high frequency command angle) in

Current reference frame i_h decomposed into i_{dhi}andi_{qhi}using Θ_{hi} (high frequency current angle)

Current lag angle $\Theta_i = \Theta_h - \Theta_{hi}$

Flux angle estimation:

Using feedback voltage (or current) to obtain the flux angle by injecting high frequency signal to control voltage signal to power a pulse width modulated inverter. High frequency signal in the high frequency phase angle and generates the first injection signal (the multiplication of the magnitude and sine of phase angle of high-frequency), other two phase signals are delayed by 120°, the first, the second and the third phase are shifted separately, then add to a separateone of the three-Voltage command Signals. Derivative of flux angle position (used to overcome problems not including speed sensors in it) used as estimated stator frequency (function of motor parameters). Hence estimated rotor speed calculated by deducting the slip frequency from the estimated frequency of the stator.

Stator resistance varies with temperature, tempreture of Stator winding is average winding current, which increases current feedback from through analog to digital converter, these current feedbacks are converted to two-phase Rectangular currents using current angle of High-frequency, offer the close-loop system in the first of two-phase Rectangular currents to be driven, the first of two-phase Rectangular

IOP Conf. Series: Materials Science and Engineering **906** (2020) 012003 doi:10.1088/1757-899X/906/1/012003

currentsto Zero using the Proportional Integral controller, as the angle of lag in current can be estimated using the PI controller's output and, The current lag angle can be used to estimate as an estimate of the lag angle between voltage-current generate an estimate of the stator resistance. Estimating the stator resistance of three-phase motor is defined by an inductance of Stator leakage and powered by a pulse width modulation scheme connected to the motor through three supply lines. A system operating method that gets a voltage signal having high frequency and causes high-frequency voltage injected at high frequency in each of the three-phases of motor.

$$R_s = ctg(\varphi_d)(\omega_h)(L_{\sigma s}) \tag{29}$$

Where,
ω_h is the injected high frequency (rad/sec)
$L_{\sigma s}$, Stator leakage inductance,
I_h high frequency current,
V_h high frequency Voltage
Φ_d voltage-current lag angle
Current vectors are present in the current reference frame include a components direct-axis and quadrature-axis. After that, the current vector of the q-axis is defined negatively and step up with the proportional-integral controller to produce angle between voltage and current. Estimation in the current Φi' which is equal to the lag angle estimation between voltage and current Φd'. Estimation, $\Phi_i' = \Phi_d'$ is now again deducted by command angle Θ_h, to create a close-loop. With addition to that, the estimated Φ_d' is also put into Equation for angle $\Phi_d$to obtain resistance estimate R_s'.

$$R_s' = \tan(90° - \varphi_d')(\omega_h)(L_s') \tag{30}$$

Rotor resistance estimation

The estimate of the lag angle used to generate estimate value of resistance of rotor. For the case, step to use lag angle estimation between voltage and current to obtain the estimation a resistance of rotor in few embodiment including a step of identification, storage of an initial lag angle estimation of voltage current, Classifying the initial value of rotor side resistance estimation, division of the initial voltage-current lag angle estimate by the estimation of voltage, current to generate a stator resistance ratio and multiplying the initial value of rotor resistance estimated with the help of Stator resistance ratio hence create an estimate of rotor resistance.

The resistance of the rotor Rr varies in proportional to the change in the estimation of a Φd', Thus, if an the rotor initial resistance Rro is determined and Stored. Initial voltage-current lag angle estimate Φdo' is calculated, related to resistance estimation done initially. Rro and Stored, these values can be estimated together with the instantaneous lag angle estimate Φd', which is used for estimation of the instantaneous resistance value of rotor Rr'. To this end, instantaneous estimate Rr'.

$$R_r' = (R_{ro})(\varphi_{do}'/\varphi_d') = (R_{ro})\tan(90° - \varphi_{do}')/\tan(90° - \varphi_d') \tag{31}$$

A Voltage Vh, is injected in between the axis of d_{hv}, and q_{hv} and the high-frequency current I_h results. In this case, angle between voltage and current is Φ_d' can be obtained using

$$\varphi_d' = abs\{abs(\varphi_i) - abs(\varphi_v)\} \tag{32}$$

If the high-frequency signals were injected between direct and quadrature-axis, instead of identifying a system of co-ordinate where the current Ih coincides with the d-axis, a new system of co-ordinates can be identified where the current Ih coincides with the q-axis. Using a following equation, a lag angle estimation Φq' can be obtained.

IOP Conf. Series: Materials Science and Engineering **906** (2020) 012003 doi:10.1088/1757-899X/906/1/012003

$$\varphi_q' = abs\{abs(\varphi_i) - abs(\varphi_v)\} \quad (33)$$

The steps involved in the Rockwell voltage current angle lag method to estimate the parameters of induction motor mentioned below.

Step 1: Identify and store the initially estimated Current Lag angle, an initial estimation of rotor resistance and inductance of stator leakage.

Step 2: Identify High Frequency Command Angle.

Step 3: A lag angle of voltage and current then Subtracted from a High-Frequency Command Angle Estimateto obtain a current angle of High-Frequency.

Step 4: Obtain the feedback currents from sensors.

Step 5: Convert feedback currents to two-phase currentsto generate lag angle of voltage and current.

*Step 6:*A mathematicaly combine angle estimation, leakage inductance of stator and command signal with high-frequency to generate resistance of stator.

Step 7:A Mathematicaly combine the initial angle estimate, the estimation of lag angle between voltage and current and the estimation of initial rotor resistance to obtain the estimate of rotor resistance.

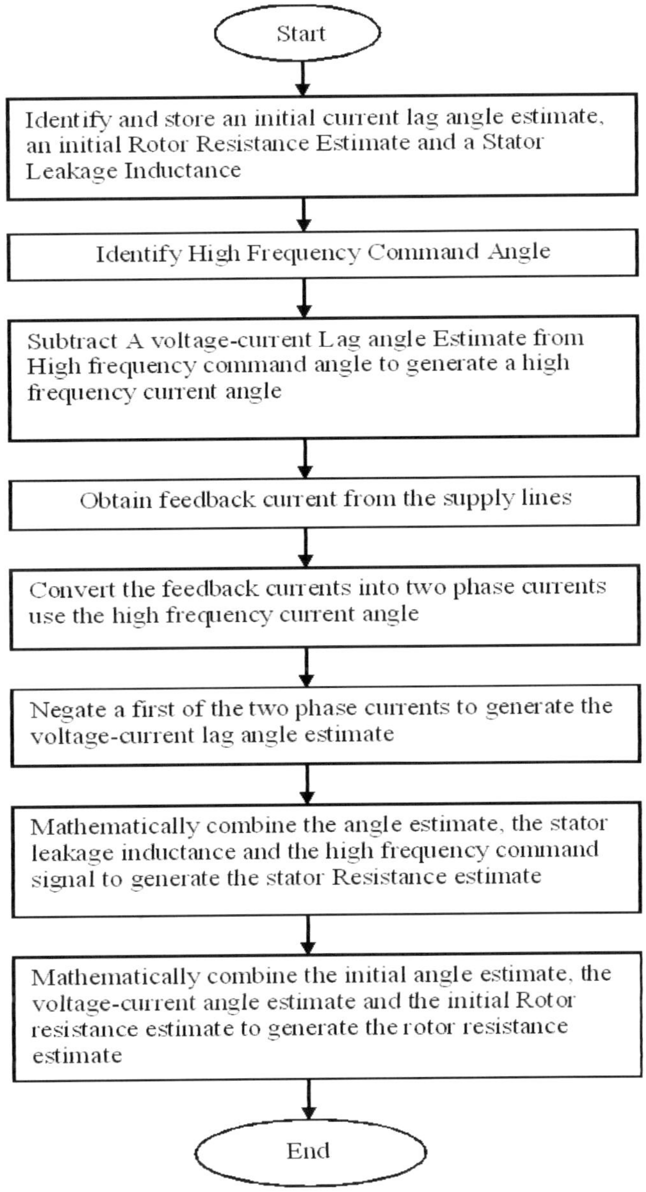

Figure 3.Flowchart of computation of the electric parameter estimations using angle between voltage and current.

3. Experimentation

The comparison between the Danfoss electric parameter estimation method and Rockwell electric parameter estimation methodwith the "real" IM electrical paremeters are determined from conventional No-load and Locked-rotor test in this paper. Which are shown in table 2 and ensure the reliability and accuracy of the IM electrical papameter estimation methods given by Danfoss and Rockwell.

Figure 4. Danfoss Drive and Rockwell Automated Drive from (left to right) with a motor set up.

IOP Conf. Series: Materials Science and Engineering **906** (2020) 012003 doi:10.1088/1757-899X/906/1/012003

4. Comparison between Danfoss and Rockwell Automation Drive

Table 2. Comparison of different parameters among Danfoss and Rockwell drives

PARAMETERS	ROCKWELL AUTOMATION	DANFOSS
Overview	<u>**Power Flex 525 drive**</u> Voltage Rating: 200V,400V 15 HP/11 kW Overload: 150% for 60 seconds frequency up to 16 kHz	<u>**VLT Automation Drive FC 302**</u> The input power required : 90-710kw Voltage=380-690V
Topology	a) Parks Transformation (3Ø to 2Ø conversion) b)Inverse Parks Transformation (2Ø to 3Ø conversion) c)Voltage reference frame d)Current reference frame	The rotor resistance (Rs) of an Induction Motor is measured automatically by means of an inverter while being acted upon by a non-rotating field
Equivalent circuit	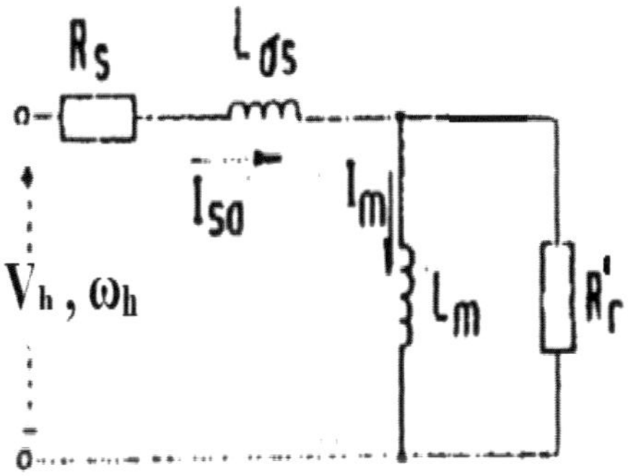 Voltage(V_h) of High frequency is injected across the stator winding of motor rotates with Θ_h , high frequency(ω_h in rad/sec) generates high frequency current (I_h)	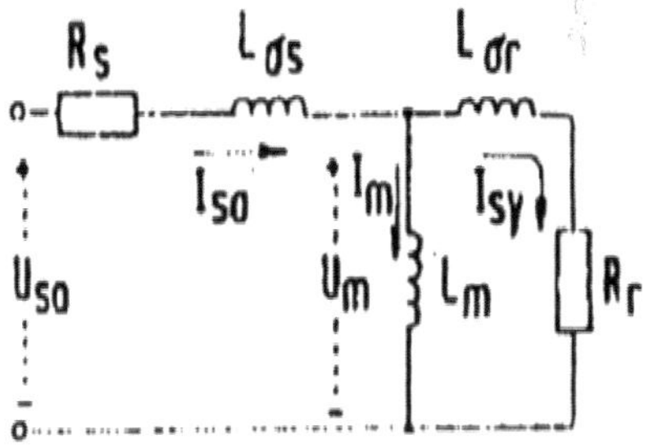 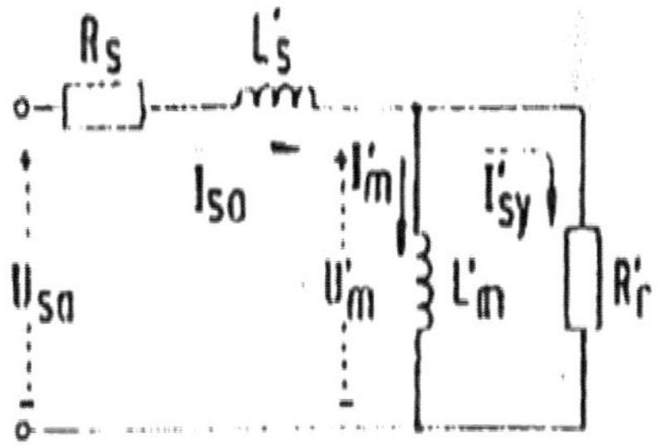
Number of sensors	• 3 voltage sensors • 3 current sensors	• 3 current sensors
Type of filter	➢ Band Pass Filter ➢ Notch Filter ➢ Low pass filter	• Sine wave filter • Harmonic filter • dU/dt Filters • common-mode filter

Parameter estimation	1. Resistance of Stator (Rs) 2. Rotor Resistance of motor (Rr) 3. Leakage inductance of stator of motor (L_{sl})	1. Ohmic stator resistance of motor(Rs) 2. Ohmic resistance of Rotor (Rr) 3. leakage inductance of rotor ($L_{\sigma r}$) 4. leakage inductance of Stator ($L_{\sigma s}$) 5. main static inductance of the Induction Motor(L_m) 6. main Dynamic inductance(L_{Dm})
Accuracy	More	Low

5. Conclusion

In this paper modelling of induction motor is done for three phase squirrel cage induction motor, to obtain some important relation between parameters of induction motor and also online parameter estimation which is also called self-commissioning of induction motor is explained in detail and compare from two different manufacturers, Danfoss and Rockwell automation drives are compared for self-commissioning of the motor. Danfoss uses the frequency injection method which is comparatively less accurate than method used by Rockwell automation which is voltage current lag angle method. Also Danfoss drives uses only current sensor, numbers of estimated parameters are more so somewhat more calculative burden will be there which will in turn increases the complexity and also calculation time.

6. Reference

[1]. "*Power Electronics and Variable frequency drives technology and Application*", Bimal.K.Bose

[2]. Henke, M.; Narjes, G.; Hoffmann, J.; Wohlers, C.; Urbanek, S.; Heister, C.; Steinbrink, J.; Canders, W.-R.;Ponick, B. *Challenges and Opportunities of Very Light High-Performance Electric Drives for Aviation.*
Energies 2018, **11**, 344. [CrossRef]

[3].www.vdfs.org

[4].www.pumpsandsystems.com/new-advances-variable-frequencydrives

[5].*United States Industrial Electric Motor Systems Market Opportunities Assessment* December 2002

[6].Nordin, K. B., D. W. Novotny, and D. S. Zinger, "*The influence of motor parameter deviations in feed forward field orientation drive systems*," IEEE-IAS Trans., Vol. IA-**21**, no. 4, pp. 1009-1015, July/August 1985.

[7].Luo.H.Liu, J.Wan,"*Offline Identification of Induction motor Parameters*", Electrical Drives 2006, **8, 16**-21.

[8].Jing Tang, Yongheng Yang, FredeBlaabjerg, Jiechen, LijunDiao and ZhigangLiu,"*parameter Identification of inverter-Fed Induction Motors: A review*", Energies 2018, no.11, pp.2194.

[9].C.Attainanese, G. Tomasso, A. Damiano, I. Marogiu, and A. Perfetto, "*A novel approach to speed and parameters estimation in induction motor drives*," IEEE Trans. Energy Convers., vol. **14**, no. 4, pp. 939–945, Dec. 1999

[10].O. Scaglione, M. Markovic, Y. Perriard, "*On-line Parameter Estimation for Improved Sensorless Control of Synchronous Motors*", IEEE-2012.

[11].Jens Godbersen, "*Method for Automated Measurement of the Ohmic Rotor Resistance Of An asynchronous Machine*", united states patent. Aug. 10, 2004.

[12].SemoyanRoyak, Mark M Harbaugh, Russel J. Kerkman, "*Stator and Rotor Resistance Identifier Using High Frequency Injection*", United States patent, Oct. 21, 2003.

[13].Jens Godbersen, "*Method for Automated Measurement of the Ohmic Rotor Resistance Of An asynchronous Machine*", united states patent, Aug. 10, 2004.

[14].SemoyanRoyak, Mark M Harbaugh, Russel J. Kerkman, "*Stator and Rotor Resistance Identifier Using High Frequency Injection*", united states patent, Oct. 21, 2003.

IOP Conf. Series: Materials Science and Engineering **906** (2020) 012004 doi:10.1088/1757-899X/906/1/012004

IOT based Smart Energy Manager for Low Voltage Consumers with Remote Monitoring and Control

S Banumathi[1*], P Preethika[2], S Shangeetha[3], A Anusuya[4]

Department of Electrical and Electronics Engineering, M.Kumarasamy College of Engineering, Karur- 639 113, India.

bbanumathi1974@gmail.com

Abstract. Electricity consumption at houses and companies are calculated using digital energy meters installed in the consumer's premises. The consumer will aware of the energy consumption rate only at the end of the period and get worried about the high electricity bill. This IoT based smart energy manager is capable of monitoring the electricity usage and send SMS/E-mail to the consumer when the consumed energy reaches beyond the threshold value. In addition to the above it aims to save energy by turning off the appliances from remote place when energy consumption gets exceeded. Remote monitoring and control of electricity is obtained with the help of Arduino, ESP-12E Wi-Fi module. The current sensor is used to measure the current consumption and gives input signal to the Arduino Uno, which is programmed to calculate the electricity bill. With the help of CritaC App IoT, IFTTT, it sends SMS/E-mail about the total energy consumption and electricity bill. The prototype model is implemented and test results are discussed in this paper.

1. Introduction

India is the second largest populous country and one of the fastest growing economies in the world. In 2017, according to the International Monetary Fund (IMF), the Indian economy was nominally worth \$2.6 trillion; it is almost \$10 trillion as a 3rd largest Purchasing Power Parity [1]. India is rich in natural resources but the demand on power increases continuously (Figure 1-source: 19th EPS report-National Electricity Plan, Central Electricity Authority, India) due to increasing population, in order to satisfy the power demand new technologies should be included [2]. Renewable resources such as solar, wind should be used effectively to satisfy the emerging demand [3-4].Today technology plays an important role in day to day life, it makes our life simpler and enable us to stay connected to family and friends with almost no cost. Now- a- days, home automation can be done in smart phone for controlling appliances with the help of sensors inbuilt within it [5].

Arduino plays a major role in the IOT platform and used to develop lot of projects in all areas. Arduino consists of microcontroller, various analog and digital I/O pins. The microcontroller present in Arduino can be coded by C and C++ programming language. With the help of Arduino, it is easy to combine both hardware and software. Here arduino is chosen as the major component since, it is cost effective and compact in size and it can easily work with all types of sensors that are available. When arduino is connected to the Wi-Fi module, it can be accessible by the smart phones[6]-[13].

The proposed system provides energy monitoring and control with the help of CritaC IoT App which calculates the tariff and send a mail if it exceeds the threshold value. The threshold value completely depends on the customer's maximum demand at the home. By this system it is easy to monitor and control the energy, here controlling is done with the help of relay connected to every appliance that trips the appliances when it exceeds the threshold value.

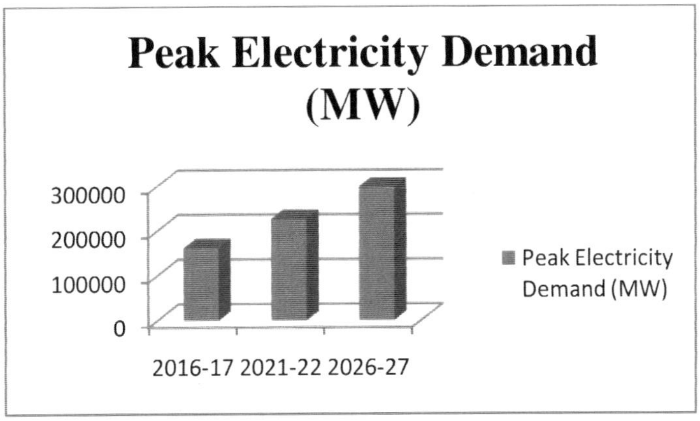

Figure 1. India's Peak Electricity Demand Source: EPS

2. Proposed System

The proposed idea gives a detailed report on energy monitoring and controlling purpose with help of Arduino, ESP8266, current sensor and relay. The Adafruit IO platform continuously fetches the data from the Wi-Fi component connected to the current sensor that measures the current consumption of the appliances. The CritaC IoT App is used to view consumption of energy for every second through smart phone and the load can be controlled with the help of relay. Thus the relay can turn off load

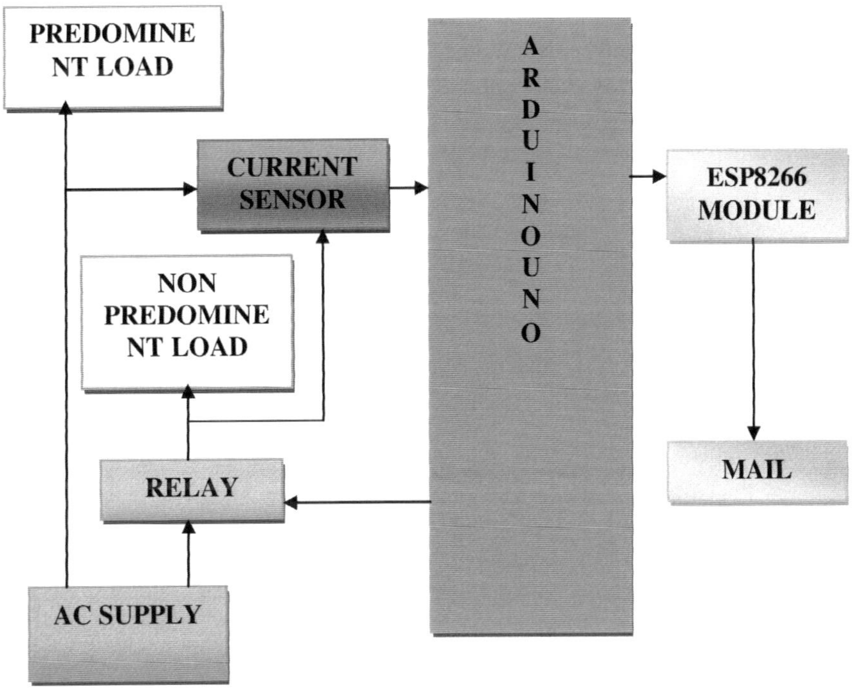

Figure 2. Block diagram of IOT based smart energy manager for low power consumers

automatically, hence it is programmed in Arduino that if it exceeds the threshold value. The main objective of the proposed project are to (i) monitor the energy consumption (ii) display energy consumption in a smart phone through CritaC IoT App (iii) display the amount of energy consumption shown as bill and (iv)to control the appliances if the energy consumption exceeds the threshold value.

The general block diagram of the proposed system is given in figure 2. The loads are classified into predominant and non- predominant loads. In this work the maximum limit is taken as 5 units. If the power consumption is beyond the maximum limit, it is not necessary to switch off all the loads. Rather than switching of all the loads only non-predominant loads are turned off with the help of relay. Here the energy monitoring and control is not enough, power saving is very much important for future. By turning on and off of the predominant and non–predominant loads at correct time is necessary and it can be done with the help of this proposed project. The comprehensive implementation details can be

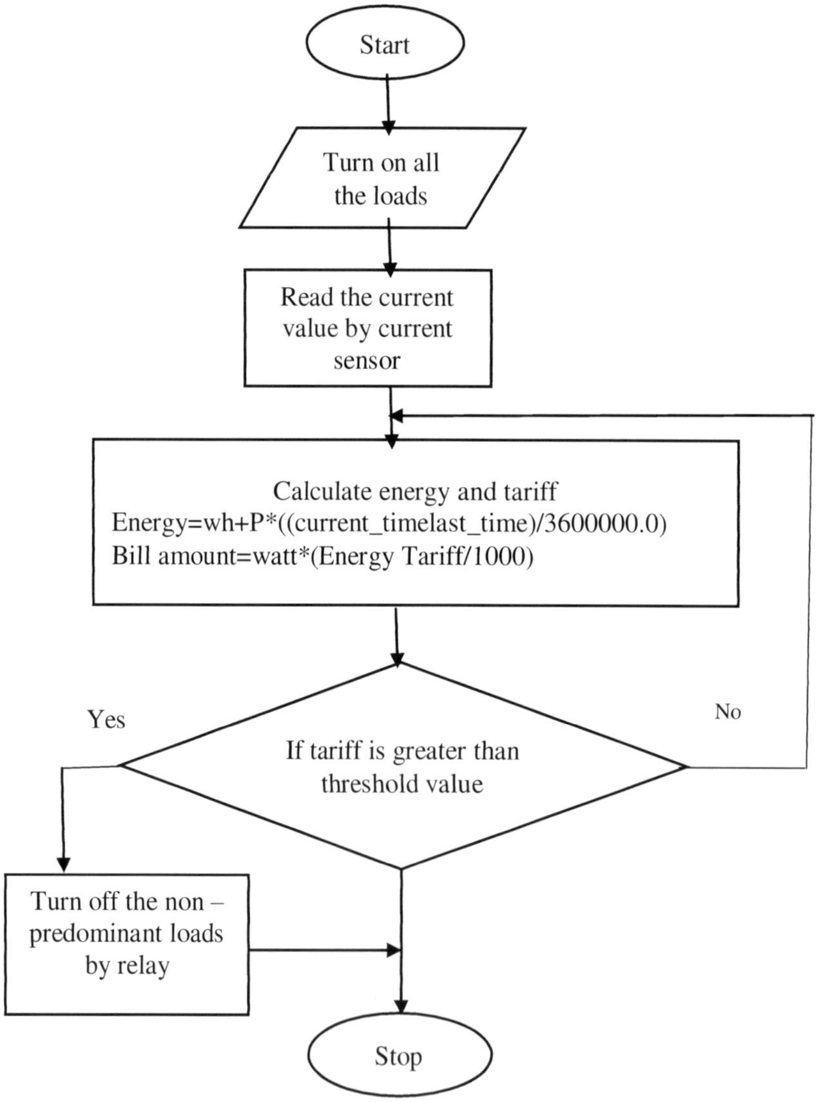

Figure 3. Flow chart for proposed system

visualized using flowchart as shown in figure 3. The flowchart describes the function of proposed prototype in which when the tariff is greater than threshold value, the non -predominant loads are turned off by the relay which is programmed by the Arduino.

2.1. Hardware setup

The hardware setup consists of Arduino Uno, relay, load, current sensor, bread board, ESP-12E module and a smart phone (Figure 4). Here the output can be viewed with the help smart phone. The loads are connected with the arduino, where ESP-12E module is used for transferring data to Adafruit Io platform. This chapter describes the components involved in this work.

The Arduino Uno is an open source Micro Controller board based on the microchip Atmega328P.The board consists of both analog and digital input & output pins that can be interfaced with other circuits. The Uno board is different from all other boards by featuring the ATmega328P microcontroller programmed as a USB to serial converter. Here, Arduino is used for serial communication with ESP12 by writing code for both transmitting and receiving data. The loads are connected to the arduinouno board as it has two analog to digital pins.

Current sensor is able to sense the flow of current in the appliances, which can detect AC or DC current. It generates a signal proportional to the current. The generated signal could be in any form and it is connected to Arduino to calculate tariff and bill using mathematical formulas (figure 3). Current sensor is the key component of the project that can able to sense up to 30A current. This module operates with Hall Effect principle.

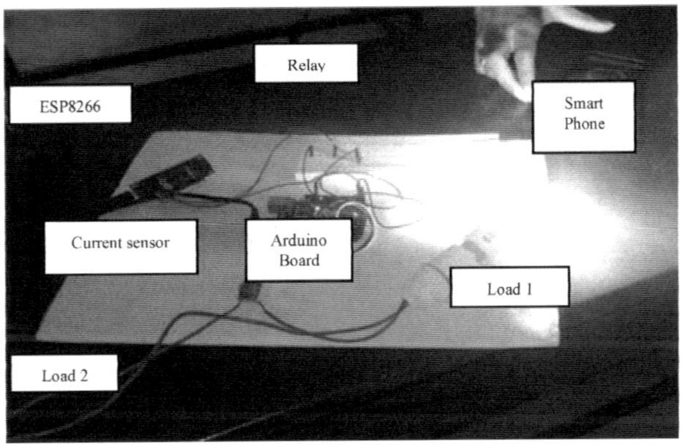

Figure 4. Experimental setup of smart energy manager

ESP-12E module is a Wi-Fi module that helps to connect the mobile phone with the sensor components. The module connected to the current sensor that collects the data from the sensor and it is transferred into mobile phone through Wi-Fi. The code written for ESP-12E module is also connected to Adafruit IO with AIO user name and AIO key that used to trigger when the electricity bill exceeds the reference value. In this project, the module is only used for transmitting and receiving data.

Features
32 bit LX106 RISC microprocessor
80 to 160 MHZ adjustable clock frequency
128 KB RAM and 4MB flash memory
802.11b/g/n HT40 Wi-Fi transceiver

2.2. Software modules

Adafruit IO platform is mainly used for storing electricity energy meter reading and communicates to CritaC IoT app. It can trigger when the bill amount exceeds the threshold value. In Adafruit IO

IOP Conf. Series: Materials Science and Engineering **906** (2020) 012004 doi:10.1088/1757-899X/906/1/012004

platform a new dashboard is created for energy meter. AIO key is generated that consists of username and password which can be used for CritaC IoT app. Then feed for power and bill are created.

CritaC IoT App is an android app that used to connect devices with mobile phone or tablet by which appliances can be easily controlled anywhere in the world. The appliances like light, fan, etc...Can be controlled with the help of this app. It helps the user by sending notification about the energy consumption, thereby it can take quick actions directly to control device by turning off.

IFTTT is both the website and mobile app which is mainly used to do more with all applications. IFTTT stands for If This Then That. IFTTT is used for triggering Gmail or SMS for energy meter when bill increased beyond the threshold value. Create the new applet to connect the Adafruit IO to the Gmail, which helps to trigger as Gmail or SMS.

3. Result and discussion

Smart energy manager works as an energy monitor and controller [6]. When all the appliances are working at a time it consumes lot of energy and it goes beyond the threshold value. In order to reduce the power consumption of the appliances the relay is used to control with the help of Arduino. The energy consumption and tariff can be viewed with the help of CritaC IoT App that monitors continuously with the help of current sensor. It continuously collects the data from the Wi-Fi module and displays in the smart phone. When the energy consumption is beyond the threshold value, the user is intimated with SMS or E-MAIL.

The figure 5, 6, 7 describes how to create new dashboard for energy monitor and control in Adafruit IO. After creating a dashboard it is ready to link with Gmail through IFTTT (Figure 8).In order to trigger it is necessary to monitor a feed on Arduino IO as shown in the figure 9. The complete action field (figure 10) will send a Gmail and review and finish work is shown in the figure 11.

The output in the experimental setup (figure 4) shows the current consumed by the loads and the maximum limit or threshold value fixed in the IOT server. It also display the bill amount for the power consumed totally, if it exceeds the threshold value the non-predominant loads are turned off with the help of relay automatically that programmed bythe Arduino .In the proposed prototype experiment kit only two incandescent lamps are connected as loads.When the lamps consume power more than the maximum limit then one lamp is automatically tripped off with the help of relay. The power consumption and tariff is continuously viewed by CritaC Iot app in the smart phone. Based on the data retrieved by sensor, the chart is drawn by android app as shown in the figure 12.

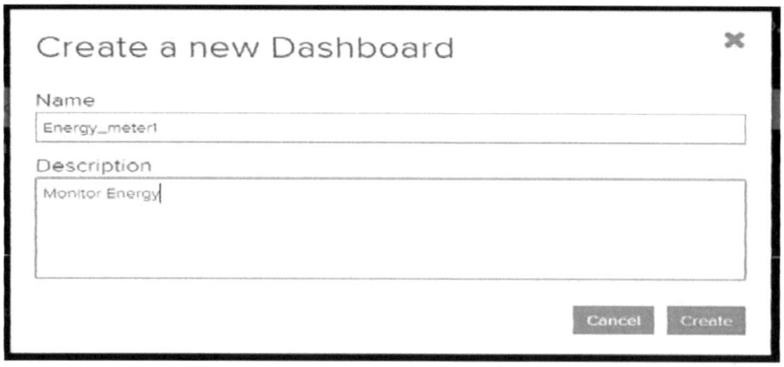

Figure 5- Creating new dashboard in Adafruit IO

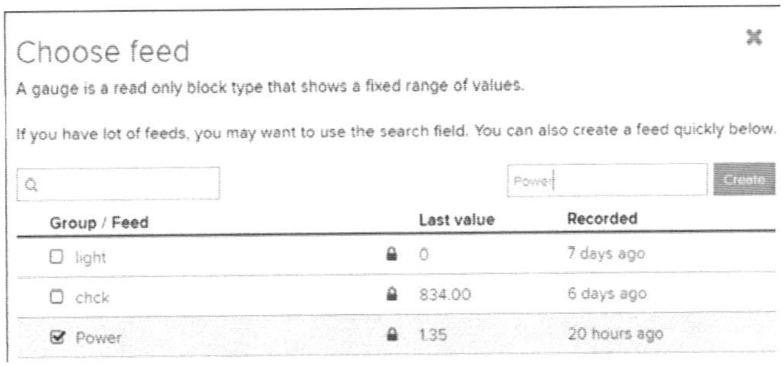

Figure 6- Choosing the feed

Figure 7. Setting minimum and maximum values

Figure 8. Linking Adafruit to SMS/E-mailusing IFTTT

IOP Conf. Series: Materials Science and Engineering **906** (2020) 012004 doi:10.1088/1757-899X/906/1/012004

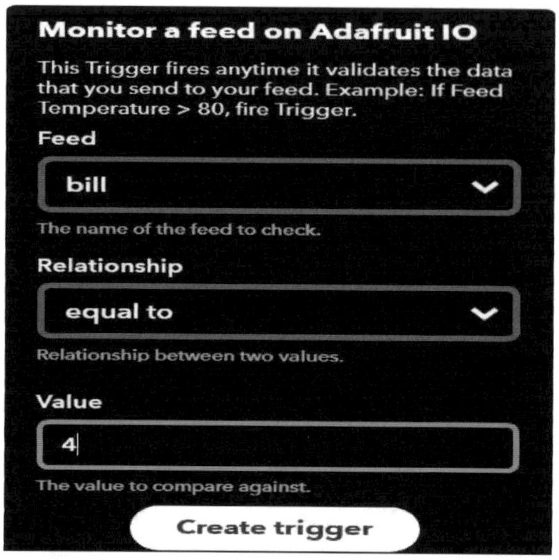

Figure 9. Monitor a Feed on Adafruit IO

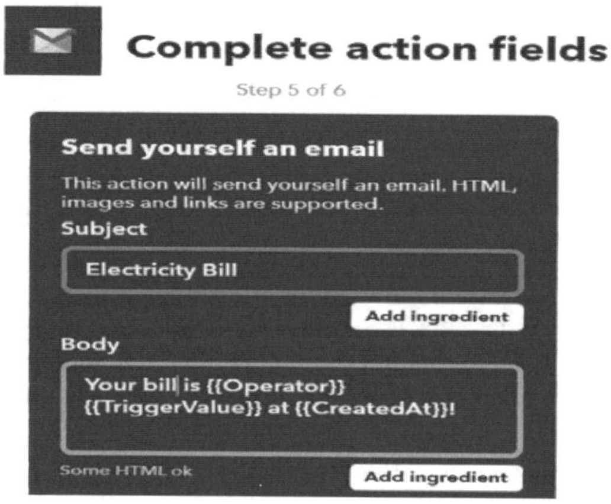

Figure 10. Complete action field-Gmail

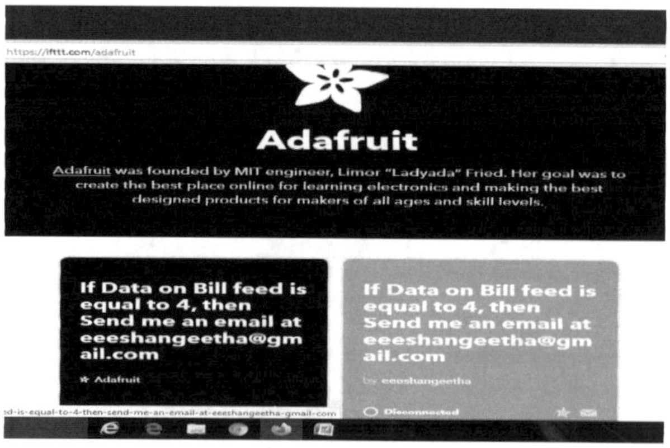

Figure 11. Review and finish

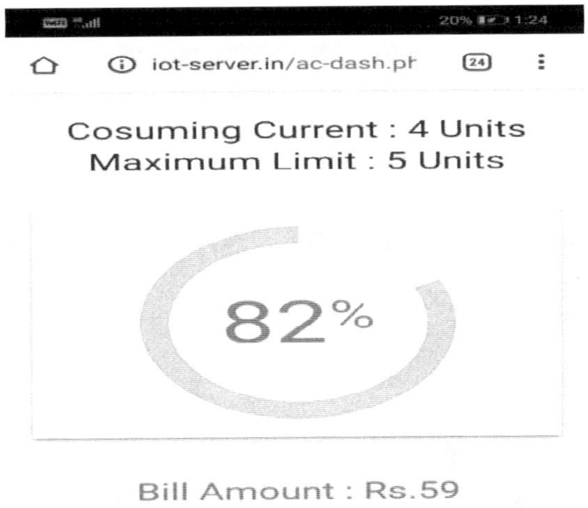

Figure 12. Current consumption and bill shown in mobile phone

4. Conclusion

The energy monitoring and control has become more vital in the present scenario and power saving is very much important for future. By turning on and off of the predominant and non–predominant loads at correct time with the help of this proposed project the energy can be saved.

This paper presented about energy monitoring and control with the help of smart phone that comprises with platform such as Adafruit IO, CritaC IoT and IFTTT. The experimental analysis of the proposed system clearly showed that it is applicable for domestic and industrial sectors and it is easy to install in both sectors.

References

[1] Ravi Kishore Kodali , Subbachary Yerroju .2018.Energy Efficient Home Automation Using IoT, International Conference on Communication, Computing and Internet of Things, February 15-17,Chennai,India.A reference.

[2] BalamuruganS,Saravanakamalam D.2017.Energy Monitoring and Management using Internet of Things, International Conference on Power and Embedded Drive Control, March 16-18,Chennai,India.

[3] Hariprabhu M, Sundararaju K.2019.IoT based Fault detection in Solar Panel using Arduino UNO wirth Wi-Fi Module ESP 8266 ,International Journal of Recent Technology and Engineering (IJRTE) vol 8 no 4 pp2630-2633.

[4] Firas A. Hadi, SamahShyaaOudah, Rafa A. Al-Baldawi.2020.An Economic Study of a Wind Energy Project Using Different Sources of Wind Data,Iraqi Journal of Science 2020, Vol. 61, No. 2, pp: 322-332.

[5] Sumi L, RangaV.2016.Sensor enabled internet of things for smart cities,Fourth International Conference on Parallel, Distributed and Grid Computing,December22-24,Waknaghat, India.

[6] Lazar Berbakov, Marko Batić , Nikola Tomašević.2019.Smart Energy Manager for Energy Efficient Buildings, International Conference on Smart Technologies, July 1-4, Novi sad , Serbia.

[7] SitiSyaidatulSyazlinaMohdSoleh , MohamadMdSom ,MohdHelmyAbdWahab, Aida Mustapha, Nurul Ain Othman, MohdZainuri Saringat.2018.Arduino-based Wireless Motion Detecting System, IEEE Conference on Open Systems, November 21-22,Langkawi Island, Malaysia.

[8] Barret,S.F.2013.Arduino Microcontroller Processing for Everyone.Third edition, 8(4),pp:26-28.

[9] Saikat Mukherjee, ArpitaGhosh ,Subir Kumar Sarkar.2019. Arduino based Wireless Heart-rate Monitoring system with Automatic SOS Message and/or Call facility using SIM900A GSM Module, International Conference on vision Towards Emerging Trends in Communication and

Networking, March 30-31, Vellore, India.
[10] Apurva Singh , Harsh Mehta , AnujNawal , Gnana O.V,Swathika.2018.Arduino Based Home Automation Control Powered by Photovoltaic Cells, SecondInternational conference on Computing Methodologies and Communication, February 15-16, Erode, India.
[11] Swati Dhingra, RajasekharaBabuMadda, Amir H. Gandomi, Senior Member, RizwanPatan, Mahmoud Daneshmand.2019.Internet of things Mobile: Air pollution Monitoring system.IEEE Internet of Things Journal,1-1, volume 6, issue 3,pp:5577-5584.
[12] AyadGhanyIsmaeel,Kitab ,Mohammed Qasim Kamal .2107.Arduino IoT Home Automation System for IR Devices,International Conference on Current Research in Computer Science and Information Technology, April 26-27,Slemani,Iraq.
[13] Dinesh kumar S , VengateshR,Subiramani K, ThangamaniP,RajeshwaranM.2018.Effectiveness and utilization of energy is compute by auditing system using IOT device in college demesne, International journal of pure and applied mathematics,Volume 118,issue 20, pp:2115-2125.

IOP Conf. Series: Materials Science and Engineering **906** (2020) 012005 doi:10.1088/1757-899X/906/1/012005

Energy Saving on Industrial Drive Technologies - Past, Present, and Future Perspective

M. Baranidharan[1], R. Rajesh Kanna[1], R. Raja Singh[2]

[1] School of Electrical Engineering, Vellore Institute of Technology, Vellore, India.
[2] Advanced Drives Laboratory, Department of Energy and Power Electronics, Vellore Institute of Technology, Vellore, India.
Correspondence: rrajasingh@gmail.com

Abstract. Power is a crucial factor in sustainable economic development and growth, the world is facing the obstacles to pass towards a cleaner and more accessible route for utilization, a recent review of new developments in the energy-efficient sectors, concentrating on selected innovations instead. Efficiency of energy for various application, enhancing the productivity, performance and precision. At the same time, using the energy efficiently on any industrial applications reduces the operation cost, therefore adopting suitable energy efficiently to the system in a conventional profit. Also, there is no limit for innovation of new energy saving technologies. These technology is defined by in terms of their energy efficiency, economy and performance in the environment. This paper provides an overview of energy savings on industrial drive technologies, sustainable economy and the environment, are going to be running out of technologies in the future.

1. Introduction

Energy is the primary source of power to our lives. Power has an influence on the economic and social development of the nation in modern society. Due to instant availability, controllability, flexibility and cleanliness, electric power plays a prominent role in [1-2]. Over the last few decades, electrical energy has led to human well-being, growth and technological advances. Adequate generation of energy and proper use of electrical energy improves industrial investment in a country that results in gross domestic product (GDP) production [3-6]. The motor drives of modern global industry are the backbone that is well covered, but also play a major role in industries which often improve the world with our environment more sustainable. Electrical energy produced at any time must be used by the consumer within a small period of time. In figure 1 the sector wise energy consumption is mentioned, the Electric power transmits through grid, and current distribution is centered on line impedance [7-10].

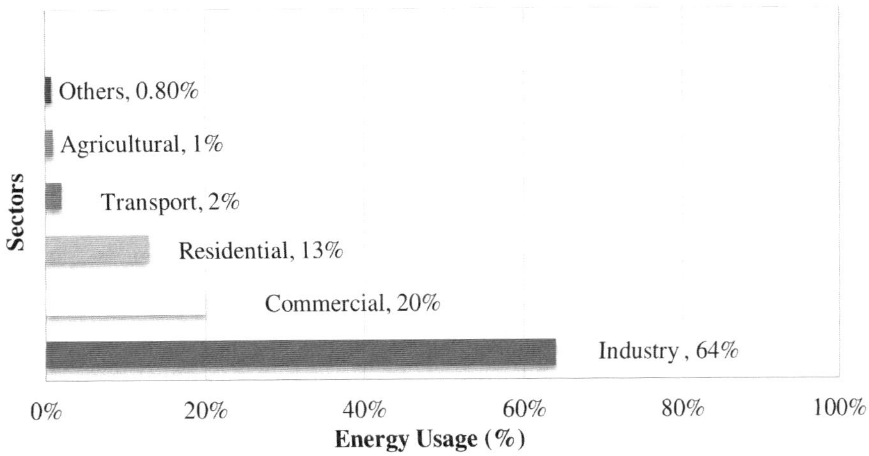

Figure 1. Sector wise energy consumption

IOP Conf. Series: Materials Science and Engineering **906** (2020) 012005 doi:10.1088/1757-899X/906/1/012005

In general, drives are commonly used for various application especially, fan, pump, compressor [11-12]. Similarly, the drives are adapted to perform any one of the function, speed control, process control in AC drives, high velocity drives, continuous frequency drives, VSD, frequency converters, power converters and inverters, By varying the frequencies and voltage of the power supply to an electric motor [13], drives can control the speed, enabling process control to be enhanced, reducing energy utilization and producing energy reliably or controlling the function of industrial applications based on electric motors [14-15].

1.1. Electric Drives for Industrial Application

Electrical drives are widely used in manufacturing industries and process industry to regulate the speed and to enhance the efficiency of the system. Such specifications are used to assess the demands for horsepower and torque. Owing to low starting torque requirements reduced voltage starters can be used [20].

Table 1. Industrial Drives Application

Industry	Application of Drives	Motor Type	Year & Citation
Petrochemical industry	Energy savings and process control optimization.	Three phase induction motor	2002 [5]
Pharmaceutical industry	Minor alterations in the Design stage and improve the energy efficiency	Brushless DC, AC squirrel cage, wound rotor, stepper and servo motors.	2003 [13]
Textile industry	Controllers in energy saving.	Induction motor	2007 [1][17]
Steel industry	Reduced maintenance costs and saved a considerable amount of electricity.	AC Synchronous motor	2008[2][18][19]
Petrochemical industry	Reduced Total Cost of Ownership and improve energy efficiency.	Three phase induction motor	2008 [4][21]
Cement plants	Compare the operational characteristics of the induction, synchronous and DC motors.	Squirrel cage induction motor	2009 [9]
Mineral products industry	Converting DC systems to AC systems	AC synchronous motor	2014 [8]
Garments industry	Reduce emissions for different energy-saving schemes.	Brushless DC/ induction motors	2016 [14][26]
Iron and steel industry	Improvement of performance and more energy-efficient	AC Synchronous motor	2017 [3]
Oil, petrochemical, pulp industry	Fans are using air exchange, drying / cleaning purposes.	Induction motors	2018 [6][22]
Paper industry	Reduce the energy input and improving energy efficiency	Induction motors	2019 [7][23]
Mining industry	Improving its sustainable and efficient industrial operating.	Induction motor	2019 [12][25]

1.2. Industrial Revolution

Industrial Revolution, in modern history, the trend of moving with an agriculture and economy driven by manufacturing industry and machinery. That process started in the 18th century in Britain and expanded to other regions of the world from everywhere.

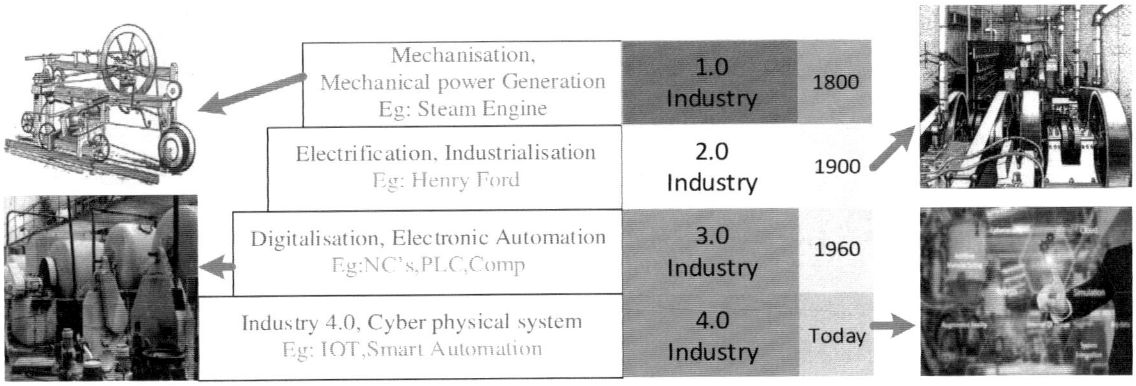

Figure 2. Stages of industrial revolutions

The main features of a four industrial revolutions are the first industrial revolution are transformed our economic and lives from the agricultural field and manufacturing economics to a production and machinery-dominated economy. The initial manufacturing revolution introduced in 1760 when the steam machine was invented. The second manufacturing revolution began in 1900 when the inner burning engine was invented. In 1960 the third industrial revolution began to be characterized by the use of microchip technology and IT for automating manufacture. The Fourth Industrial Revolution describes a future in the people turn from virtual worlds to offline reality using connected technologies to make life possible and controlled and also it includes computer-generated product design and 3D printing which can build up successive layers of materials to produce solid objects are shown in figure 2 [16].

Over the past few decades, public awareness of environmental concerns has led to a strong focus on topics such as resource-efficient, and energy-efficient development in both research and industrial applications. Numerous studies were performed, evaluating the energy usage of manufacturing environments from different perspectives. Among other things, the results showed substantial energy consumption during periods of non-value adding, i.e. during times when production equipment is stopped or idled, for almost every considerable way of classifying machinery and in almost all major manufacturing industries.

A method was designed to provide a standardized collection of possible energy-saving states for the production equipment, clustering these states according to the commonly applicable equipment behaviour. Improving energy efficiency and increasing demand for energy are commonly seen as the most promising, fastest, cheapest and safest means to mitigate climate change. Most incentives tend to be cost-effective at current levels of energy, and can provide additional benefits such as increased fuel poverty reduction and increased economic productivity. As a result, the International Energy Agency (IEA) and other organizations are placing their priority on reducing energy demand, the European Commission has set long-term targets to reduce energy demand, and countries around the world are implementing a range of policies to achieve those reductions.

1.3 Energy Conservation on Industrial Drives

Energy saving means by reducing energy demand by means of a smaller amount of an energy used. Although energy saving reductions vitality properties, it can contribute to improved quality of the climate, national security, individual financial protection and advanced savings. Energy saving is a significant element of vitality policy on a broader scale [24]. Vitality competence is also greatest economical approach to power losses. Energy saving is certainly of great position to us all, because its depend on energy for all the day [1]. Energy is generated – this is equal to 2 units of energy created its save one unit of energy. Saving energy to reduce pollution – generating and utilizing energy accounts for a large amount of air pollution and extra 83% of greenhouse gas emission productions [26-28]. If the highest of the ladder for renewable energy. It also reducing the energy costs and potential depletion of resources. Resource can be conserved by reducing waste and losses, increasing efficiency by technological changes, and improving the operation.

2. Energy Efficient Strategies on Electric Motor Drives

The manufacturing sector consumes more energy for all other end-use areas and this industry actually absorbs nearly 20% of the universe total energy supply. Energy is used in the manufacturing area by a diverse business population, including manufacturing, forestry, mining and construction, and for a wide range of activities, including production and assembly, room conditioning, and lighting [30]. The various energy efficient strategies are discussed in figure 3.

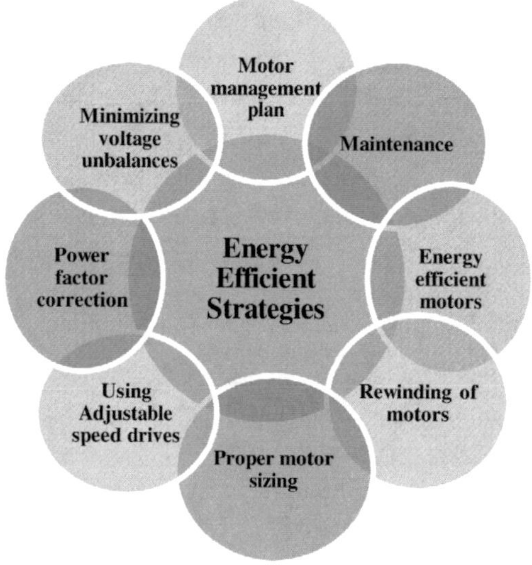

Figure 3. Energy efficient strategies

2.1. Management Plan of the Motor

An engine controlling system is the important share of a company's energy managing plan. The implementation of an engine management plan would help companies achieve long-term energy savings in the engine system, which will ensure efficient and cost-effective control of motor failures.

2.2. Maintenance

Motor preservation objectives remain to extend motor lifetime and anticipate a malfunction in the engine. Therefore, motor maintenance steps may be considered as either predictive or preventive. Preventive steps contain minimization of power inequity, awareness of weight, alignment of the motor, lubrication and ventilation of the engine.

2.3 Energy-Efficient Motors

Efficient power engines minimize the energy consumption by improved construction, improved products, tighter tolerances and advanced production techniques. Energy-efficient motors can also stay warmer with proper installation, can also decreasing heat loads on the facility and have extreme duty numbers, longer containing life, longer life span of insulation and much less noise.

2.4. Rewinding of Motors
For certain situations, rewinding a current, energy-efficient engine can be cost-effective, rather than purchasing a new engine. When repairing or rewinding an engine, it is critical to select an engine facility center that meets best preparation motor reversing standards to reduce possible losses to performance. The rewinding is applied efficiency losses are typically reduced less than 1%.

2.5. Proper Size of the Motor
Its common misconception that huge motors, particularly engines working less than 50 percent of the valued load, are not effective and must be directly changed by energy-efficient components of sufficient size. In fact, it takes many pieces of information to complete an accurate evaluation of energy saving.

2.6. Adjustable Speed Drives
Its ideally suited to load supplies for engine processes, thereby ensuring the engine energy usage for customized by particular uses. Because is motors' vitality consumption is about comparative flow rate cube, relatively small movement decreases, proportional to the pump speed, still result in significant energy savings.

2.7. Power Factor Correction
Energy metric is really the proportion of power delivered and work-time. That tests how power generation is being used properly. A high-power factor signifies a successful use of electric energy while a low-power factor implies inefficient use of electrical energy.

2.8. Minimizing Unbalances in Voltage
An imbalance in voltage reduces the presentation of three-phase engines and shortens their life. A voltage imbalance produces a current imbalance, resulting in rotation pulses, improved pulsation and mechanical strain, augmented losses, and motorized warmness, which container minimize the lifetime of the engine's winding protection.

3. Trends in Energy Saving
Energy efficiency has massive potential to boost financial growth and reduce greenhouse gas emissions, but a trend which has major impacts on people, companies and the environment is slowing the global rate of progress. The energy sector continues to turn its attention to energy efficiency with rising regulations. Industrial energy use remained roughly constant at EU level and since then has fallen rapidly twice as fast as industrial activity in figure 4[3].

Figure 4. Global renewable energy trends (source: https:www.deloitte.com)

3.1. Past Trends in Energy Saving
An analysis of these remarkable achievements helps explain both the factors its inspiring the combined motor drives and contemporary expansions in power electronics expertise that complete those induction motor drives (IMDs) possible. Such advancements are listed in a roughly sequential instruction to type it informal to classify the evolution of the expertise that has complete building more advanced IMDs technologically feasible [11].

3.1.1. Automotive Alternators
Moreover, the industrial inverter chosen for its first instance may not classify as an incorporated drive system but are included in this segment as it provides an essential reference to the following IMDs. Additional precisely, in an unregulated full-wave bridge configuration, the power electronics consists of 6 (or more) diodes short of any measured control buttons required to create variable-frequency ac power waveforms. The close connection with automotive generators and IMDs relates to the fact that the diode devices are mounted inside the alternator framework in areas close to the Lundell claw-pole synchronous alternator assembly, introducing both to a higher heat and vibration factors to follow this charging design.

3.1.2. Blower Fan Drive Electronically Commuted Motor (ECM)
Some of the initial main IMD devices designed for a residential usage in applications for boiler, ventilation and air preparing (HVAC), the Electronically-Commutated Engine, primary published as a creation in 1987. The ECM was intended aimed at usage in housing and bright profitable HVAC devices such as heater blowers anywhere they offered a way to improve the overall efficiency and comfort of the furnace system by in case adjustable-speed air flow to the active places rather than traditional on-off thermostat created regulator through a fixed-speed line.

3.1.3. Submersible Water Pump
Installing the motor and drive in the underground submersible pump unit enabled the control cabling essential in the well shaft to be reduced although using the impelled water as an abundant then efficient refrigerant for the motor and electronic drive.

3.1.4. Industrial-Grade Integrated Motor Drives
Modern motor drive manufacturers have produced combined motor drive product lines that are designed for overall industrial uses such as pumps and fans / blowers that result from adjustable speed operation in performance improvements and energy cost savings [31].

3.2. Present Trends in Energy Saving
By integrating solar energy systems for hot water and producing electricity, we achieve a further reduction in energy consumption. Recently there has been an increasing trend in the integration of solar photovoltaics in buildings.

3.2.1. Design of Energy Efficient Motors
Efficiency has a direct impact on energy consumption and the conservation of energy. Motor charging, relative to full load, affects the motor's efficiency and overall performance. The efficiency can be increased by reducing losses which can be a combined effect of decreasing the number of turns, increasing the size of the conductor and increasing the length of the core causing an increase in the efficiency [32].

3.2.2. The Generation IMMD with GaN-Based Power Electronics
Its benefit of GaN power devices benefit the third-generation IMMD prototype systems to reduce inverter form and capacity, with advanced changing frequencies then lesser conductive damages. The

IOP Conf. Series: Materials Science and Engineering **906** (2020) 012005 doi:10.1088/1757-899X/906/1/012005

prototype IMMD unit also uses a 6-phase ac system like the aforementioned 2nd generation IMMD, except it is an induction system in this case. Another distinction between the two models is that the 3rd generation con- figures the sequence of 6 pole-drive elements into binary full-bridge 3-phase inverters, rather than parallel ones. It allows GaN devices to be used with lower voltage levels, raising the system's on-state resistance and conduction losses. Since this energy electronics package is placed on the outer external of the finish bell of the motor designed for easy watching short of disassembly, the energy electronics has remained suitably reduced towards be compatible by the inserting in the finish bell cavity of the motor short of at all alteration in the measurements of the housing an evaluation of the 2nd and 3rd group IMMD prototype elements therefore delivers substantial indication of the optimistic effect of elegant WBG established energy electronics on the forecasts of attaining fixed energy electronics in upcoming IMDs.

3.2.3. Current Source Inverters enabled by WBG

A stated in Unit V.A, new WBG control semiconductors made from silicon carbide(SiC) and gallium nitride (GaN) are attractive accessible which can turn additional 10 times earlier than their silicone equivalents. The research communal is highly optimistic that these new WBG control switches that bring significant benefits to the uses such as IMDs, which urgently need breakthroughs leading to smaller , lighter, and more powerful electronics.

3.3. Future Trends in Energy Saving

Evaluates or findings that suggest an operator change and to analyse driver patterns to have a sense of where the industry is going, EPRI will track developments on a routine basis, finding significant changes that may suggest a major effect on scenarios, drivers, global perspectives, or the Interconnected Energy Network.

3.3.1 Ultra-efficient Heat Pumps

Stepping up energy efficiency will be needed as global demand for energy continues to rise. The technology chances are endless and possible real investments, but so far customers and values have remained measured to capitalize in the maximum economical, energy-efficient machineries. In-use houses, electrical equipment, and devices' energy efficiency is well short of what is theoretically achievable.

3.3.2 Variable Speed Drives

The AC induction motors translate the electrical energy into mechanical and other types of usage. Therefore, about two-thirds of the electrical energy is fed to the world's motors for this reason. And the

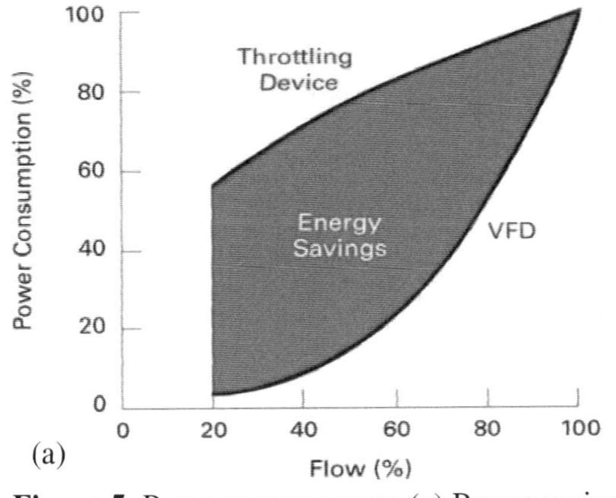

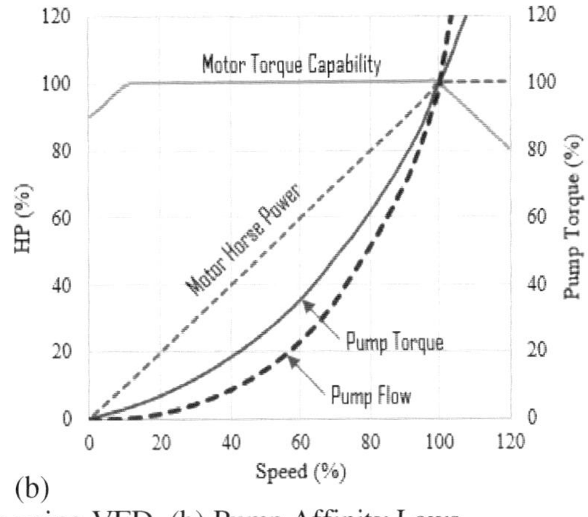

Figure 5. Pump energy curves (a) Power savings using VFD, (b) Pump Affinity Laws

big restriction is energy efficiency for motors. And this energy saving is accomplished largely by the use of variable frequency power. In general, three types of major load applications are classified as variable torque type load, constant torque type load and constant Horsepower type load [29]. VSD control, using inverter technology, it allows the regulation of starting inrush currents, the management of specific fixed speeds, quick change of speed and the reversal of AC motors. Such types of electric motors are responsible for a significant part of energy usage in manufacturing, operating in combination with fluid-moving machinery such as pumps, fans, compressors etc [33]. Most variable torque drives have Proportional Integral Differential (PID) capability for fan and pump an application, which allows the drive to hold the set point based on actual feedback from the process, rather than relying on estimation. High levels of accuracy for other applications can also be achieved through drives that offer closed-loop operation. Closed-loop operation can be accomplished with either a field-oriented vector drive, or a sensor less vector drive. The field-oriented vector drive obtains process feedback from an encoder, which measures and transmits to the drive the speed and/or rate of the process, such as a conveyor, machine tool, or extruder. The drive then adjusts itself accordingly to sustain the programmed speed, rate, torque, and/or position Variable speed drives, on the other hand, gradually ramp the motor up to operating speed to lessen mechanical and electrical stress, reducing maintenance and repair costs, and extending the life of the motor and the driven equipment[6].In variable torque applications, the torque required varies with the square of the speed, and the horsepower required varies with the cube of the speed, resulting in a large reduction of horsepower for even a small reduction in speed[7] [8].The motor will consume only 12.5% as much energy at 50% speed than at 100% speed as shown in Figure 5 (a) and (b).

3.3.3 Energy Efficient Motors

IE3 & IE4 Motors have high efficiency at any ambient temperature. Hence these are costly than IE2 Motors. The materials cost of the motor is increased by a few percent. While trying to reduce copper losses, we end up increasing core loss. Hence the starting current of motor is high, it may increase fault levels and in turn cable size. The disadvantages are overcome as the payback period for the customer can be as little as six months for a continuously loaded motor. Through improving architecture, improving materials and improving production techniques, the energy-efficient electrical motors decrease energy losses. Replacing an engine may be defensible based on the fuel cost savings resulting from an energy-efficient model. It is correct if the engine is constantly running, the power levels are great, the engine is overweight for the mission, or damage or past rewinds have reduced its nominal performance.

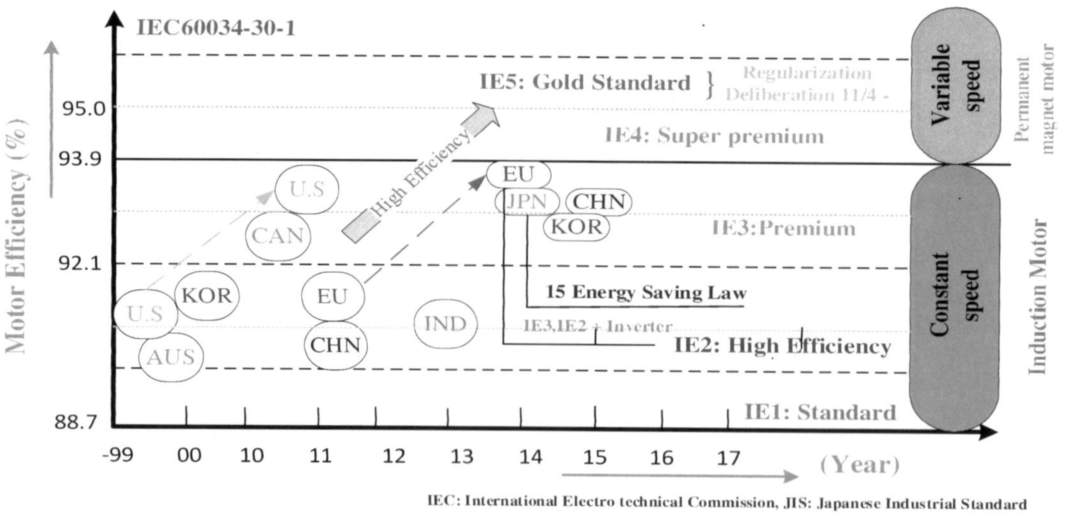

Figure 6. Movement of regulation of motor efficiency

It also has good performance as under loaded as compared with the IM, and has higher torque density and overload capacities the movement of regulation of motor efficiency are shown in figure 6. When less heat is produced inside, it is easier to refrigerate. Syncrel has another important benefit when it is required to substitute the normal motors with energy efficient ones. Since the high efficiency IMs at the same parameters are larger than the regular ones, the high efficiency SyncRels have the same height of the centerline shaft as those to be replaced. The upgrade is therefore simple and quick, without the need for mechanical modifications [35]. SynRel and PMSM engines it can be provided as a possible additional to the induction engines in applications where speed variation is required, realizing in consistency and outstanding in performance. Such machines run at synchronous speed, minimizing rotor losses and are thus able to achieve very high performance levels, often reaching IE5 standards. The better assets of these machineries be able to similarly remain used to greatly decrease the complete size and mass of the engine, meaning that the design of the engine needs less materials.

3.3.4 Soft Starter with Energy Savers

The Soft Starter offers a consistent and cost-effective solution to these difficulties by supplying the motor with a measured power issue, resulting in phase smooth, less acceleration and deceleration. As damage to the windings and rollers decreases, motor life will increase. In 3 step units, Soft Start & Soft Stop offers guided start and stop with a range of ramp times and current limit settings to match all applications [34].

3.3.5 Bidirectional Converter for Energy Recovery

Most applications driven by electric drive, such as lifts, tooling machines, packaging machines, etc., are branded by frequent stops / starts with erratic duty rounds; thus, significant energy-saving capacity exists through the recovery of extreme kinetic energy during braking intermissions. Though, modern variable frequency drives do not support regenerative braking, since they are normally provided by bridge diode rectifiers. The renewed kinetic energy is thus degenerate in case of assistance of decelerating devices. It can result in a considerable amount of vitality wastages, particularly once there are frequent stops / starts.

3.3.6 Internet of Things (IOT)

Energy management system for commercial and industrial spaces through the Internet of Things. Using the data that comes from a network of on-site sensors and meters, the system offers saving steps to minimize energy usage and increase productivity. IoT can help to improve the various processes to be more quantifiable and measurable by the large amount of data collected and processed. IoT could potentially improve quality of life in various spheres including medical services, smart cities, construction, Agriculture, water, and the energy sector [36].

3.3.7 Digital Twin Technology

Digital twins are used in the energy sector to optimize the functioning and maintenance of physical assets, systems, and manufacturing processes. The digital twin is used through the visualization of change over time to track construction and identify problems. This capability helps the team to immediately respond to changes on-site.

Table 2. Comparison of past, present and future energy saving techniques of industrial drive

Components	Past	Present	Future
Motor	Normal motor	Energy efficiency motor	Light weight motor
Converter	Power electronics switches	Advanced power electronics switches	Silicon carbide switch
Drives	Industrial-Grade IMD	Adjustable speed drive	VSD with digital twin technology
Load	Normal load	Pump and fan load	IOT based connected load

4. Conclusion

Most developed countries have adopted legislation close to those of the European Union on energy efficiency. Motors representing 40–50 per cent of total global energy usage are becoming the largest appliance requiring high-efficiency standards. Currently the most common program in the motor industry is the research and technological production of DC brushless motors. Certain R&D products include refining motor design, enhancing silicon steel sheets electromagnetically, encouraging the advantages of permanent magnetic fields, adjusting wire-winding technology, and developing the electronic control system. The emerging energy saving and carbon cuts issues as well as the high-energy price age approach have led to the rapid production and popularization of high-efficiency engines. There is a trend towards increasing efficiency inside the electrical machine itself, but the greatest system-level improvements occur when the machine is coupled with an electronic power converter to produce a variable speed drive. The key drawbacks to this lie in the initial expense of a variable-speed motor, but the payback period is very limited in many situations. Future technological developments will reduce the cost of capital of the drive and thus existing markets will expand and new markets will open up[10].The motor is the sector's most energy-using part. Promoting high-efficiency motors is conducive not only to improving industrial productivity but also to reducing greenhouse gases. Several countries have thus been discussing the enhancement of motor-efficiency requirements in their energy and industrial policies in recent years.

References

[1] B. C. Mecrow and A. G. Jack, "Efficiency trends in electric machines and drives," Energy Policy, vol. 36, no. 12, pp. 4336–4341, 2008.
[2] H. Gavrila, V. Manescu, G. Paltanea, G. Scutaru, and I. Peter, "New Trends in Energy Efficient Electrical Machines," Procedia Eng., vol. 181, pp. 568–574, 2017,
[3] T. M. Jahns and H. Dai, "The Past, Present, and Future of Power Electronics Integration Technology in Motor Drives," CPSS Trans. Power Electron. Appl., vol. 2, no. 3, pp. 197–216, 2017.
[4] R. D. Cepoi, F. F. Jaşcău, and L. Szabó, "Current trends in energy efficient electrical machines," J. Electr. Electron. Eng., vol. 10, no. 2, pp. 13–18, 2017.
[5] A. De Almeida, J. Fong, C. U. Brunner, R. Werle, and M. Van Werkhoven, "New technology trends and policy needs in energy efficient motor systems - A major opportunity for energy and carbon savings," Renew. Sustain. Energy Rev., vol. 115, no. September, p. 109384, 2019.
[6] Bhukya, P., & Basak, D. (2014). Energy Saving Technologies in Industries-An overview. International Journal of Scientific and Research Publications, 4(1), 2250–3153. www.ijsrp.org
[7] D. Maheswaran, V. Rangaraj, K. K. J. Kailas, and W. A. Kumar, "Energy efficiency in electrical systems," in 2012 IEEE International Conference on Power Electronics, Drives and Energy Systems (PEDES), Bengaluru, Karnataka, India, 2012, pp. 1–6.
[8] Matsuse, K., & Matsuhashi, D. (2019). New technical trends on adjustable speed AC motor drives. Chinese Journal of Electrical Engineering, 3(1), 1–9.
[9] I. Karatzaferis, E. C. Tatakis, and N. Papanikolaou, "Investigation of Energy Savings on Industrial Motor Drives Using Bidirectional Converters," IEEE Access, vol. 5, pp. 17952–17961, 2017,
[10] S.-M. Lu, "A review of high-efficiency motors: Specification, policy, and technology," Renew. Sustain. Energy Rev., vol. 59, pp. 1–12, Jun. 2016 .
[11] Peter Love*, "The Past, Present and Future of Energy Conservation in Ontario" Energy Regulation Quarterly article., June 2015 – Volume 3, issue 2 2015
[12] Nadel, S., Rainer, L., Shepard, M., Suozzo, M., and Thorne, J., 2019. "Emerging Energy-Saving Technologies and Practices for the Buildings Sector," Washington, DC: ACEEE.
[13] Mine K. Yücel ,The energy industry: past, present and future Article · February 2003 Source: Biotechnology Conference Volume.
[14] Anglani, N., Consoli, A., & Petrecca, G. (2008). Energy Efficiency technologies for industry and tertiary sectors: the european experience and perspective for the future. IEEE Energy 2030 Conference, ENERGY 2008.
[15] Malico, I., Nepomuceno Pereira, R., Gonçalves, A. C., & Sousa, A. M. O. Current status and future perspectives for energy production from solid biomass in the European industry. In Renewable and

Sustainable Energy Reviews in 2019.
[16] Vuksanović, D., Ugarak, J., & Korčok, D. (2016). Industry 4.0: the Future Concepts and New Visions of Factory of the Future Development. October, 293–298. sinteza-2016.
[17] C. T. Raj, P. Agarwal, and S. P. Srivastava, "Particle swarm optimized induction motor for a textile mill load diagram," IET Semin. Dig., vol. 2007, no. 2, pp. 379–383, 2007.
[18] Paul, A. K., Banerjee, I., Santra, B. K., Neogi, N., & Vvvf, A. (2008). Application of AC motors and drives in Steel Industries. December, 159–163.
[19] K. He and L. Wang, "A review of energy use and energy-efficient technologies for the iron and steel industry," Renew. Sustain. Energy Rev., vol. 70, no. June 2015, pp. 1022–1039, 2017.
[20] M. J. Melfi, S. D. Rogers, S. Evon, and B. Martin, "Permanent-magnet motors for energy savings in industrial applications," IEEE Trans. Ind. Appl., vol. 44, no. 5, pp. 1360–1366, 2008.
[21] B. Lockley, "Adjustable speed drives in the petrochemical industry," Conf. Rec. - IAS Annu. Meet. (IEEE Ind. Appl. Soc., vol. 2, pp. 1240–1243, 2002.
[22] O. Eseosa and A. Christian, "Energy Efficiency Optimization of Three Phase Induction Motor Drives for Industrial Applications," Int. J. Eng. Appl. Sci., vol. 5, no. 8, 2018.
[23] S. Haider, M. S. Danish, and R. Sharma, "Assessing energy efficiency of Indian paper industry and influencing factors: A slack-based firm-level analysis," Energy Econ., vol. 81, pp. 454–464, 2019.
[24] D. Ma, R. Fei, and Y. Yu, "How government regulation impacts on energy and CO2 emissions performance in China's miningindustry," Resour. Policy, vol. 62, no. September 2018, pp. 651–663, 2019.
[25] M. A. Habib, M. Hasanuzzaman, M. Hosenuzzaman, A. Salman, and M. R. Mehadi, "Energy consumption, energy saving and emission reduction of a garment industrial building in Bangladesh," Energy, vol. 112, pp. 91–100, 2016.
[26] M. M. Elkholy, "Optimal energy saving for variable speed single phase induction motor drives," 2017 19th Int. Middle-East Power Syst. Conf. MEPCON 2017 - Proc., vol. 2018-Febru, no. December, pp. 754–763, 2018.
[27] O. Muravleva and O. Muravlev, "Induction motor improvement for energy saving technologies," 7th Korea-Russia Int. Symp. Sci. Technol. Proc. KORUS 2003. (IEEE Cat. No.03EX737), vol. 2, pp. 17–19, 2003.
[28] O. Muravleva and O. Muravlev, "Induction variable speed drive as the basis of efficient energy saving," 8th Korea-Russia Int. Symp. Sci. Technol. - Proc. KORUS 2004, vol. 1, pp. 264–267, 2004.
[29] B. Sergelen et al., "Energy saving by using variable speed drives with AC induction motors in industries," Proc. IFOST-2008 - 3rd Int. Forum Strateg. Technol., pp. 564–567, 2008.
[30] N. Heryana, F. Sasongko, and P. A. Dahono, "Study and implementation of energy conservation in a small cottage," Proc. 2009 Int. Conf. Electr. Eng. Informatics, ICEEI 2009, vol. 2, no. August, pp. 715–719, 2009.
[31] F. J. T. E. Ferreira and A. T. De Almeida, "Overview on energy saving opportunities in electric motor driven systems - Part 2: Regeneration and output power reduction," Conf. Rec. - Ind. Commer. Power Syst. Tech. Conf., vol. 2016-June, pp. 1–8, 2016.
[32] O. Muravleva and O. Muravlev, "Power effective induction motors for energy saving," Proc. - 9th Russ. Int. Symp. Sci. Technol. KORUS-2005, vol. 1, pp. 358–362, 2005.
[33] S. Halász and I. Kádár, "Power saving of PWM rectifier-VSI fed induction machines," 7th Internatonal Conf. Power Electron. ICPE'07, pp. 1098–1102, 2008.
[34] C. D. Pitis and M. W. Zeller, "Power savings obtained from supply voltage variation on squirrel cage induction motors," 2008 IEEE Electr. Power Energy Conf. - Energy Innov., pp. 1–3, 2008.
[35] N. H. Motlagh, M. Mohammadrezaei, J. Hunt, and B. Zakeri, "Internet of things (IoT) and the energy sector," Energies, vol. 13, no. 2, pp. 1–27, 2020.

IOP Conf. Series: Materials Science and Engineering **906** (2020) 012006 doi:10.1088/1757-899X/906/1/012006

Review of Power Factor Correction (PFC) AC/DC-DC Power Electronic Converters for Electric Vehicle Applications

***M.Bharathidasan* [1*] *and V.Indragandhi* [2]**

[1] Research scholar, School of Electrical Engineering, VIT, Vellore, India.

[2] Associate Professor, School of Electrical Engineering, VIT, Vellore, India.

[*]Corresponding author E-mail: bharathidasan.mohan@vit.ac.in

Abstract: This article fully discusses the general review on power factor correction of single-phase AC/DC-DC power electronic converters for electric vehicle applications. To magnify power quality in the declaration of Power Factor Correction (PFC) make the proposal of assorted for demand of solid-state DC-DC converters. These converters are look forward to minimize the Total Harmonic Distortion (THD) at the both input and output side and then to synchronize the DC voltage in DC-DC converters representation of both Uni/Bi- Directional power flow are analyzed. A complete review of Improved Power Quality Converter (IPQC) configuration, design feature, and specific value of selection are presented in this paper. The battery charger necessity to have a converter flow of power ability, a solution needs for an on/off-board charger of an Electric Vehicles (EVs).Due to high power density and efficiency, a DC-DC converters is preferred for medium power and high voltage applications, the most attractive feature of DC-DC converters that makes it suitable for handling medium power supplies is its soft-switching capability. This paper is aimed to research, design and to improve the working applications on clean energy and smart grid field.

Keywords: DC-DC converter, Uni/Bidirectional converter, Power Factor Correction (PFC), Total Harmonic Distortion (THD), Electric Vehicles (EVs)

1. Introduction

Over the last years, research on power factor correction AC/DC power electronic converters techniques have been thoroughly rising. These rectified power electronics converters are generally utilized in various applications like telecommunication equipment, battery charger, medical, industrial machinery, military, etc. The charging is mostly encouraged by unidirectional charging of the battery that keeps up a single flow of power mode grid to vehicle. Conventional electric vehicles are analyzed only as a load, still now they consume grid power to EVs for recharge the battery. The power factor correction division can be successfully achieved by applying a boost inductor and diode with a transformer and power electronic switch specifically. When the grid line is provided, the PFC step employ as a half-bridge converter which operates [2] likes the conventional boost PFC converter.

Two PFC switches share the inductor input during both the half period or cycle, which results in minimized switching and conduction losses and then minimize the charger size and cost [6]. The grid collects the power from both renewable and non-renewable energy sources. The technique state helps to connect the vehicle to the grid network which intern helps the grid to connect with the battery of EVs and permit bidirectional power flow between them. During flow of power in each direction, a soft switching is achieved and it is keep all over operation range, without any additional component [7].Consistently, a bidirectional link having the ability to match with variable level of voltage is needed between the source and load (EVs), the battery will be charging while producing low harmonic current and to consume the power of battery reverse to the grid under act. Here, converter of power electronics employ as voltage of grid to rechargeable dc voltage and inversely. The representation of Bi/Uni-directional converter has shown in figs. 1& 2 accordingly. However, using unidirectional converters will result in increased by system weight, volume and cost. With the modern development in solid-state topology, bidirectional converters only favour for electric vehicle application.

Nomenclature

AC	Alternating Current	HRPWM	Hybrid Resonant Pulse width modulation
ACM	Average Current Mode	IPQC	Improved Power Quality Converter
ANN	Artificial Neural Network	HUT	Hold-Up Time
A-PWM	Asynchronous Pulse Width Modulation	IVFF	Input Voltage Feed Forward
ASLFR	Adaptive Sliding mode based Loss Free Resistor	LLC	Inductor –Inductor- Capacitor
BADC	Bidirectional AC-DC converter	LMS	Least Mean Square
BHB	Bridgeless boost Half-Bridge	MBLPOL	Modified Bridgeless Positive Output Luo
BL	Bridgeless	MC	Matrix Converter
CC	Constant Current	MLC	Multilevel Converter
CCM	Continuous Conduction Mode	MOSFET	Metal Oxide Semiconductor Field Effect Transistor
CICM	Continuous Inductor Conduction Mode	MPPT	Maximum Power Point Tracking
CV	Constant Voltage	OBC	On-Board Charger
CVM	Constant Voltage Mode	PEI	Power Electronics Interface
DAB	Dual Active Bridge	PEV	Plug-in Electric Vehicles
DBR	Diode Bridge Rectifier	PFC	Power Factor Correction
DC	Direct Current	PFM	Pulse Frequency Modulation
DCM	Discontinuous Conduction Mode	PI	Proportional- Integral
DM	Differential Mode	PLL	Phase-Locked Loop
EMI	Electromagnetic Interference	PP	Propulsion
ESS	Energy Storage System	PV	Photovoltaic
EVBC	Electric Vehicle Battery Charger	PWM	Pulse Width Modulation
EVs	Electric Vehicles	RB	Regenerative Braking
FB	Full-Bridge	SEPIC	Single Ended Primary Inductance Converter
FC	Fuel Cell	SiC	Silicon Carbide
G2V	Grid –to-vehicle	Si	Silicon
HB	Half Bridge	SPV-SVC	Solar Photovoltaic Standalone Voltage Control
HESS	Hybrid Energy Source System	THD	Total Harmonic Distortion
HF	High frequency	V2G	Vehicle-to-grid
HFT	High frequency transformer	ZCS	Zero Current Switching
		ZVS	Zero Voltage Switching

This paper reviews the different converters of power electronic techniques that have been reported for electric vehicle applications. Different methods of charging/discharging systems have been developed. All of this order has been viewed in this paper along with the current concepts of EVs charger /discharger. This paper is sort out as follows: segment 2 reviews the converter techniques of Uni/Bi-directional, Segment 3 discusses the different Power factor corrections techniques, Segment 4 studies the different methods of charger and discharger systems, Segment 5 presents the design of DC-DC converters for EVs and Segment 6 Remarks of this paper is concluded.

2. Uni/Bi-directional converter

The initial stage of the converter, needed for smooth bidirectional AC-DC power electronic converter and power factor correction. Although more than a few recent researches focus on applying a single rectifier to reduce the losses, component count and cost of the system and also to increase the efficiency. A large number of literatures have been reported with help of AC/DC-DC converter for application of EV.Generally, a bidirectional converter is employed at two methods of current control. In grid side, a sinusoidal Pulse Width Modulation (PWM) is created with help of the indirect current control method of PWM control. The closed current control loop is agonizing from quiet performance of transient and dc component showing in the grid side. The indirect current control method has some disadvantages can be eliminated. Discontinuous Conduction Mode (DCM) that involve of two closed control system – the inner current loop is used for to control the inductor current and the dc voltage is stabilized in a outer voltage loop. Indirect control method is treated for to increase the system performance (Static and dynamic) in [1]

2.1 Bidirectional AC-DC converter

The Bidirectional AC-DC converter (BADC) employ as a rectifier and inverter mode as required by the system. In AC side, the total harmonic distortion and power factor correction of BADC is achieved with IEEE standard, dc voltage Bidirectional converter in between two levels, with the ability of bidirectional power flow on modulation topology and control algorithm to secure soft-switching each operating span of the converter in[7].The important function of BADC can be precisely stated as follows: 1) Primary side switches by use of Zero current turn-offs and 2) Secondary side devices by use of Zero current turn-on, each operation without any extra components for battery charging application. For EVs charger existing of half-bridge converter of current fed on the input side and full-bridge converter on the output side, with transformer of high frequency in the middle of supplying galvanic isolation as mentioned in [7]. As a result, current fed technology supply a high current profile on the AC side and IEEE 519 standards of Total Harmonic Distortion (THD) is achieved.
The advantage of bidirectional DC-DC converter has an easy circuit with less component count that provides a wide range of voltage gain, a command ground and less stress of voltage. In extension the synchronous AC-DC converter permit ZVS turn on and turn off without anyone additional tools needed and the converter efficiency is developed in [4]. As a Hybrid Energy Source System (HESS) is a combination of a supercapacitor and battery, an electric vehicle is considered as the best way to increase the lifespan of battery and general efficiency of the vehicle. For greater power density, more life period and good charge/discharge efficiency, the supercapacitor with the high voltage dc bus, keep away from any enhancement in current from the battery is used.
The energy storage and converter components are used as perfect, [4] and Continuous Conduction Mode (CCM) is employed in this converter and each the capacitors are more enough, so we consider all capacitor's voltage is fixed over all switching cycle or state. Regenerative braking and acceleration of supply power with the battery meets the requirements for long-range operation of energy storage density. As a result, a very easy circuit, a number of component counts is reduced, gain of maximum voltage, a common ground, minimum stress of voltage, and are achieved.
The schematic diagram of the bidirectional converter show in Fig.1.

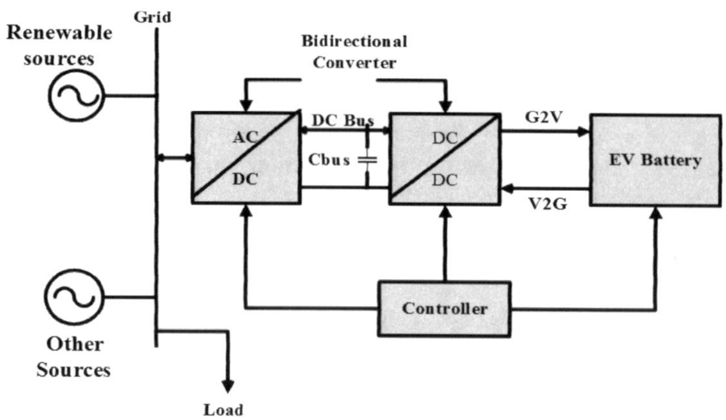

Fig.1. Bidirectional converter for vehicle to grid

To draw sinusoidal grid currents by use of modulated converter, Single-Step Three- Phase AC-DC

Converter in a bidirectional Soft-Switched DAB was presented in [1]. For the application of maximum voltage and medium power, the DAB converter is required, results in: 1) low-loss switching: in ac side ZCS and ZVS at dc side. 2) Without dc-side load current sensing, a power factor correction of open loop input is achieved and 3) the control input or phase shift is proportional to the shifted active power in closed loop control. This uses a strategy of space vector modulation and rectifier followed by converter of DAB, [10] control loops are needed in [27] bidirectional T-type Multilevel Converter (MLC). When the average of current inductor is greater than its ripple, the converter of boost is worked at Continuous Conduction Mode (CCM).To minimize the voltage stress and more-frequency period - by- period voltage balance at power switches and originate a maximum output voltage magnitude make use of power electronic switches with half of the maximum opposite voltage. More switches and capacitors are required in T-type MLC, to balance the voltage and also increase the switching and conduction losses.

Calculation of high frequency inductor, the total losses in the transformer and the conductor losses in primary and secondary bridge are has been presented in [10] due to bidirectional power flow ability , loss-less switching, and high power density, the present solution may list out a favorable solution for vehicle to grid application. Increased power density by use of the switching frequency and loss-less switching of the converter.

2.2 Unidirectional converter

The unidirectional converter circuit, shown in fig.2. The charging pattern encouraged by battery chargers of unidirectional that maintains up a single power flow way source to load, the electric vehicles are analyzed only as a load on the power system, still now battery is recharged from a grid power.

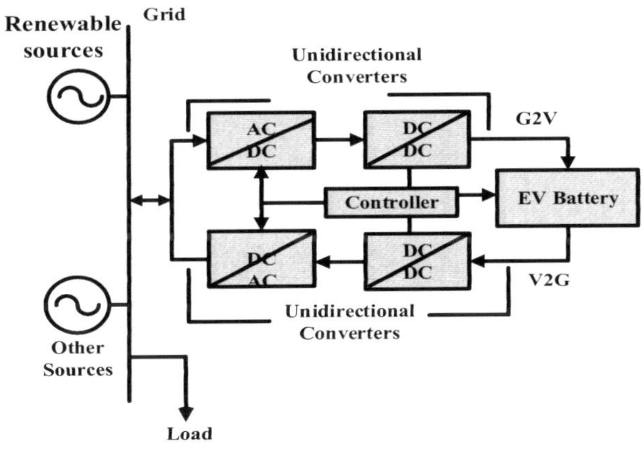

Fig.2. Unidirectional converter for electric vehicle

3. Power Factor Correction (PFC)

The converter of in active step up PFC use as the efficient HB converter during the Hold-Up Time (HUT), it can be operated normal state only, [2]. High current stress, high voltage, and Electromagnetic interference (EMI) noise due to the reverse recovery problem of Diode are reduced by using ZCS. The converter of HB-LLC is obtained normal dc-link voltage at the normal state PFC, during Hold-Up Time (HUT), converter of bridgeless boost PFC is inactive, which help to cover maximum voltage range. The converter of HB-LLC operates at normal state is designed with variation of limited switching frequency and large magnetizing inductance and without the aforementioned drawbacks. To decrease the conduction loss by use of PFC step boost transformer, the efficiency also slightly improved by using this converter.

3.1 Hybrid Resonant PWM Bridgeless AC–DC Converter

For sensing positive and negative input ac cycle operation, the converter does not required extra circuitry due to sharing a gate signal to all PWM switches. The operation mode of hybrid resonant boosts the problem of reverse recovery for the PWM switches. To improve the reliability of the system by limiting inrush current and improve the efficiency. A bridgeless rectifier PFC converter of a hybrid resonant pulse width modulation is similar to a boost converter. By eliminating the problem of reverse recovery with PWM switches of body diode, principle operation, design, and analysis of the converter is explained in [5]. To achieve power factor correction by use of boost converter of single – step working in a Continuous Conduction Mode (CCM), the system efficiency is improved and minimizes the conduction losses by the use of CCM. The PFC is reached by the use of average current control mode, which enables surge protection and allows simple implementation of lightning of system.

3.2 Hold-Up Time operation of DC-DC full-bridge converter

As a requirement, DC-DC Full-Bridge converter of ZVS must also take care of PFC by interleaved boost converter and total harmonic injection is obtained from the grid. An important aspect to be considered of this converter is to reliably expose the possibility grid zero voltage switching. This can conduct to voltage waveform, current waveform, efficiency, and Hold-Up Time (HUT) operating waveforms as in [8]. In rated input, converter is controlled with help of the variable frequency and the phase shift control and operation of HUT. DC-DC Full-Bridge converter of ZVS is operated under variable control methods were analyzed and described in detail in [8].

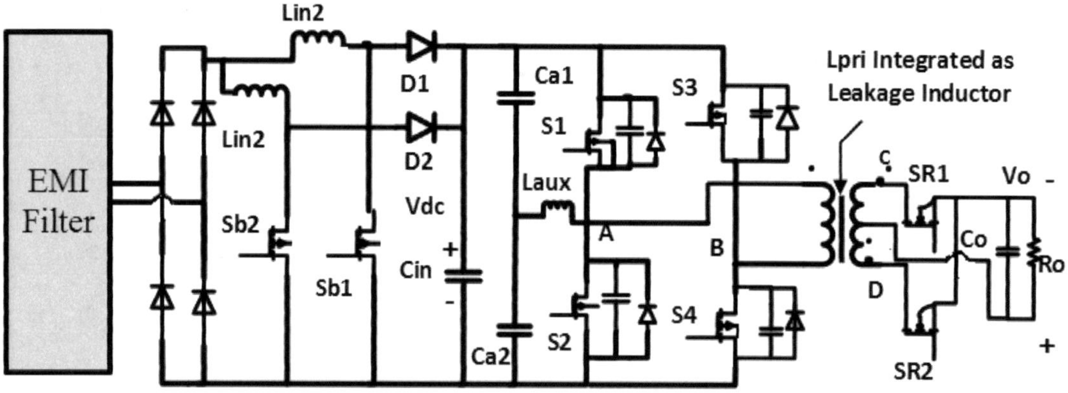

Fig.3. Complete implementation topology of Hold-Up Time operation of DC-DC full-bridge converter [8]

The DC-DC converter with HUT ability, reducing the losses and dc-link storage capacitance, thereby increasing the reliability of the system. .Complete implementation topology ZVS converter of DC-DC Full-Bridge with Hold-Up Time is shown in fig.3.

3.3 A Soft-Switching Bridgeless AC–DC Converter

A soft -switching, bridgeless boost converter for PFC for power supply and battery charging application is presented in [3]. The integrated converter functions as an in pulse width modulation and resonant mode one and all switching cycles that make use of average current control. Design analysis and working principles of ZVS HRPWM converter have been reported in [3].A continuous conduction mode and the achieved ZVS is used for minimizing losses of switching and system efficiency is increased To increasing reliability of the system due to minimize the PWM switches turn-off losses and neglecting switching losses that also reverse recovery losses is minimized by using diode output employ with controlled di/dt turn off and THD, and battery heating is limited in the operation of power quality.

3.4 Modified Bridgeless Positive Output Luo Converter

In output side, defined DC voltage is maintained and power quality is increased at the input side, it can be reached by design control of a modified bridgeless positive output Luo (MBLPOL). In input side a Discontinuous Conduction Mode (DCM) is applied for PFC with a minimum riddle, a single voltage sensor with a circuit of feedback control and to manage the DC load voltage for supply voltage and various load condition is implemented for PFC. A design, working principles, and testing point are discussed in [13]. Unity power factor is maintained with rated load and supply current total harmonics distortion is reduced. This can lead to voltage and current waveform, THD is in [13].

3.5 Single-Phase/-Step Buck/Boost converter

Rectifier technique is obtained for the plug-in electric vehicle application of on-board battery charging, assumed from converter of buck-boost pulse width modulation. The power factor correction (PEC) is designed for the following factor (a) Dc link voltages ripple (b) Damped input filter less influences (c) inductive current ripple is converter operation. A differential mode (DM) filter is used to reach a great converter performance as it takes on less output impedance with high –frequency attenuation. A two fast current control loop is used to manage PEC. The design and implementation a single-step/phase PWM PFC rectifier of control stragies and battery charger topology have been discussed in [20] to present maximum efficiency, power density, and power factor. The schematic structure of single- step/phase AC-DC unidirectional buck-boost topology shown in fig.4.
As a result, in full load condition, to achieve, maximum power factor and efficiency, the switching and conduction losses are analyzed in [20].In single state topologies with employed power processing, to minimize area and weight, because it is a transformerless techniques and minimum number of modules are employed. An operation of PFC is an analysis of the worked rectifier has been performed [20], and more conduction step, operation section, control strategy and explanation of the modulation structure.

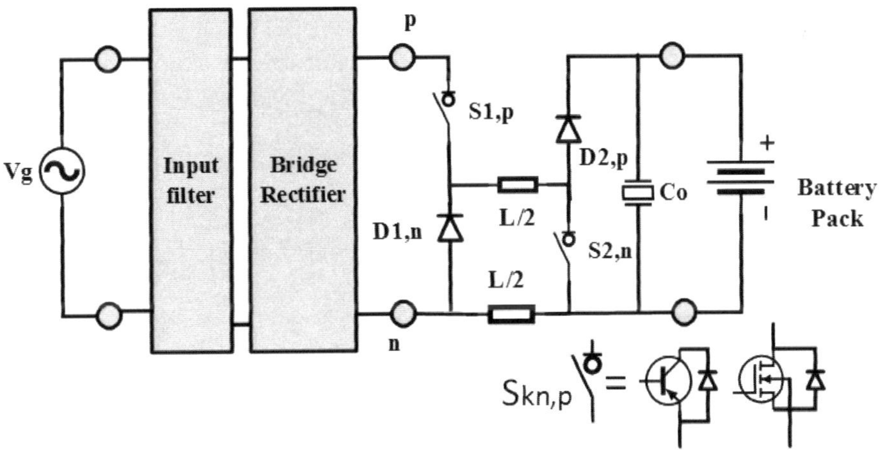

Fig.4. Schematic diagram of single-step/phase AC-DC unidirectional buck-boost topology [20]

3.6 Power Electronic Interface

The Power Electronics Interface (PEI) converter obtained from a secondary ended primary inductance isolated converter is developed to work productively in battery charging, Regenerative Braking (RB) and propulsion (PP). To optimize the power factor correction using continuous conduction mode,

MPPT is applied to take out the maximum number of power with help of this converter for battery charging from solar photovoltaic sources, this neglect the required of another DC-DC converter through operation of charger for MPPT. The solar photovoltaic and grid based converter of isolation, controller design, control algorithm, comparative analysis, and operation of PP and RB have been presented in [6] for battery charging and reducing conduction losses.

As a result, the charger has low stresses and increasing compactness of the charger and also construction, it is the greatest solution for the onboard battery charging application. Moreover, the PEI carries out buck/boost operation effectively, without any design modification for all vehicle modes. Because of this, due to availability of charging point the battery will be charged. The characteristics of the Photovoltaic (PV) panel, grid voltage and current shows in [6].The battery charging/discharging conduct to attain power and load levelling together with increase in system reliability. To encourage a uninterrupted power to the grid, hence increasing the possibility of the Battery Energy Storage (BES) hold up to the system, converter of voltage source perform as an active power filter and carry out the mitigation of harmonics with compensation of reactive power, a special control is expanded [9] due to mitigation of failure ,grid to be reconnected. To keep an uninterrupted power supply to the load with help of Phase-Locked Loop (PLL), maintaining the power balance, and increases the battery lifespan through the attainment of the operation of the system. The switches are controlled in zero crossing voltage and current. Solar Photovoltaic Standalone Voltage Control (SPV-SVC) is obtained to maintain an endless power to the loads. Instantaneous quantities of proportional –Integral and load current is taken out appling adaptive filtering along with the LMS algorithms in [9].

3.7 Bridgeless Isolated SEPIC Converter

Restriction of efficiency in an EV battery charging due to more condition losses correlated with a DBR at input side in a conventional Power Factor Correction (PEC). To diminish this issue, a great power quality with a Bridgeless (BL) Single Ended Primary Inductance Converter (SEPIC) is given in [14].During charging mode, the current input display a operation of unity power factor. To achieve greater power density and high efficiency, using the operation of Discontinuous Conduction Mode (DCM) with BL isolated SEPIC converter has been analyzed in [14]. Due to constant current /constant voltage control mode the current of battery is controlled. A control topology of PWM create the need gate pulse for the converters by emulate the current of output managing a cascaded both Proportional-Integral (PI) controller to give the response of fast dynamic loop, saw tooth carrier and minimum number of steady state error.

As a result, conduction losses and conduction of current through less number of devices are crucially reduced by eliminating BDR.To improves the efficiency and intrinsic PFC thus, enhances the complete performance of the charger. A charging technique is existence of a usual inductor input in the course of all the half period. Two switches are sharing the inductor input and as solution, minimize the battery charger size and cost.

3.8 Bridgeless / and switched inductor Cuk Converter

A DC-DC converter with common voltage input comprises of a boost converter for PFC during the two-step converters for battery charging of electric vehicle. This two-step conversions conduct to increased component count and poor efficiency, A Bridgeless (BL) Cuk converter based Electric Vehicle (EV) battery charging with great power quality, maximum efficiency is developed and implemented in [11]. By eliminating DBR, to reduce the additional conduction losses and this charger absorb a minimum number of devices running over period of one switch. Hence, it is increasing the charger's efficiency. To keep the desired charger current throughout the battery followed by the CC & CV charger regions, the instructions are synchronized by a flyback converter.

To improve efficiency, high stage down gain, less current stress, and power electronics component count is minimized in PFC converter based a single-step switched inductor Cuk converter is applied in this scheme. The design equation, operational analysis for different items of cuk converter are follow up in Continuous Current Mode (CCM) using a multiplier approach of current that has been presented in [21] concerning power quality sign like voltage THD and current THD. The DCM of action in BL cuk PFC converter is designed in [11], control of voltage follower mode is used, it is developed using PI controller no circulating current succeeding through the inductor inputs. Due to cascaded dual PI controller, converter of flyback is controlled and output of controller to supply CC charging to the each operating condition of battery. Completely charged the battery in CV mode at the end area of all period and the drawn source of current is cut down to completely charged condition of battery depicting value is less.

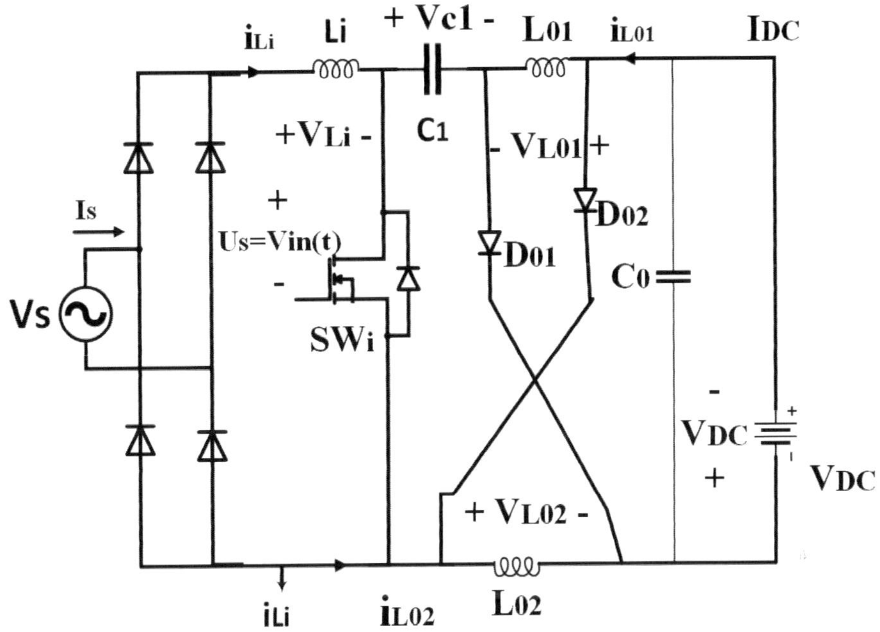

Fig.5. Circuit diagram of Switched inductor Cuk converter. [21]

The DCM, due to voltage follower approach, the scheme of control is minimized by need of sensor and converter of output voltage is sensed by quantity.BL cuk converter based charger extract a sinusoidal current from the ac main and THD in the current of supply is minimized. Undesired conduction by the diode body of the idle switch in the earlier on advanced converter of BL cuk is ignored, also unpopular coupling of capacitive loop is erased and minimize the number of components count over one switching period. This significantly increases the efficiency of charger's. To reduce voltage stress and low output voltage on devices, a transformer may be need, which is more prices, mainly if galvanic isolation is not needed. Cuk converter by dividing inductor into two equal inductors and rectifier diode into two diodes. Here frequency of supply line is lower than the frequency of switch. The total power factor is follow out with different methods of battery load and resistive load in one and the other CC and CV mode.PFC converter is successfully developed, the following assumptions are create in [21].Circuit diagram of switched inductor Cuk converter show in fig.5.

As a result, large voltage ripples in the intermediate capacitor, current stress in switch, and voltages of the battery are notice to develop during this cycle. And also to reduced conduction losses, switching losses, diode conduction losses, and input inductor losses and increased efficiency can be noted. This topology consists of two converters, first-step cuk converter for PFC by working in the converter of

LLC resonant, while the CCM of the primary inductor and the second stage the CVM and CCM control algorithms are operated for essential charging of EV battery. To minimize the turn off losses in LLC converter, the converter employ a control algorithm of frequency to actively change the dc-link voltage based on the state of the battery. The benefit of cuk converter of the control algorithm based LLC resonant and phase-shifted zero voltage switching converters are presented in [23] used for high power fast charger. As result, the input current of harmonic free is realized by pulse width modulation (PWM) control of the cuk converter utilizing a two-loop design, neglecting any battery charging risk by use of LLC controller, to minimise turn off losses & switching losses, less LLC transformer flux density and then to increase efficiency.

3.9 Three-level topology

To minimize the size of inductor filter and switching losses considerably in a [37], full-bridge (FB) BADC operating an asymmetrical three-level circuit. The FB- BADC is created by the power switches and the asymmetrical three-level converters are also created. As can be seen from the figure, to charge the battery from asymmetrical three-level converters, thus minimize the power losses and rating of the alike BDC. The BADC is employed with the help of sinusoidal PWM techniques. As a result, achieving a high efficiency either controlling methods for a more voltage selection is done.

3.10 A Modified Luo Converter

To increase current and voltage through the semiconductor in terms of single step/switch PFC a modified isolated Luo a high power AC-DC converter based EVs charger two switches and enclose clamped diodes at High-Frequency Transformer of primary isolation is used. A modified negative output Luo converter fed EV charger with minimized voltage switch is shown in fig.6. Minimize the complexity of control due to minimum number of sensors in the operation of Discontinuous Conduction Mode (DCM) at a modified Luo converter have been in [15], to obtain PF correction capability at sources side. Constant Current (CC)-Constant Voltage (CV) modes provided operate at better low ripple based charger and increase the long term battery service life.

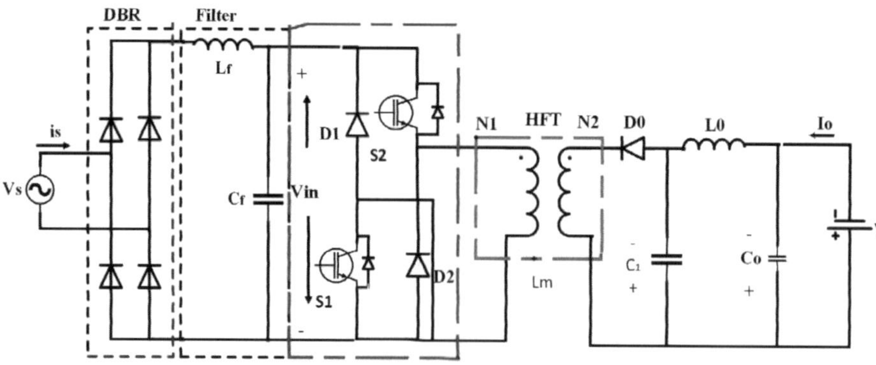

Fig.6. Modified negative output Luo converter fed EV charger with reduced switch voltage. [15

The limitations of high ripple, high component stress and high voltage stress by use of single-phase/stage PFC based EVs charger. Which in turn, result in lesser efficiency and conduction loss is very low since the conduction of line diode in all switching period are bisected. This can lead to voltage and current waveform and THD are in [15].

3.11 A Bridgeless Boost Half-Bridge DC-DC Converter

The characteristics of bridgeless, it is a single-step and galvanic isolation, converter input is permit to accurate the power factor , to encourage ZVS for all of its switches by the Asynchronous Pulse Width Modulation (A-PWM) in more to keeping a minimum semiconductors. a Bridgeless boost Half-

Bridge (BHB) DC-DC converter as a battery charging for electric vehicle is designed & operated, as well as the loop control will be presented in [18]. A control structure has been controlled by an average current of conventional, keeping a compact adjustment in [18]. As a result, obtaining a minimize the number of semiconductor and losses increase the efficiency.

3.12 Matrix Converter based topology

Matrix Converter (MC) is a combination of DC-AC and AC-DC converter; it is simply called a direct AC-AC converter and straight converting high-frequency ac voltage to power frequency ac grid and power frequency ac voltage to high frequency ac voltage. To achieve high frequency in the condition of ZVS, each power devices turn-on below ZVS state and turn –off below AVS situation. For greater reliability, greater power density capability, reasonable THD, simple control strategy, fast response, and fast charging of the EV, the BADC is recommended. The downside for the considerable problem in the configuration model needs sixteen power devices for it's in action and tolerate from less-frequency ripples attended in the battery current charger. An isolation transformer-based single phase full – bridge converter has been detailed in [38].

Current Trends

The channel resistance and less output capacitance of Silicon Carbide (SiC) power devices along with more voltage closing ability can allow maximum operation of frequency, less reverse recovery charge, high power density and good thermal characteristics [32].

3.13 A SiC-Based LLC converter

In the LLC converter can work in closeness of resonant frequency and more efficiencies above the maximum collection of battery store voltage utilized in PEV, the dc- link voltage can be actively adjusted. Conventional boost type PFC converters can't supply ultra-wide dc-link voltages, due to conventional constant dc-link strategy of control, Silicon Carbide (SiC) based onboard PEV charging is designed and controller in [22] . LLC converters of resonant are worked due to short circuit, galvanic isolation and soft switching. To keep greater efficiency, when battery voltage is lesser than it's assess voltage at a minimum step of charger it leads to increased switching losses and circulating losses. A sic based converter [22] is minimizing conductor losses & switching loss and have greater efficiency capability. By varying the air gap in the middle of the two slices of magnetic cores due to magnetizing inductance.

The circuit modeling of LLC resonant converter along with converter of SEPIC PFC, a two-step isolated charger is presented in fig. 7, the frequency operation of the LLC resonant converter is put moderately greater than resonant frequency to secure the ZVS running at dissimilar loads. To reach the required voltage gain & lower efficiency, when the battery voltage is under the rated voltage, the switch frequency turning off from the resonant frequency.

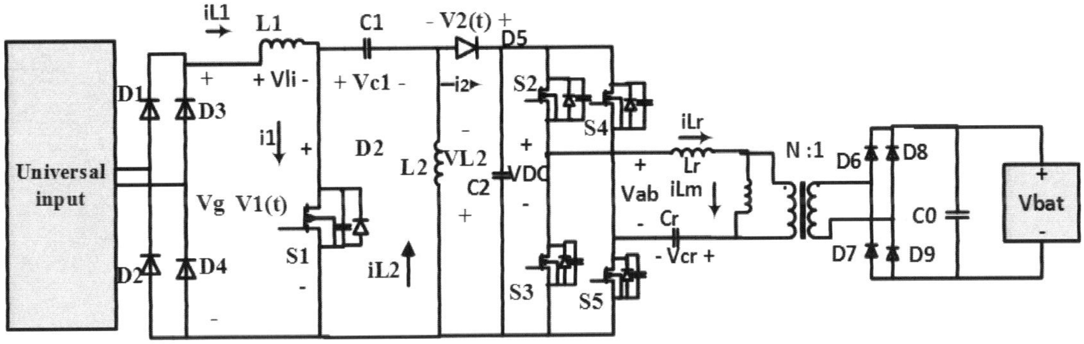

Fig.7.A two-step isolated charging based on SEPIC PFC and LLC resonant converter. [22]

The unity power factor is achieved along with low THD in a SEPIC PFE converter and an increase in overall efficiency of the charger of the rated load to full load.

3.14 A Soft Switched Boost Cascaded-by-Buck converter

A two-step converter couple the input supply of grid to the battery store whose voltage generally vary for different type of vehicle, for maintaining power factor and smooth input current, ZCS boost cascade operated by buck converter and analysis worked in application of PFC in [25].A mixing of resonant capacitor and inductor are managed to generate ZCS throughout turn-off switch. Due to less line input voltage the switching stress is more and size of the components, the boost structured PFC topologies is designed in [25]. As a result, to achieve higher efficiency and keep a power factor, the structure of control loop supply the unfixed DC-link voltage outputs at PFC with general voltage input is designed in [25].

3.15 Boost DC-DC Converter

Generally, the stress of voltage is minimized in the half of the voltage input, the absolute value of voltage gain is 1/(1-d),power inductor and switches are less than those of the more voltage gain in two-step cascaded and three-level boost DC-DC converter, but it's the result of the efficiency of all step ,more voltage stress in output step. A maximum voltage gain but stress of voltage beyond the power electronic switch is similar to the voltage output in an inductor of switched DC-DC boost converter. The single switch with diode capacitor units boost DC-DC converter is a easy structure, the principles of operation, the enlarge step, converter steady state characteristics and operation of fault-tolerant are analyzed and given in [30] and also small-signal sample is derived. The converter is operated at the state-space average mode; the capacitor is charged in parallel and discharged in series. Due to voltage loop control PWM and Proportional –Integral (PI) voltage controllers are adopted in this converter, which provides less voltage stress, less switching losses and voltage gain is more.

3.16 A Step-Up Multi-input DC-DC Converter

The input sources of Energy Storage System (ESS), Fuel Cell (FC) and Solar Photovoltaic (PV) for the step-up multi-input DC-DC converter in [12].

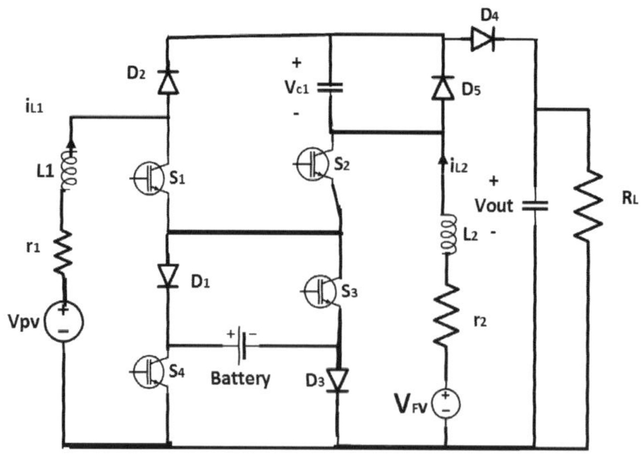

Fig.8. Three-input boost DC-DC converter [12]

IOP Conf. Series: Materials Science and Engineering **906** (2020) 012006 doi:10.1088/1757-899X/906/1/012006

The DC-DC converter has the potentiality of arranging the requested power by load in the non-attendance one or more resources. The FC is considered a main source for clean electricity generation, output current ability of high density, and to charge the battery by use of PV minimizes the fuel economy and improve the efficiency. Three-input boost DC-DC converter shown in fig.8. A continuous conduct mode to provide smooth current with minimum current ripple in a multiple-input boost DC-DC converter. The state-space sample is prepared for DC-DC converter modelling and control in [12]. The MPPT is applicable for to obtain the high power from photovoltaic and the system steady state error is neglected with help of Proportional –Integral controller (PI).As a result, the inductor current and switching losses are more in the session.

3.17 Step-Up Resonant DC-DC Converter

The ZVS in two power electronic switches and ZCS in the rectifier diode operation over the full load range to achieve maximum efficiency. To minimize the stress of the component, and also decreases the transformer requirements with the help of interleaved step-up technique. To reach ZCS and ZVS also to extended voltage control with help of series resonant tank, DC-DC converter of more voltage gain boost employing resonant technology has been presented in [16]. It gives the voltage step up through its turn's ratio but brings needed galvanic isolation between the high output voltage and low input voltage, Discontinuous Conduction Mode (DCM).To reach a high voltage gain and high switching frequency in an unregulated LLC topology cascade with a step-up topology.
The limitations of the topology in [16] to reduce copper losses, iron losses and conduction losses, easier flux design, employing an interleaved boost topology, and also in transformer stress of current is minimized and improve the efficiency. In voltage gain, uniform operation of the interleaved topology, in resonant mode component stress and power loss. For operation mode of D=1/2, since the positive and negative cycle is uniform in [16], positive cycle only analyzed.

3.18 Dual input–dual output DC-DC converter

To transfer the power between input sources to loads/EVs. A dual input-dual output DC-DC converter is developed in [19] for the incorporation of solar photovoltaic, battery, and ultra capacitor for the EV application. DC-DC converter is controlled inappropriate switches and different modes using the same architecture by [19]. This converter is used for neglecting current stress and minimizes weight and size impact on the formation of minimum cost and more efficient electrical system, and also involved conduction and switching losses and high component count.

3.19 Isolated/Non-isolated topologies

The isolated bidirectional converter is mostly used for high voltage range applications, it is used to limit the diode reverse recovery problem, voltage and current spikes, and switching transients. However, it essential a right design to increase the soft- switching range and inductor to enhance the circulating energy in such a converter. Current- fed DAB BADCs give maximum voltage gain [43].The non-isolated bidirectional converter make use of an energy transfer devices, inductor for transfer the power from source to load, and thus utilize easy controller plan. When the switch is turn on during operation of boost, the magnetic formation of inductor energy is stored, so that it can be deliver when the switch is turn off at boost operation. To realize fast transient response in a direct current control method. A full-bridge BADC coupled with two-step interleaved buck-boost bidirectional converter has been recorded in [39, 40]. The circuit keeps away from the usage of an irresponsible dc-dc capacitor as in-between storage devices, resulting in current ripple of inductor is more; the feed forward control and direct current authority has been utilized to track the quantity of power flow and direction, secure energetic response of the system and to regulate the voltage with help of PI controller.

3.20 Single-Step Three-Phase PFC Converter

A single-Step isolated cuk converter using Three-phase AC-DC converter of PFC is developed and analyzed for the CCM is modulated for charging of EV. The PFC along with compact voltage order is obtained in the control scheme of Adaptive Sliding mode based Loss Free Resistor (ASLFR). The Total Harmonic Distortion (THD) is met the grid standard in [29], Front converters operate a two-stage rectifier of a resistor emulator of the active method is free harmonic rectification and second converter for improved voltage regulation. To reduce efficiency and extra PFC units improved the number of component, complexity, and reduce the efficiency. The circuit diagram of the expansible three-phase rectifier using Cuk converter is shown in fig.9.

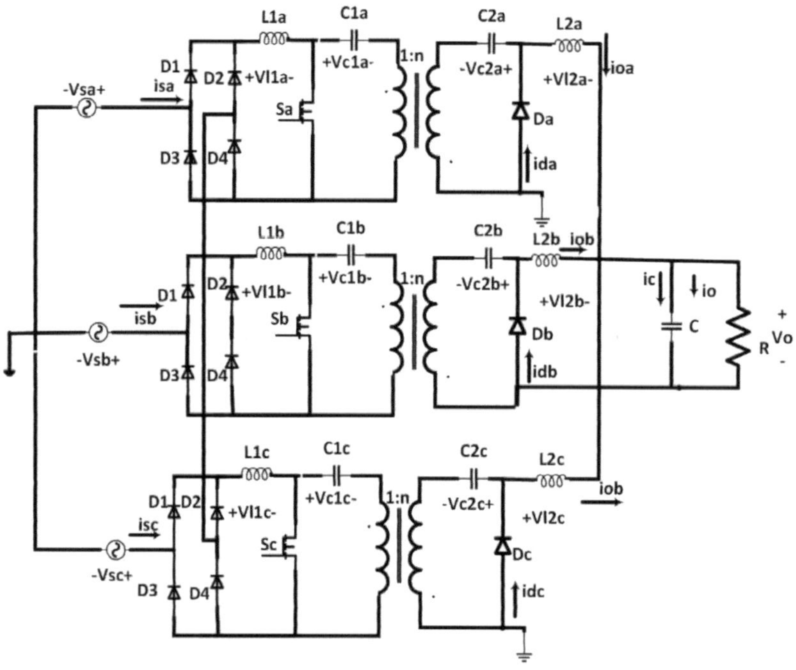

Fig.9. Circuit diagram of the expansible three-phase rectifier using Cuk converter. [29]

Single-step approach generally, high efficiency but current input is not a complete sinusoidal, besides it has a minimum size as balanced to the two-step converter, maintained unity power factor nearly to provide a trade-off, which can increase the power delivery while inscribe number of component and cost. To keep the response of fast transient even with the lower conversion steps while offering an increase the response, high efficiency, and emulating an adaptive losses free resistor.

Comparison of various BDCs

A full-bridge converter is given isolated Bidirectional DC-DC Converter (BDC) technique, but the buck/boost converter is almost all common nonisolated BDC technique. The advantages of non-isolated topologies are less weight, less cost, lower footprint, and easy control, mostly due to non-appearance of the isolation transformer. In converter, increase the component for less switching. However, they restrict the power density by reason of conduction loss created by maximum current pass through the switching scheme and further component of passive element, so that in this converter get less efficiency. For fast charging/discharging capability, the interleaving is applied in nonisolated BDCs. The switching losses are addressed in soft switching techniques, nonisolated BDCs do not apply ZVS/ZCS in the control algorithm. The scaling flexibility in case more power

IOP Conf. Series: Materials Science and Engineering **906** (2020) 012006 doi:10.1088/1757-899X/906/1/012006

transfers ability is considered in a cascaded buck/boost topology and it give the good transient performance. The good safety during a fault condition, the isolated BDC topologies are considered, but at the amount of a high-frequency transformer is required. Isolation is all-important according to international standards, for the most part, when maximum power transfer is carried out. Comparison of Power factor correction and BDCs developed for Electric Vehicle application is given in table 1.
To increase the reliability of the converter and minimize the elements, on resonant topologies is prepared. When matched to non-resonant technology, the current pass through the switching scheme is somewhat on the higher side. As a result, more conduction losses and less efficiency. As a result of more bandgap devices, maximum efficiency, high power density potentiality and good thermal quality and can display to be power tools for maximum power charging/discharging systems. If further minimizing the area and footmark of the bidirectional converter is require, but battery shame is inescapable, sinusoidal charging is considered. In-vehicle to grid applications, nonisolated BDC topology is considered.
The comparative performance characteristics of bidirectional and unidirectional converter using silicon and silicon carbide with respect to various switching frequency and efficiency of DC-DC converters is measured with help of Si & SiC technology has been shown in fig.10.Furthermore, when the more number of switching frequency in Silicon Carbide (SiC) as compared with silicon (Si), the minimum significant of reduction efficiency. Where the maximum number of losses in Si as switching frequency is varied from 20 to 80 KHz, while a minimum number of losses in SiC.

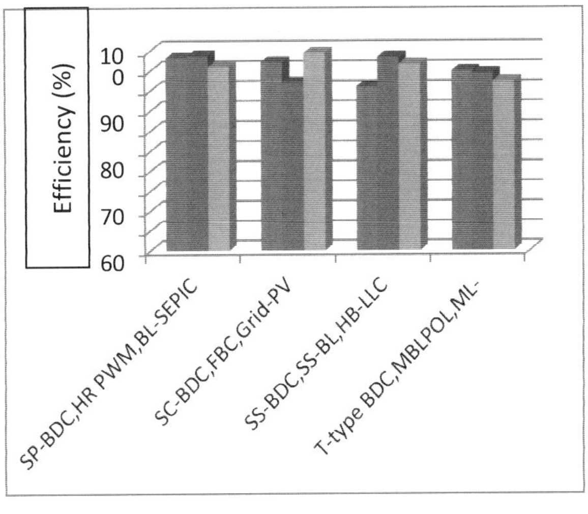

Fig .10. Measurement of efficiency in a DC-DC converter using Si and SiC technology

Table 1: Comparison of Power Factor Correction and BDCs advanced for Electric Vehicle applications as established in the literature

Topology	Rated power	Charging quantity	Range of battery voltage	No. of switching tools	Passive required	Power factor	Switching frequency (KHZ)	Isolation (Galvanic)	Efficiency (%)	Reference	Remark
Boost PFC converter is required in PSUs	480W	2	------	4	C	0.98	20	Yes	<90	2	Conduction losses copper loses & switching losses are much more high, Max. current stresses of L&D
(HRPWM) Bridgeless	650W		---	2	C	0.9	70	No	97.5	5	More complex impedances in EMI filter, grow the losses
Soft-Switching Bridgeless boost Converter (ZVS HRPWM)	650W	1	----	3	C	0.999	150	No	96.95	3	More complex impedances in EMI filter and grow the losses, maximum current at less duty cycles when the voltage input is more
BDC	1.5KVA	1	220–336 V	8	L&C	Unity	100	Yes	96.5	7	Great conduction losses, High ON state resistance by reason of less current rating
BL SEPIC converter	760W	1	48V	2	C	Unity	20	Yes	92.1	6	Maximum current stress of 28A, Maximum voltages stress of 520 V
BL cuk converter	1.2KV	2	48V	2	C	Unity	20	No	93	11	Maximum current stress
Modified Luo converter	760w	2	48V	2	L&C	Unity	20	yes	90	15	Converter bulky, high switch stresses
Boost Cascaded-by-Buck	1KW	2	-----	2	C	0.99	30	No	96.4	25	Maximum stress of voltage ,Big areas with indirect power transfer capabilities
Switched inductor cuk converter	500W	1	48V	1	C	0.99	25	No	Standard	21	Output inductor losses , ESR losses of the intermediate capacitor
LLC resonant converter with cuk converter	500	2	Upto 20Ah	3	L&C	Unity	30	Yes	Standard	23	Increase the number of components count
Soft-Switching Single-Stage Converter	3.3KW	2	360v	2	C	Unity	50	No	96	17	Poor soft switching and circulating current , Two-stage structural problem
L3C resonant converter (PV)	3 kW	2	230–430 v	4	C	---	180-400	No	97.5	33	More conduction losses, Converter efficiency decreases at light loading conditions
Half-Bridge resonant	3.3kW	1	360V	4	C	Unity	180-200 KHz	Yes	96	42	Minor footprint and More Complex

IOP Conf. Series: Materials Science and Engineering **906** (2020) 012006 doi:10.1088/1757-899X/906/1/012006

4. Methods of charging/discharging systems

For Electric Vehicle (EV) applications charging/discharging are generally classified into (a) On-board or off-board systems, (b) Integrated or Non-integrated systems, and (c) Inductive or capacitive. All of the above mentioned methods have been discussed in the following sub-segments. An inductive charging/discharging system is one of a non- integrated design system, where one scrap of the system is assigned over on-board of the electric vehicle and further scrap over the off-board of the EVs.

4.1 On-board and off-board charging/discharging

Accommodation to charging/discharging the vehicle in on-board systems, wherever an electric power is available at outlet, as well the EV battery voltages match from an onboard system. The voltage from grid is normally constant, whereas the voltage of battery is not and it differs from V2V. The on-board systems are specifically fabricated by EV construct for all vehicles in order to meet the particular voltage need of the grid and battery. The same doesn't suits with off-board systems that are developed in order to keep a large collection of battery voltages. However, design parameters and schematic diagram of the on- board charger/discharger system [25]. Fast charging/discharging is achieved by a current focus, which request maximum power heavy converter system and battery profile shown in [22]. It can be concluded that the EV operation should incur the cost of both charging / discharging but its reduced switching and conduction losses increase the efficiency of the system due to low diode conduction losses. The characteristics of on-board systems as [21].

4.2 Integrated and Non-Integrated charging / discharging

In a non-integrated method of charging/discharging, the system drive is free of the charger/discharger [35], this demands a committed converter system to look after the charging/discharging of the EV battery. A Non-integrated type is preferred for charging/discharging of the system. However, if grid to vehicle and verse visa operation and tracking are shown in [27]. Such an integrated charger/discharger is called an on-board system. An integrated system minimizes the voltage and current stress, conduction losses of an inductor, neglect the sub-harmonic oscillation in voltage output unexpected load changes, and the effect of thermal ageing is been eliminated. Some other integrated charging/discharging systems has been discussed in [32] that use an on-board three-phase full-bridge BADC for propulsion as well as to the regenerative braking and the three-phase permanent magnet synchronous motor windings to promote the power flow in both directions.

To better power quality an isolated LLC resonant of cuk converters for an Electric Vehicle Battery Charger (EVBC) is planned definitely to charging a small Electric Vehicle (EV). The technique contain a two converters comprising of the first step cuk converter for PFC by operating in the LLC resonant converter, while the CCM of the primary inductor and the second stage the CVM and CCM control algorithms. To minimize the turn off losses in LLC converter switching devices, the converter make use of a frequency control algorithm to sincerely vary the in-between dc-link voltage based upon the state of the battery. The lead- acid batteries are charged by the C/10 ratio due to the battery thermal characteristics the keep up charging time limits their time of EV driving and but lithium batteries solve these issues under fast charging time (C/5,1C) minimum charging duration with travel long distance.

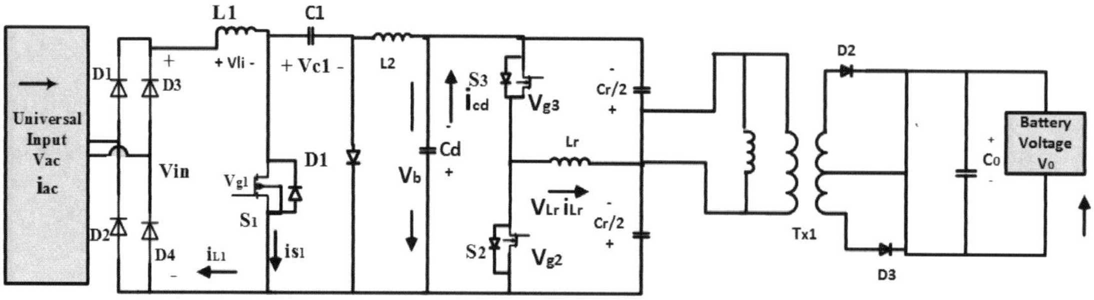

Fig.11. Configuration of PFC-Cuk converter fed isolated HB-LLC converter. [28]

The benefit control algorithm of the cuk converter based LLC resonant and phase- shifted zero voltage switching converters are presented in [28] used for high power fast charger. As result, the input current harmonic less is achieved by Pulse Width Modulation (PWM) a two loop structure of cuk converter is controlled, neglecting any battery charging risk by use of LLC controller to reduce turn off & switching losses, with less flux density of the LLC transformer and then to increase efficiency. Configuration of the PFC-Cuk converter fed isolated HB-LLC converter is shown in fig.11.The On-Board Charger (OBC) supplies ZVS of each switches and minimize the reverse recovery problems by ZCS of the diode outputs. Two interleaved cells sharing a full-bridge diode rectifier for OBC in [17], all cells are composed of a series resonant and active clamp circuits.
The interleaved technique helps to keep the actual power quantity with maximum efficiency and the control algorithm permit power quality by the OBC without need of auxiliary circuit and the current input is balanced in between the two cells and also converter make sure compact size, a long lifetime and need no electrolytic dc-link capacitor. For low power application (less than 1kW) single stage topologies are applied. Single-stage, single cell OBC is shown in fig.12, the control algorithms and development of a soft-switching single-stage converter, and then OBC transformer transforms the ac-voltage input into the desired dc-voltage output in a single stage is presented in [17], to achieve desired power factor and maximum efficiency. This converter is comprises of full bridge diode rectifier, along with a maximum-frequency DC-DC converter combined with a full-bridge series resonant circuit and an active clamp circuit. The main advantages of this converter is absorbed, the simple structure and maximum efficiency was conferred by soft switching of all elements. Moreover, to enable sharing power distribution between the interleaved cells, there needs an auxiliary PFC circuit. To achieve unity power factor at ac mains, a PFC step up converter based half-bridge LLC resonant converter which utilizes as a front end converter is operated in CICM for Electric Vehicle Battery Charging (EVBC) application.

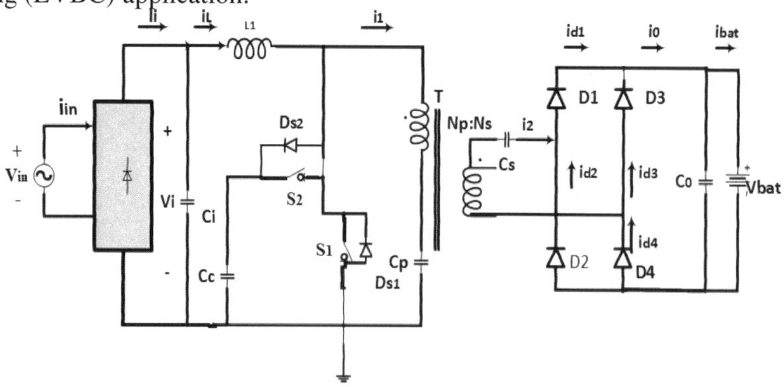

Fig.12. Single-stage, single cell OBC [17]

LLC converter uses a Pulse Frequency Modulation (PFM), whereas a step up converter is works in PWM technique at a constant frequency to keep the power delivered to the battery. To measure the low value of inductor current, boost inductor that works in CCM and Average Current Mode (ACM) control is operated for less power loss.PFC boost converter is developed below CICM mode using ACM control at a desired switching frequency is described in [31]. A smooth rectified sine template even below the poor voltage quality control for the current generation is obtained in the phase-locked loop technique. DC link voltage is sensed by a less-cost possible divider that forms supply a reference current for inner current loop and outer voltage control loop. As a result, less ripple current throughout the inductor less the stress of current in semiconductor, high efficiency, and reduced charging time with less-cost developed power quality EVBC. An Asymmetrical Pulse Width Modulation (APWM) technique is used for controlling the output voltage and foremost switches, resulting in maximum efficiency.
The design and implementation of the controller and control technique are introduced in [24], which minimize the switching current stress , as well as the conduction losses and the circulating losses during the freewheeling session compared with the auxiliary inductor current is controlled by use of conventional phase shift modulation method, so that can increase the system efficiency. In full-bridge DC-DC converter topology, maximum power density, high reliability of the system, and isolation ability. A bidirectional DC-DC converter of switched-capacitor is shown in fig.13, reduces the number of component count, and gets a large range of voltage gain along with less stress of voltage and a common ground. Additionally the synchronous rectifier permits to achieve ZVS turn on and turns off without need anyone additional hardware and converter efficiency is better in [26]. As a Hybrid Energy Source System (HESS) is a combination of a super capacitor and battery, an electric vehicle is considered as the best way to increase the battery life and overall efficiency of the vehicle. For maximum power density, greater cycle life and high efficiency, the supercapacitor with the high voltage dc bus which keep away from any change in step in current from the battery is used. Regenerative braking and acceleration of supply power with the battery meets the requirements for long-range operation of energy storage density.

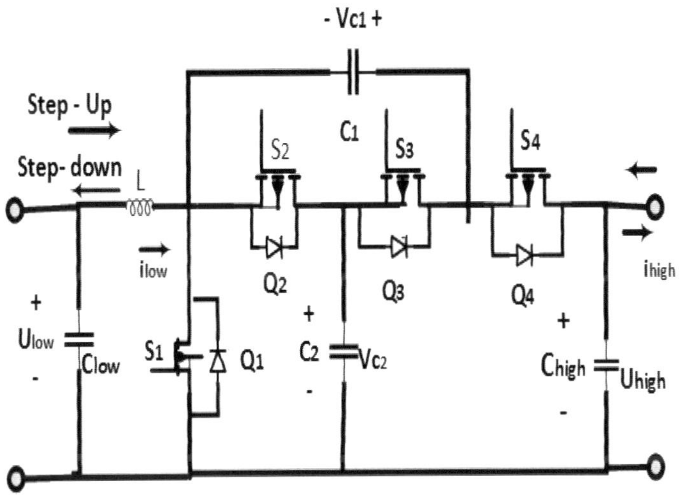

Fig.13.A bidirectional DC-DC converter of switched-capacitor [26]

As a result, a very easy circuit of minimized number of components count of greater voltage gain, a less stress of voltage, and a common ground, but conduction and switching losses are high. A switched inductor-capacitor, coupled inductor, Walton multiplier, boosts converters with greater voltage gain ability as the nonisolated DC-DC power conversion step for Photovoltaic to applications of grid. Successfully, DC-DC boost converter helps the Photovoltaic (PV) voltage to reach the maximum-performance level, but they are not suggested for EV applications since they need a mandatory galvanic isolation in the middle of the PV panel and the greater voltage battery store. These proposed converters prosperous handle the PV voltage variations and MPPT demands to respond to the different battery charging state, but it doesn't address the pair of input and output variations simultaneously.

From complete discharged condition to the charged floating voltage phase, the charger must operate in CVM and CCM or constant power modes based on the battery charge state. A schematic structure of full-bridge L3C resonant converter with capacitive output filter is shown in fig.14, applicable for PV to more-voltage battery applications. The challenge of the converter is to combine both variables PV input voltage and an energy system utmost voltage gain variations from input to output at different current and voltage quantities. A high-efficiency fourth-order L3C resonant converter has been presented in [33], resonant converter all the time operate in ZVS for different load control and have less noise output voltage. L3C converters not only come up with protect the full-bridge inverter form over current and short circuit control but also charging the battery pack at constant current.

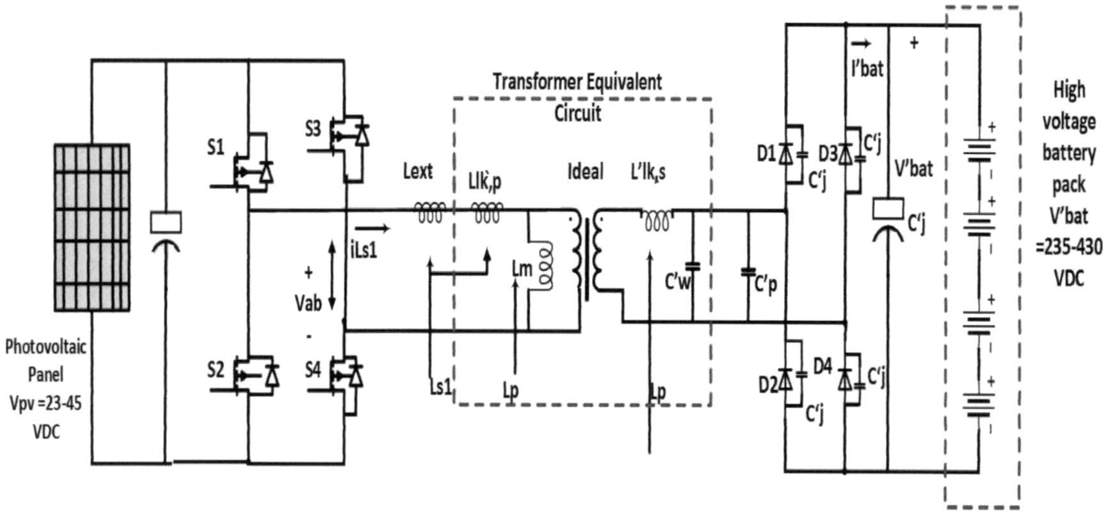

Fig.14 A schematic diagram of L3C resonant converter of full bridge function with capacitive output filter [33]

To neglect the maximum voltage constructed by the reverse recovery current of diodes and to drop the noise level in the receiving end voltage, a function of rectifier diode output in a Zero Current Switching (ZCS). In order to charge the Li iron battery at a lesser price with an extended time of charge without lesser the life and allowed maximum current charger with the help of constant voltage and constant current mode. As a result, show the high-quality voltage output, the output current is limited, during the CCM a battery current is controlled, over current protection, maximize the converter efficiency, and during light loading conditions a converter efficiency is reduced and it's having more conduction losses. Fig.15 shows the Single-step, two-way AC/DC converters topology for OBC, the output voltage always great or less than the input voltage, and smooth transitions between the modes are operated in CCM and performance of the converter is expressed in [34].

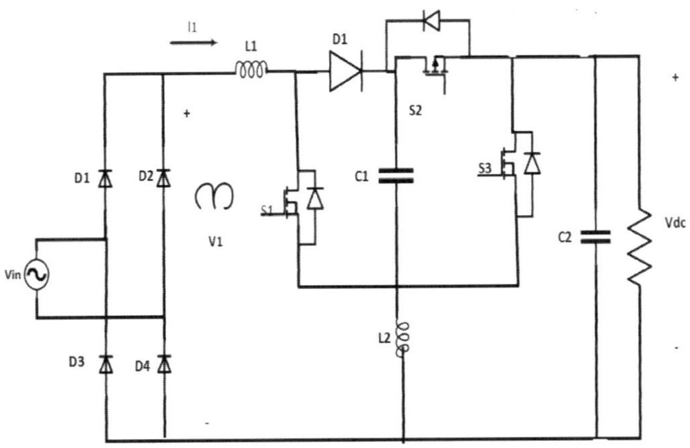

Fig.15 Single-step, two- way AC/DC converters topology for OBC [34]

The input current as the inner loop and voltage output as an outer loop is implemented in an Average Current Mode (ACM). To take the variations of input voltage throughout both step up and step down modes and the Input Voltage Feed Forward (IVFF) terms is as well attached with gain. Voltage and current controller depends on carrier generation block and Pulse Width Modulation (PWM). The simplicity of design and robustness is well-established for all applications, the Proportional -Integral controllers are used. A single-stage, two-mode AC/DC converter is implemented for low voltage and current stress, inductor conduction losses, neglect the subharmonic oscillations in voltage output unexpected change of load the effect of thermal ageing is been eliminated. In the front step, due to a less number of active elements they have maximum power, maximum efficiency and reliability of the system with lower application cost and low harmonics of input current. The control of Average Current Mode (ACM) rectifier step up PFC type designed and analyzed with maximum efficiency and to reach unity power factor.

In the second stage, maximum power factor and less THD for extended battery lifespan, maximum efficiency and permit EVs to get extra miles per charge. An Artificial Neural Network (ANN) is designed to control the half-bridge LLC resonant converter. ANN is applied to synchronize the LLC resonant converter load voltage with feed-forward back propagation technique and Levenberg Marquardt (LM) activation algorithm for fixed voltage battery charge methods in [36]. The control method of current mode is favour for PFC with a pulse width modulation based PI controller. LLC resonant converter is operated under maximum voltage and power conditions, a minimum number of voltage stresses on the diode output, and the potentiality to employ the ZVS with a broad load range. To increase power quality, efficiency, battery lifespan, and then reduce in harmonics of the input current is achieved by use of this topology.

4.3 Inductive / conductive charger / discharger

Conductive charging/discharging systems are any one integrated or non-integrated systems and apply a direct conductive approach, normally a cable with Z connector to permit power flow in the middle of the grid and the EV. Such systems can supply a fast charging station to the EV [41]. However flexibility and convenience is compromised since the EV has to plug in the wire all time. Inductive charging/discharging systems are all the time non- integrated method, where the pick-up side is placed on-board the EV and the primary side off-board the EV.Uneven in between the primary and pick-up side inductive plan of action is the major of the problem of the inductive charging/discharging. Which small change the magnetic coupling between each sides leads to power transfer level and variable resonant frequency. When power is transferred inductively, perceptive

configuration is important. As a result of bidirectional power flow, when the pick-up side LCL network of every one Electric Vehicle is tuned to the frequency of the track current and the load side converters are employed either in rectification or inverting method depending on the regulation of the power flow.
A fallback phase angle solution in power flow from the source to load side, while leading phase angle outcomes in power flow load to the source side. The direction of power flow depends on the corresponding phase angle and magnitude between the voltages at primary to load side. A phase-modulated square wave voltage is generated from a PI controller and a triangular wave is generated at source or load side control system. The output voltage and current harmonics are reduced with the help of the LCL network.

5. Design of DC-DC converters for EVs

A design condition of DC-DC converters for EV is represented in this segment. To determine the duty cycle, inductor, capacitor, average input voltage and resonant frequency value of DC-DC converters with help of design equations. Each parameter is taken from both simulation and hardware models discussed in the above literature. Based on value of design and parameters, the prototype test is designed, and performance characteristics of dc-dc converter are analyzed.

5.1 Bridgeless Isolated SEPIC DC-DC Converter

There are three operating principles in bridgeless isolated SEPIC DC-DC Converter (i) ON mode switch (ii) OFF mode switch (iii) Switch OFF .When switch is turned ON, the inductor current linearly rises as it energy is stored from the input supply. An energy stored in magnetizing inductor, the energy transfer capacitor is discharged from it. During this condition the output diode is reverse bias and the dc-link capacitor supply the battery form battery current. During turned OFF condition, the two inductor currents are force to conduct the output diode. Releasing the energy from inductors to the voltage output, when both inductor currents are decreasing. In mode iii, primary inductance of HFT energy is completed depleted and still in switch is OFF. The current inductor remains on and off for the remain of the switching period, in this condition, there is no energy moving between the HFT primary to secondary, as a diode output is reverse biased and the need power of charging is delivered by the dc-link capacitor.

When switch S1 is ON

$$i_{L_i}(t) = I_{L_i}(t_0) + \frac{V_{in}}{L_i}(t - t_0) \quad \text{For } t_0 \leq t \leq DT_s \tag{1}$$

$$i_{L_{m1}}(t) = I_{L_{m1}}(t_0) + \frac{V_{c\,1}}{L_{m1}}(t - t_0) \tag{2}$$

When switch S1 is OFF

$$V_{in}(t)D + \left(V_{in}(t) - V_{c1} - nV_{dc}\right)D_1 = 0 \tag{3}$$

$$i_{L_i}(t) = I_{L_i}(t_0) + \frac{V_{in}}{L_i}DT_s - \frac{D}{D_1}\frac{V_{in}}{L_i}(t - t_1) \tag{4}$$

$$i_{L_{m1}}(t) = I_{L_{m1}}(t_0) + \frac{V_{in}}{L_{m1}}DT_s - \frac{D}{D_1}\frac{V_{in}}{L_{m1}}(t - t_1) \tag{5}$$

When switch S1 is still OFF

$$i_{L_i}(t) = I_{L_i}(t_0) + \frac{V_{in}}{L_i} DT_s + \frac{V_{in}-V_{c1}}{L_i}(t - t_2) \tag{6}$$

$$i_{L_{m1}}(t)=0$$

$$V_{sav} = \frac{2\sqrt{2}\ X\ V_s}{\pi} \tag{7}$$

The input and magnetic inductances are

$$L_i = \frac{V_s D}{2 f_s \Delta_{i_i}} \tag{8}$$

$$L_{mc} = \left(\frac{N_1}{N_2}\right)^2 \frac{V_{dc}(1-D)^2}{2D f_s I_{dc}} \tag{9}$$

The resonant frequency

$$f_r = \frac{1}{2\pi\sqrt{L_i + L_m C_1}} \tag{10}$$

The design equations of capacitors are

$$C_{1,2} = \frac{N_2}{N_1} \frac{V_{dc} D}{(\Delta V_{c1,2} f_s R_{dc})} \tag{11}$$

$$C_{dc} = \frac{I_{dc}}{2\omega \Delta V_{dc}} \tag{12}$$

The specification value of bridgeless isolated SEPIC DC-DC converter is given in [6]

5.2 Single-phase Single-Stage Isolated AC–DC Converter

In this converter, a current-fed bidirectional based half bridge converter on primary side and full-bridge converter interlinking battery on secondary side. A four quadrant operation of voltage and current also is performed in [7]. In first mode of this converter, when the input voltages is positive half period and transfer the power from grid to battery, primary switches are worked at high frequency with modulation of duty ratio and 180 degree phase shifted in a gate pulse. In second mode, two switches of S1a & S2a are maintained at ON step and switches of S1b & S2b are employed at HT with modulation of duty ratio is greater than 0.5 and 180 degree phase shift are maintained. S1b & S2b switches are maintained ON and switches of S1a & S2a are kept OFF stage, negative direction of current is maintained in third mode of this converter. Phase angle should be differing from 0 to 0.5 and voltage output also varied. In fourth mode of this converter, a switching state of primary devices are maintained based on the direction of current input.S1b and s2b switches are ON-stage and current flow through s1a and s2a of diode.

The output voltage is

$$V_0 = \frac{n\ XV_{in}}{(1-d_1)} \tag{13}$$

Transformer turns ratio

$$n = \frac{V_{0,min} \cdot (1-d_{1,min})}{V_{in,max}} \tag{14}$$

$$L_1 = L_2 = \frac{V_{in,max} \cdot d_{1,min}}{\Delta I_L f_s} \tag{15}$$

$$C_{dc} = \frac{2P_0}{\Delta V_0 . \omega . V_{0,min}} \tag{16}$$

5.3 Improved bridgeless Cuk Converter

A flyback converter is worked as constant voltage and constant current charging and a improved bridgeless Cuk converter is recommend outstanding PFC characteristics of Discontinuous Conduction Mode (DCM) and there are three modes of operating: (i) ON mode switch (ii) OFF mode switch (iii) DCM mode. When first mode is activated, its works at positive half cycle, capacitor voltage being decreasing through the S1 switch and inductor output is supply the need load current to fly back converter. Due to polarity of capacitor voltage, the diode output is reverse bias in this condition. When switch is OFF, the output diode is conducted and decreasing the stored energy of input inductor as the capacitor voltage being to increasing. As the addition of input and output current of inductor becomes zero and output diode current also zero, a fixed current pass through the circuit in the mode of DCM.
The regulation of dc-link voltage and the multiplier approach of current to supply a basic wave model characteristic in a Continuous Conduction Mode (CCM). More sensors are need in a CCM and the sensing precondition of the input current and voltage as well as output voltage. However, due to voltage follower approach, the need sensors are minimized as the control structure and voltage output is only sensed in this converter. The input and output inductance, magnetizing inductance, output capacitor and duty cycle equations for Improved bridgeless Cuk Converter design and value are [11].

$$L_{i1,2} = \frac{V_s X D_{Cuk}}{\Delta_{i_{Li1,2}} X f_s} \tag{17}$$

$$L_{01,2} \leq \left(\frac{L_{eq} L_{i1,2}}{L_{eq} - L_{i1,2}}\right) \tag{18}$$

$$C_{batt} = \frac{D_{fb} \, X \, V_{batt}}{f_{sf}\left({V_{batt}}^2/_P\right) \gamma V_{batt}} \tag{19}$$

$$D_{fb} = \frac{V_{batt}}{\left({n_s}/{n_p}\right) V_{Cuk} + V_{batt}} \tag{20}$$

5.4 A modified isolated Luo converter

To reduce the stress of voltage across the switch and voltage input is clamped in a modified isolated Luo converter with clamping input diodes at the primary side of an isolated HFT and two switches.PFC ability to achieve in a discontinuous conduction mode of this converter , minimize the number of sensor required and minimize the complexity of control, Three mode of operation is discussed in a converter.Mode-1: Two switches are ON, the supply current increase linearly as energy stored in a magnetizing inductor. Voltage of capacitor is reducing through inductor L0 as inductor being charging through in-between capacitor C1.Mode-II: The diode will be conduction when the switches are OFF. During this condition, the magnetizing inductance supply the energy to diode output, capacitor voltage vC1 is being rising through the magnetizing inductor energy at the secondary winding of HFT, with help inductor output the energy is supplied to capacitor of output and the need current of charger is transferred to the battery. Mode-III: The two switches are OFF and two diodes are reverse biased. The magnetizing inductor store the energy in DCM operation of this converter, supply the battery energy is need by a capacitance of dc-link through inductor and the energy provide capacitor. In this converter, expression of filling efficiency, relation between input and output voltage, magnetizing inductance , maximum capacitor filter are analyzed in [15]

$$m = \frac{1}{\zeta_m} = \frac{M^2}{D^2 \frac{R}{2fL_m}} \tag{21}$$

$$M = \frac{V_0}{V_{in}} = \frac{D}{n(1-D)} \tag{22}$$

$$L_{m\,min} = \frac{Rn^2(1-D)^2}{2Df_s} \tag{23}$$

$$C_{f\,max} = \frac{I_{s\,max}}{\omega V_{s\,max}} \tag{24}$$

5.5 Single –step converter for OBC

A single-cell, Soft switching converter for on-board battery charging consists of full- bridge diode rectifier with two interleaved cells, which is keep the desired power level with maximum efficiency. A series resonant and active clamp circuits are relaxed in all cells, this converter supply ZVS of each switches and diode output of Zero-Current Switching is used for minimize the reverse recovery problem. When ZVS is ON, transformer leakage inductance and magnetizing inductance to store the recycle energy and main switch of surge voltage is clamped in this circuit. The resonant capacitor is composed in this series resonant circuit, a full-bridge diode and the leakage inductance of the transformer. Resonance intermediate the leakage inductance and resonance capacitor supply ZCF is OFF. The three mode of operation is presented in [17]. During ZVS –ON, the negative current of main switch is flow in an initial time interval. Input current of rectifier is calculated, during one grid time interval with assumption of the average output power distributed equally and null power losses. When the current is reaches zero, diode of ZCS turn OFF.
The expression of duty ratio, primary inductance, relation of input voltage and battery voltage and battery voltages are given below

$$D = 1 - \frac{nV_i}{2V_{bat,.ref}} + L_1 \frac{n}{2V_{batt.re}T_s} \Delta i_1 \tag{25}$$

$$D_n = 1 - \frac{nV_i}{2V_{bat,.ref}} \tag{26}$$

$$L_1 \frac{\Delta i_1}{T_s} = V_i D + \left(V_i - V_{C_c}\right)(1 - D) \tag{27}$$

$$\frac{V_{batt}}{V_{in}} = \frac{n}{2(1-D)} \tag{28}$$

$$V_{batt} = \frac{nDV_i}{1-D} - V_{cs} \tag{29}$$

$$\Delta D = L_1 \frac{n}{2V_{batt.ref}T_s} \Delta i_1 \tag{30}$$

The different mode of principle operation is presented in [17]

5.6 Single –step switched inductor Cuk Converter

In this converter is designed for maximum efficiency, less current stress, maximum gain of step-down with minimum number of component current and converter is operated with continuous conduction mode. Two mode of operation is presented in this converter, during charging mode of inductor, switch is ON and diodes are turn OFF by capacitor average voltage is free ripple.
The instantaneous value of inductor input, output capacitor and duty ratio are expressed as.

The average value of inductor current input is

$$I_{Li} = I_{dc} \frac{V_{dc}}{V_{in}} \tag{31}$$

Output capacitor is carried out by effective series resistance

$$\Delta V_{dc} = \frac{\Delta I_{C1}.T_s}{8C_0} \tag{32}$$

Based on input voltage, a duty ratio is expressed as

$$D = \frac{2V_{dc}}{2V_{dc}+V_{in}} \tag{33}$$

The intermediate capacitor is

$$C_1 = \frac{DT_s}{2\frac{\Delta V_{C1}}{V_{dc}}.R_0} \tag{34}$$

5.7 LLC Resonant Converter

Four mode operation of Half-bridge LLC converter with Cuk converter is operated in this converter, dc-link voltage regulated for to keep sinusoidal current input and converter is operates in CICM. During mode 1, both inductors current is increased linearly and energy stored in input inductor only. While first capacitor energy is discharged to the capacitor dc- link through switch S1.The stored energy is discharged from L2 through capacitor dc-link. The LLC converter work under frequency of resonant, thus keep away from series resonant during high voltage condition of ac mains. The LLC transformer work at fixed flux density all over variation of ac voltage.

$$L_1 = L_2 = \frac{1}{\Delta i_{L1\Delta}\, X\, f_s} \left(\frac{{V_s}^2}{P_{max}}\right) \left(\frac{V_{dc}}{V_{dc}+V_{s_{peak}}}\right) \tag{35}$$

$$C_1 = \frac{1}{\omega_r{}^2 \, X \, (L_1+L_2)} \tag{36}$$

$$C_{dc} = \frac{P_{max}}{2 \, X \omega_L \times \delta \times V_{dc}} \tag{37}$$

$$C_r = \frac{i_{Lr_{peak}}}{2\times\pi\times f_s\times 0.5\times V_{dc}} \tag{38}$$

$$L_r = \frac{1}{2\times\pi\times f_r{}^2\times C_r} \tag{39}$$

$$L_m = 6 \times L_r$$

Different value of capacitor, inductor and transformer ratio is expressed in [28, 31]

The following equations for losses evaluation of DC-DC converter

MOSFET losses during the ON/OFF-state

$$P_{son} = \frac{1}{2} f_s V_{ds} I_d (t_{12} + t_{23}) \tag{40}$$

Switching loss is

$$P_{sw} = 2P_{son} \tag{41}$$

MOSFET of conduction loss is

$$P_{R_{ds}} = f_s \int_{t_4}^{t_5} i_{ds} R_{ds} dt \approx D I_d{}^2 R_{ds(on)} \tag{42}$$

MOSFET of Reverse Recover loss is

$$P_{bdrr} = Q_{rr} V_{ds} f_s \tag{43}$$

MOSFET of driving loss is

$$P_{drive} = Q_g V_{gs} f_s \tag{44}$$

Core loss is

$$P_C = W_t\left(6.5 f_{sw}{}^{1.51} B_{ac}{}^{1.74}\right) \tag{45}$$

Conduction loss is

$$P_R = R_L I_{L,rms}{}^2 + R_C I_{L,rms}{}^2 \tag{46}$$

Diode loss is

$$P_D = I_{Drms}{}^2 r_f + V_{F0} I_D + \left(\frac{V_0}{V_{cc}}\right)^{0.6} E_{rr} \left(\frac{I_{d,rms}}{I_c}\right)^{0.6} f_{sw} \tag{47}$$

Semiconductor loss is

$$P_T = I_{s,rms}{}^2 r_{CE} + V_{CE} I_s + \left(\frac{V_0}{V_{CC}}\right)^{1.2} \left[E_{off}\left(\frac{I_{s,rms}}{I_C}\right) + E_{on}\left(\frac{I_{s,rms}}{I_c}\right)\right] f_{sw} \tag{48}$$

Total losses of converter is

$$P_{con} = N_{ph}(P_T + P_D + P_R + P_C) \tag{49}$$

Efficiency of converter is

$$\eta = \frac{P_{out}}{P_{out}+P_{con}} \times 100 \tag{50}$$

6. Conclusion

Grid to vehicle and in reverse is a favorable technology that permits EV to utilize as circulated resources, which can distribute or store energy at a suitable time, thus allowing an power exchange between the grid and EV and vice- versa, this boost reliability, stability, efficiency of the grid and expand the overall power generation capacity. To transfer power between grid and EVs, a bidirectional converter is normally used. Different classification of bidirectional converters have been reported and effectively developed in electric vehicle applications. This paper reviewed the different bidirectional and unidirectional topologies are applied to the electric vehicle applications. Moreover, the method of charger/discharger systems implemented for EVs application was studied. Formed on the review, it is possible to recommended that a standard power electronics converters and Uni/Bi-directional c h a r g e r /discharger system planed for improved quality based EVs. Interrelated must meet the following requirement: 1) To minimize the conduction and switching losses, 2) high power density potentiality for fast charging, 3) Rule of grid and EV battery voltage, 4) Fast response, 5) Stability of power factor correction and THD, 6) High reliability, 7) To meet the interrelated requirement and standards, 8) soft-switching of the power devices reduce the stress, EMI and losses, 9) simple and efficient control system, 10) to minimize the component count and system price (cost) and 11) The efficiency of the system will be upgraded. These devices can take switching power into such region of capacity with high efficiency when compared to nowadays performance levels. Furthermore, the recent research address smart charging / discharging of Electric Vehicles can inflate the working and behaviour of the upcoming renewable incorporated grid bodywork.

References

[1] F.Li,F.Ji,H.Guo ,H ,Li ,Z wang, research on integrated bidirectional control of EV charging station for vehicle to grid ,2017 2ndInt.Conf.Power renewable energy (ICPRE) (2017) 833-838.

[2] Jae IlBaek, Jae-kukkim, Jae-Bum, Han-shin young, Gun-woo moon, A boost PFC stage utilized as half-bridge converter for high-efficiency DC-DC stage in power supply unit, IEEE Transactions on power electronics,Vol.32.No.10,(Oct 2017) 7449-7487.

[3] Muntasiralam, Wilson Eberle, Deepak S, Gautam, Chris Botting, A soft-switching bridgeless AC-DC power factor correction converter, IEEE Transactional on power electronics, Vol.32, No.10,(Oct 2017) ,7716-7726.

[4] YunZhang,YongpingGao, Lei Zhou, Mark Sumner, A switched –capacitor bidirectional DC-DC converter with wide voltage gain range for electric vehicle with hybrid energy source,IEEE Transactional on power electronics, Vol.33,No.11 ,(Nov 2018), 9459-9469.

[5] Muntasir Alam,WilsonEberle, Deepak S. Gautam, Chris Botting, Nicholas Dohmeier, and Fariborz Musavi, A Hybrid Resonant Pulse-Width Modulation Bridgeless AC–DC Power Factor Correction Converter,IEEE Transactions on Industry applications ,Vol.53,No.2, (March / April 2017), 1406-1415.

[6] Bhim Singh and RadhaKushwaha , A PFC Based EV Battery Charger Using a Bridgeless Isolated SEPIC Converter, IEEE Transactions on Industry applications ,Vol.56,No.1, (JANUARY/FEBRUARY 2020), 477-487.

[7] Udupi R. Prasanna, Anant Kumar Singh and Kaushik Rajashekara, Novel Bidirectional Single-phase Single-Stage Isolated AC–DC Converter With PFC for Charging of Electric Vehicles, IEEE Transactions On Transportation Electrification, Vol. 3, No. 3, (September 2017),536-544.

[8] RamprakashKathiresan,Pritam Das, Thomas Reindl, and Sanjib Kumar Panda, A Novel ZVS DC–DC Full-Bridge Converter With Hold-Up Time Operation, IEEE Transactions on Industrial Electronics, Vol. 64, No. 6, (June 2017) , 4491 -4500.

[9] NupurSaxena,IkhlaqHussain, Bhim Singh, and AnoopLalVyas , Implementation of a Grid-Integrated PV-Battery System for Residential and Electrical Vehicle Applications, IEEE Transactions On Industrial Electronics, Vol. 65, No. 8, (August 2018),6592-6601.

[10] Dibakar Das , Nathan Weise,KaushikBasu, RohitBaranwal,and Ned Mohan, A Bidirectional Soft-Switched DAB-Based Single-Stage Three-Phase AC–DC Converter for V2G Application,IEEE Transactions on Transportation Electrification, Vol. 5, No. 1, (March 2019) , 186-199.

[11] RadhaKushwaha, and BhimSingh,A Power Quality Improved EV Charger With Bridgeless Cuk Converter,IEEE Transactions on Industry Applications, Vol. 55, No. 5, (September/October 2019) ,5190-5203.

[12] Rouzbeh Reza Ahrabi, HosseinArdi, MahdiElmi, and Ali Ajami , A Novel Step-Up Multiinput DC–DC Converter for Hybrid Electric Vehicles Application , IEEE Transactions on Power Electronics, Vol. 32, No. 5, (May 2017) 3549-3561.

[13] J. Gnanavadivel, P. Yogalakshmi, NatarajanSenthil Kumar, K.S. Krishna Veni,Design and development of single phase AC– DC discontinuous conduction mode modified bridgeless positive output Luo converter for power quality improvement,IET Power Electronics, ISSN 1755-4535 (2019) 1-9.

[14] BhimSingh,and RadhaKushwaha,A PFC Based EV Battery Charger Using a Bridgeless Isolated SEPIC Converter,IEEE Transactions on Industry Applications, Vol. 56, No. 1, (January/February 2020),477 -487.

[15] RadhaKushwaha , and Bhim Singh, A Modified Luo Converter-Based Electric Vehicle Battery Charger With Power Quality Improvement, IEEE Transactions on Transportation Electrification, Vol. 5, No. 4, (December 2019) , 1087-1096

[16] Fei Shang, GengNiu, and Mahesh Krishnamurthy, Design and Analysis of a High-Voltage-Gain Step-Up Resonant DC–DC Converter for Transportation Applications,IEEE Transactions on Transportation Electrification, Vol. 3, No. 1, (March 2017), 157 -167.

[17] Seo-GwangJeong ,Yeong-SeungJeong , Jung-Min Kwon ,and Bong-Hwan Kwon, A Soft-Switching Single-Stage Converter With High Efficiency for a 3.3-kW On-Board Charger ,IEEE Transactions on Industrial Electronics, Vol. 66, No. 9, (September 2019) , 6959 - 6967

[18] Vitor A. G. Cunha, Antonio O. Costa Neto, Gustavo B. Lima, Luiz C. G. Freitas , A Bridgeless Boost Half Bridge DC-DC Converter for Electrical and Hybrid Vehicle Applications ,978-1-5386-8218-0/19/$31.00 ©2019 IEEE.

[19] Sundaramoorthy kumaravel ,Ravishankar Achathuparambil Narayanankutty, VemparalaSeshagiriRao , Ashok sankar , Dual input–dual output DC–DC converter for solar PV/battery/ultra-capacitor powered electric vehicle application, IET Power Electronics,ISSN 1755-4535 (2019) , 3351-3358.

[20] Francisco J. B. BritoJr, Sergio Daher, Ren´e P. T. Bascop´ e, Breno B. Chaves, Debora P. Damasceno and Bruno R. de Almeida,Design and Control for a Single-Phase/-Stage Buck/Boost-Type PWM PFC Rectifier , 2018 13th IEEE International Conference on Industry Applications , 530-536.

[21] B. R. Ananthapadmanabha,RakeshMauryaandSabha Raj Arya,Improved Power Quality Switched Inductor Cuk Converter for Battery Charging Applications , IEEE Transactions on Power Electronics, Vol. 33, No. 11, (November 2018) , 9412- 9423.

[22] Chuan Shi, Haoyu Wang, Serkan Dumez, and Alireza Khaligh,A SiC-Based High-Efficiency Isolated Onboard PEV Charger With Ultra wide DC-Link Voltage Range , IEEE Transactions on Industry Applications, Vol. 53, No. 1, (January/February 2017)501 – 511.

[23] RahulPandey and Bhim Singh, a Power-Factor-Corrected LLC Resonant Converter for Electric Vehicle Charger Using Cuk Converter,IEEE Transactions on Industry Applications, Vol. 55, No. 6, (November/December 2019) ,6278-6286.

[24] K. v. Ravi Kishore, B. F. Wang, K. N. Kumar and P. L.,A New ZVS Full-Bridge DC-DC Converter for Battery Charging with Reduced Losses over Full-Load Range ,978-1-4673-6540-6115/$31.00 ©2015 IEEE.

[25] A.V.J.S.Praneeth, and Sheldon S Williamson, A Soft Switched Boost Cascaded-by-Buck Power Factor Correction Converter for On-board Battery Charger Application,978-1-5386-8330-9/19/$31.00 ©2019 IEEE 567-572.

[26] Yun Zhang , YongpingGao, Lei Zhou , and Mark Sumner, A Switched-Capacitor Bidirectional DC–DC Converter With Wide Voltage Gain Range for Electric Vehicles With Hybrid Energy Sources, IEEE Transactions on Power Electronics, Vol. 33, No. 11, (November 2018) 9459-9469.

[27] Ahmed Sheir , Mohamed Z. Youssef , and Mohamed Orabi , A Novel Bidirectional T-Type Multilevel Inverter for Electric Vehicle Applications, IEEE Transactions on Power Electronics, Vol. 34, No. 7, (July 2019) 6648-6658.

[28] RahulPandey and Bhim Singh, A Power Factor Corrected LLC Resonant Converter for Electric Vehicle Charger Using Cuk Converter, 978-1-5386-9316-2/18/$31.00 ©2018 IEEE.

[29] NupurRathore,Sivanagaraju Gangavarapu ,Akshay Kumar Rathore , and Deepak Fulwani, Emulation of Loss Free Resistor for Single-Stage Three-Phase PFC Converter in Electric Vehicle Charging Application, IEEE Transactions On Transportation Electrification, Vol. 6, No. 1,(March 2020), 334-345.

[30] Yun Zhang , Lei Zhou , Mark Sumner, and Ping Wang, Single-Switch, Wide Voltage-Gain Range, Boost DC–DC Converter for Fuel Cell Vehicles, IEEE Transactions On Vehicular Technology, Vol. 67, No. 1, (January 2018) 134-145.

[31] RahulPandey, Bhim Singh, A Power Factor Corrected Electric Vehicle Battery Charger Using Boost Converter, 978-1-5386-4996-1/18/$31.00 ©2018 IEEE.

[32] F.Zhang,X.Wang,Q.Liao,C.G.Auth,Highlyintegratedbidirectionalvehicle-to- grid (V2G) for electric vehicles based on open winding permanent magnet synchronous motor, 201720th Int. Conf. Elect. Mach. Syst. (ICEMS) (2017) 1–4.

[33] NavidShafiei, Martin Ordonez , Mohammad Ali SaketTokaldani ,and SeyedAli Arefifar, PV Battery Charger Using an L3C Resonant Converter for Electric Vehicle Applications,IEEE Transactions on Transportation Electrification, Vol. 4, No. 1, (March 2018), 108-121.

[34] Jaya SaiPraneeth Ammanamanchi Venkata, Sheldon S. Williamson , Analysis and design of single-stage, two- mode AC/DC converters for on-board battery charging applications, IET Power Electronics,ISSN 1755-4535,(2019), 830-843.

[35] M. Kwon, S. Choi, An electrolytic capacitor less bidirectional EV charger for V2G and V2H applications, IEEE Trans. Power Electron. 32 (September9) (2017) 6792–6799.

[36] UnalYilmaz, OmerTutksoy, AhmetTeke, Intelligent control of high energy efficiency two –stage battery charger topology for electric vehicles. Energy (Elsevier) (2019)

[37] L.Zhu, et al ,An symmetrical three level dual input bidirectional dc/ac converter with improved converter efficiency for vehicle to grid application,2017 IEEE Applications Power electron ,conf.Expo (APEC) (2017) 2062-2067.

[38] D. Varajao, R.E. Araujo, L.M. Miranda, J.P. Lopes, N.D.Weise, Control of an isolated single-phase bidirectional AC-DC matrix converter for V2G applications, Elect. Power Syst. Res. 149 (August) (2017) 9–29.

[39] T.peng.P.Yang.H.Dan, H.Wang, H.han.J yang et al., A single phase bidirectional AC/DC converter for V2G applications, Energies 10 (Jan 7) (21017) 881.

[40] P.Yang, T.Peng, H.Wang, H.Han, J.Yang, H.Wang, A single phase current source bidirectional converter for V2G application,20147 IEEE 3rdInt.Future energy electron,Conf. and ECCE Asia (IFEEC 2017-ECCE Asia) (2017) 704-709.

[41] S.Ebrahimi, R.Akbari, F.Tahami, H.Oraee, An isolated bidirectional integrated plug in

hybrid electric vehicle battery charger with resonant converters, Elect. Power components syst.44 (July 12) (2016) 1371-1383.

[42] Y. Tang, J. Lu, B. Wu, S. Zou,W. Ding, A. Khaligh, An integrated dual-output isolated converter for plug-in electric vehicles, IEEE Trans. Veh. Technol. 67 (February 2) (2018) 966–976.

[43] Salehidobakhshari, J.Milimonfared, M.Taheri, H.Moradisizkoohi, A quasi resonant current fed converter with minimum switching losses, IEEE Trans.Power Electron.32 (Jan 1) (2017) 353-362.

IOP Conf. Series: Materials Science and Engineering **906** (2020) 012007 doi:10.1088/1757-899X/906/1/012007

Power Monitoring and Control System for Medium Voltage Smart Grid Using IoT

Challa Babu[1], D. Dinesh Kumar[1*], Kumar K[2], K Jyotheeshwara Reddy[3], R Sudha[4]

[1] Department of electrical and electronics engineering, Audisankara college of engineering and technology, Gudur, A.P, India

[2] Associate Professor, SV College of Engineering, Tirupathi, AP, India

[3]Associate Professor, Sree Vidayanikethan Engineering College, Tirupathi, A.P., India

[4] Division of Automation, Department of Production Systems, Högskolan Väst, Trollhättan, Sweden-46131

Abstract. This paper presents a power monitoring and control system for a medium voltage smart grid system. The smart grid interconnects the power sources between solar PV panel and 220V distribution network. This system consisting major components of inverters, measuring meters, solar charge controllers, relays, Arduino NANO and Raspberry Pi. The Current, power and energy readings are duly recorded. The Internet of Things (IoT) plays the vital role in the data communication between the sensors and electric power system. The voltage and current sensor data is used for the protection of power system network. In the traditional systems only the communication is uni-directional. The reliability of power supply is increased by used the bi directional network communication medium such as IoT. The tabulated results of voltage levels between 203.5V up to 212.8V shows the feasibility and effectiveness of proposed design. The proposed IoT model demonstrated the bidirectional communication from the sensors to the control unit and vice versa.

1. Introduction

Electric energy generated is transmitted and distributed through over head lines and underground cables at different voltage levels. In traditional system the power transactions were done via one way communication due to which generation of power at user side is not possible. This was overcome by the concepts of microgrid and smart grid. Smart grid uses the latest communicating technologies like smart metering systems and Internet of Things (IoT) which makes the tracking and controlling of power transactions more effective, flexible and transparent.

IoT is the most emerging technology where the devices communicate with each other through the cloud. IoT embedded devices will provide enhanced functionality without exposing the end user to the complexities of conventional computer based systems. These systems interface with the real world through sensors, and actuators. IoT devices are implemented in both hardware modules like Arduino, Raspberry Pi and software components. Smart metering technology with IoT makes the power transactions between the utility and consumers [1-6].

The smart grid includes hybrid power generation [7-10] integrating solar, wind and fuel cell energy resources. A Perturb and Observe (P&O) algorithm for Maximum Power Point Tracking (MPPT) is used for increasing power extraction efficiency [11-14]. In order to integrate the solar power generation with conventional grid fly back power processing converter is adopted [15]. Several monitoring techniques such as M2M low PAN wireless sensor network, integrating of renewable resources and traffic optimizing in smart grids are proposed in [16-18]. The main objective is to design the power monitoring and control system for a medium voltage smart grid using IoT.

IOP Conf. Series: Materials Science and Engineering **906** (2020) 012007 doi:10.1088/1757-899X/906/1/012007

2. Power Monitoring and Control system

The architecture of on grid solar power generation consisting of solar PV panels, MPPT solar charge controller, on grid inverter, storage batteries and three relays are shown in figure 1. The relay-1 shown in figure 1 is operated to determine the grid connected mode or islanded mode operation depending upon the supply from utility grid and charge condition of the batteries. If the batteries are fully charged the power generated by the solar panels is fed directly to the utility grid otherwise the power generated is used to charge the battery bank.

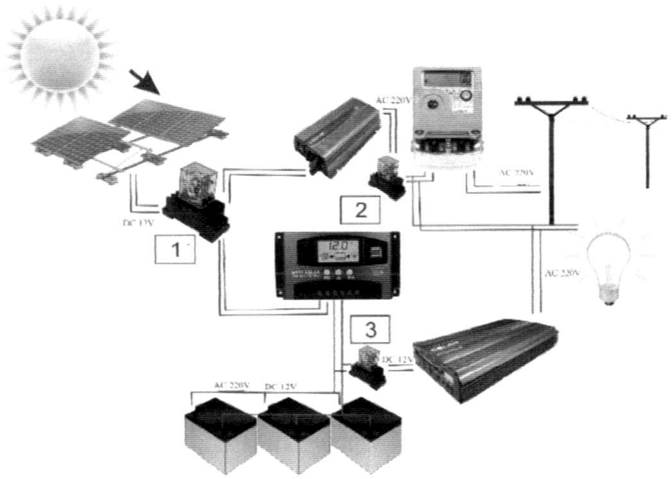

Figure 1: Architecture of on grid solar power generation.

The relay-2 is to connect and disconnect power supply from the solar PV panels while sending signal to the relay-1. If there is no power flow from the grid, the realy-2 disconnects the power supply form solar PV panels to the utility grid. In case of power flow from the grid was detected, the output power from the inverter is fed to grid.

The relay-3 connects or disconnects the battery from delivery the power to the load. This function is based on the incoming signal from the relay-2. If there is no incoming power from the utility grid relay-2 will send signal to relay-3 such that it connects battery bank to the load. In the same way if there is incoming power from the grid relay-3 disconnects the battery bank from the load.

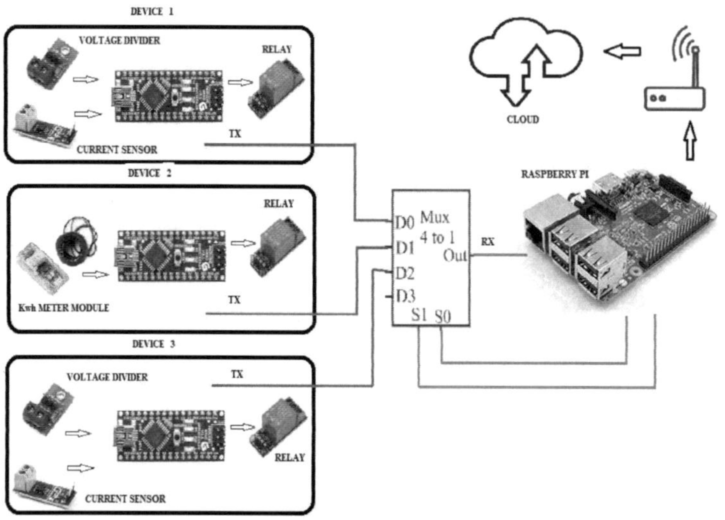

Figure 2: power monitoring and control scheme using Arduino nano and Raspberry Pi

The power monitoring and control scheme in medium voltage solar power generation is shown in figure 2. The Device 1 consisting of Arduino nano, voltage sensor, current sensor and a relay circuit which is used to regulate the power supply of the solar PV panels which feeds to the utility grid or to the battery bank depending upon the setting parameters.

The device-2 consisting of Arduino nano, Kwh meter module and a relay circuit which is used to connect and disconnect from the utility grid by sensing the incoming or outgoing current at the output side of the on grid inverter of solar PV panels. The device-3 consisting of Arduino nano, voltage sensor, current sensor and a relay circuit which is used to connect and disconnect the battery banks from the load. The central controlling unit which is Raspberry pi communicates with all the three devices. The central controlling unit receives data and sends data to the Arduino nano devices through the cloud which can be remotely accessible with smart mobile or computer.

3. Results and Observations

The monitoring system parameters voltage, current, power and energy are obtained in intervals of time via cloud which is shown in figure 3.

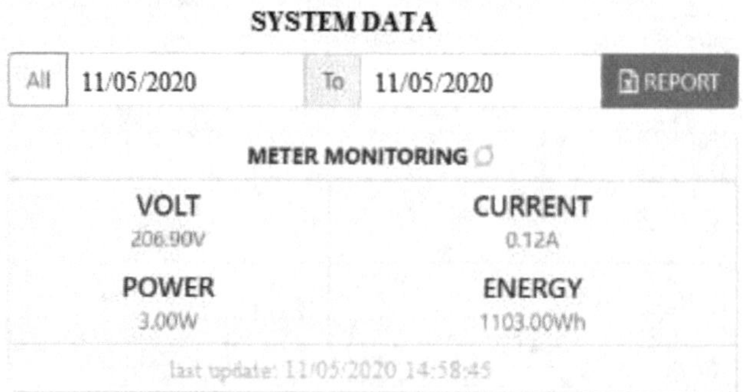

Figure 3: Display of parameters from the cloud.

A 16W lighting load is taken to monitor and control the proposed system and the testing results are tabulated in table 1.

TABLE 1: Monitoring Events at Different Instant of Time

Event No	Time (ms)	Voltage (V)	Current (A)	Power (w)	Energy (Kwh)
1	50	207.3	0.07	14.51	0.0522
2	100	207.8	0.07	14.546	0.0523
3	152	208.2	0.07	14.544	0.5240
4	200.5	209.1	0.07	14.637	0.0526
5	277	209.4	0.08	16.752	0.0603
6	310.2	209.4	0.07	14.658	0.0527
7	352	209.4	0.07	14.658	0.0527
8	400.5	209.9	0.07	14.693	0.0528

The voltage samples recorded at different instant of time are shown in figure 4. It was observed that the voltage values are in between 203.3V to 209.9V. The average value of all the events is 207.7 and standard deviation is 0.9 indicates that the voltage is monitored and controlled effectively.

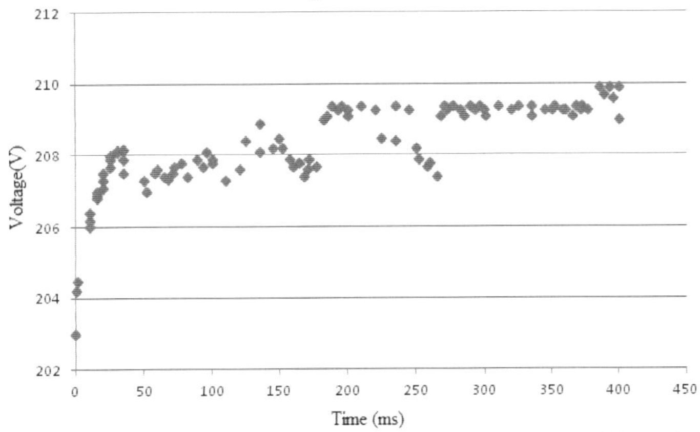

Figure 4: Voltage sample points at different instant of time (ms).

The current samples recorded at different instant of time are shown in figure 5. The average value for all the events is obtained as 0.07A. The load is disconnected at 400ms and therefore current becomes zero as shown. The energy consumed by the load in Kwh is shown in figure 6. The energy consumed was almost constant on an average at 0.05 of all the events and that becomes zero after disconnecting the load at 400ms

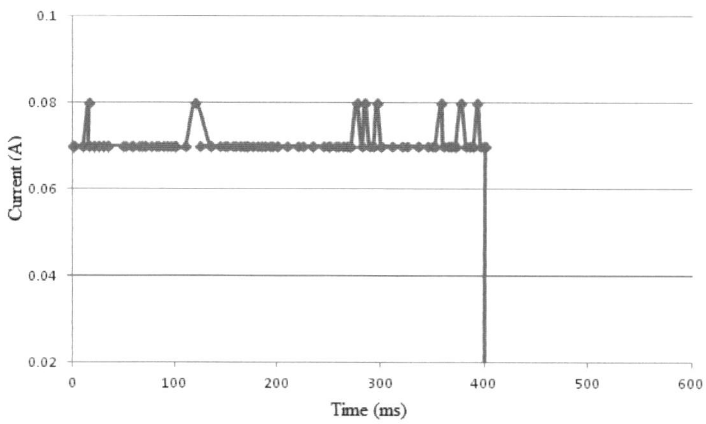

Figure 5: Current sample points at different instant of time (ms).

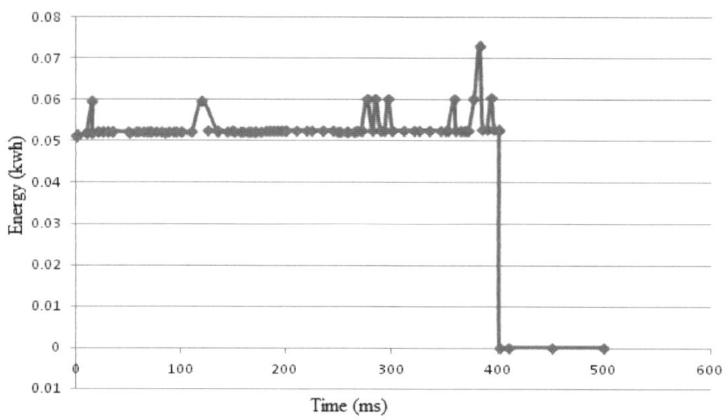

Figure 6: Energy consumed by load at different instant of time (ms).

4. Conclusion

The smart grid technology is the effective solution to perform the power transactions between the utility and the end users. In order to control distribution generating units of a microgrid, and monitor power transactions by different consumers effectively the Communication system play very important role. Therefore, A Power Monitoring and Control System for Medium Voltage Smart Grid Using IoT is developed. The voltage is controlled with a standard deviation of 0.9. The current values and energy consumed are constant at 0.07A and 0.05 respectively. The results obtained show the effectiveness of proposed scheme with and without load conditions.

5. References

[1] Challa Babu, D. Dinesh Kumar, K.Jyotheeshwara Reddy, Kumar K., R. Sudha, 'IoT Enabled Interactive Advanced Metering Infrastructure Technology in Smart Grids', International Journal of Recent Technology and Engineering, ISSN: 2277-3878 (Online), Volume-8 Issue-5, January 2020. Page No. 3483-3488.

[2] Raut, M.M., Sable, R.R. Toraskar, S.R., Internet of Things(IoT) Based Smart Grid, International Journal of Engineering Trends and Technology (IJETT), Vol. 34, No. 1, April 2016, ISSN 2231-5381, pp.15-20.

[3] Camacho, E.F., Samad, T., Sanz, M.G., Hiskens, I., Control for Renewable Energy and Smart Grids, The Impact of Control Technology, 2011.

[4] Natsheh, E.M., Albarbar, A., Yazdani, J., Modeling, and control for smart grid integration of solar/wind energy conversion system, 2011 2nd IEEE PES International Conference and Exhibition on Innovative Smart Grid Technologies, Manchester, UK, 5-7 Dec. 2011.

[5] D. Dinesh Kumar, Challa Babu, K Jyotheeswara Reddy, Kumar K, N K Kumar, "An improved P&O MPPT control algorithm for increasing power extraction efficiency of solar PV module". In IOP Conference Series: Materials Science and Engineering 2019 Oct (Vol. 623, No. 1, p. 012020). IOP Publishing.

[6] D. Dinesh Kumar, Challa Babu, P. Ponnambalam, "A Flyback Partial Power Processing Converter based Solar PV Systems Interfacing the Grid" IEEE conference, Innovations in Power and Advanced Computing (i-PACT-2019) 10.1109/i-PACT44901.2019.8960045.

[7] Shrihariprasath, B., Rathinasabapathy, V., Design and Implementation of Solar PV Smart Grid System Using IoT and M2M Integrated 6LowPAN Wireless Sensor Network, International Conference on Advances in Information Technology and Networking, Feb. 2017.

[8] Foleya, A., Connollycd, D., Leahya, P., Mathiesend, B.V., Lunde, H., Leahycd, M., McKeogha, E., Electrical Energy Storage & Smart Grid Technologies to Integrate the next generation of Renewable Power Systems, SEEP2010, Conference Proceedings, June 29th-July 2d, 2010,Bari, Italy

[9] Habault, G., Lefranc¸ois, M., Lemercier, F., Montavont, N.Chatzimisios, P., Papadopoulos, G.Z., Monitoring Traffic Optimization in Smart Grid, IEEE Transactions on Industrial Informatics, Jan. 2017, pp.1-9.

[10] Babu Challa, and P. Ponnambalam. "The role of thermoelectric generators in the hybrid PV/T systems: A review." Energy Conversion and Management 151 (2017): 368-385.

[11] Babu, Challa, and P. Ponnambalam. "The theoretical performance evaluation of hybrid PV-TEG system." Energy Conversion and Management 173 (2018): 450-460.

[12] Kumar DD, Babu C, Reddy KJ, Kumar K, Kumar NK. An improved P&O MPPT control algorithm for increasing power extraction efficiency of solar PV module. In IOP Conference Series: Materials Science and Engineering 2019 Oct (Vol. 623, No. 1, p. 012020). IOP Publishing.

[13] Babu, Challa, and Ponnambalam Pathipooranam. "PV Module Temperature Estimation by Using ANFIS." Soft Computing for Problem Solving. Springer, Singapore, 2020. 311-318.

[14] Ponnambalam, P., Praveenkumar, M., Babu, C., & Raj, P. D. (2017). Analysis of Stacked Multicell Converter with Fuzzy Controller. In Proceedings of Sixth International Conference on Soft Computing for Problem Solving (pp. 318-330). Springer, Singapore.

[15] Leela Madhav.k , Challa Babu, Dr.P.Ponnambalam, Ashutos Mahapatra., " Fuzzy Logic Controller for Nine Level Multi-Level Inverter with Reduced Number of Switches" IEEE conference i-PACT-2018, DOI: 10.1109/IPACT.2017.8244938.

[16] S. Gupta, N. Sinha, R. Sudha and C. Babu, "Breast Cancer Detection Using Image Processing Techniques," *2019 Innovations in Power and Advanced Computing Technologies (i-PACT)*, Vellore, India, 2019, pp. 1-6, doi: 10.1109/i-PACT44901.2019.8960233.

[17] Ponnambalam, P., K. Muralikumar, P. Vasundhara, S. Sreejith, and Babu Challa. "Fuzzy Controlled Switched Capacitor Boost Inverter." Energy Procedia 117 (2017): 909-916.

[18] Anipireddy, P., and C. Babu. "Modeling and Simulation of Three Level Piezoelectric Transformer Converters." Adv Robot Autom 3.120 (2014): 2.

Integrated Multi-Port Converter For Off - Grid Photovoltaic System

Divya M*, Banumathi S

Department of Electrical and Electronics Engineering,
M.Kumarasamy College of Engineering,
Karur-639113, Tamil Nadu, India
Email: divyamuralib.e@gmail.com
Email: banumathis.eee@mkce.ac.in
*Corresponding author

Abstract. The multi-port converter proposed in this work is used for the integration of many converters in a single system. The system is the solar dc-dc standalone converter with the battery storage. The proposed integrated multiport converter consists of three separate converters they are, bidirectional PWM converter, switched capacitor converter, series resonant converter. The numbers of switches are reduced and thus the loss in the power electronics circuit is also reduced. A 300 watts prototype model of the system is implemented and the hardware results obtained is discussed and the same system is simulated using matlab software and the results are matched.

Keywords: multi-port; PV standalone; PWM converter; reduced switching; PID controller.

1. Introduction

Recently, control gadgets examine has concentrated on usage of sustainable power sources to supplant petroleum product, on account of its harmful natural impacts (Debnath et al., 2015). Because of the ecological issue, e.g., environmental change, and political and monetary reasons, e.g., less reliance on the remote vitality import, high oil cost, sustainable power source has pulled in colossal consideration. As extraordinary development in sustainable power source has been made, the cost for power delivered with sustainable power source drops. The national normal power levy cost is around 5 rupees for every kWh. The cost for particular sorts of sustainable power source is lower than 5 rupees for each kWh. As the expense of ordinary vitality assets is expanding each year, the retreating pattern in the expense of sustainable innovations is empowering, making sustainable power source an efficient methods for control age sooner rather than later. Sustainable power source advances not exclusively can tackle the environmental change and decrease the reliance on remote vitality import, it is likewise appropriate for dispersed power age. In remote territories, where there are no transmission lines or the expense of building new transmission lines is high, sustainable power source can give control without costly and muddled framework foundation. Multi-port segregated converter topologies positively affect utilizing different sustainable power sources, and can accomplish higher

power and voltages with improved productivity because of develop medium power semiconductor innovation, they are starting to stand out for control hardware(Su.G et al ., 2008)
Switched capacitor converters are minimal effort and smaller and are fit for accomplishing efficiencies more prominent than 90%. Clearly, the present yield is restricted by the current conveying limit of the switches and the size of the capacitors. Switched capacitor voltage converters don't keep up high productivity for a wide scope of proportions of contribution to yield voltages, not at all like their exchanging controller partners. Since the contribution to yield current proportion is scaled by the essential voltage transformation paying little mind to whether guideline is utilized to lessen the multiplied or reversed voltage, any yield voltage size under 2VIN for a doubler or not exactly |VIN| for an inverter will bring about extra power dispersal inside the converter, and productivity will be debased relatively.
Bidirectional PWM converter has increased voltage transformation proportion and programmed current adjusting capacity for single-cell battery control frameworks. By adding just two capacitors to the ordinary interleaved three-stage PWM converter the voltage transformation proportion at a given obligation cycle is significantly increased, henceforth counteracting extraordinary obligation cycle activities. Inductor flows can be naturally adjusted gratitude to the additional capacitors, and no present sensor is essential for inductor current adjusting. Moreover, the additional capacitors lessen voltage worries of switches and charged-released energies of inductors, adding to effective power transformation and diminished circuit volume.
The series-resonant converter (SRC) are utilized to infer shut structure answers for the yield qualities and enduring state control attributes, to decide working mode limits, and to discover top segment stresses. There are two ways: Zero current exchanging and Zero voltage exchanging.
Turn-on or turn-off changes of semiconductor gadgets can happen at zero intersections of tank voltage or current waveforms, along these lines lessening or killing a portion of the exchanging misfortune instruments. Thus converters can work at higher frequencies than PWM converters. Zero-voltage exchanging likewise lessens converter-created EMI Zero-current exchanging can be utilized to commutate SCRs. In particular applications, thunderous systems might be unavoidable. High voltage converters: critical transformer spillage inductance and winding capacitance prompts resounding system.

2. Block Description

The block diagram of switched converter multi-port system switched capacitor multi-port converter is the integration of three separate converters into single converter (Sun.X et ., 2015).the three converters are Series resonant converter it consist of two switches S1, S2. bidirectional PWM converter S3,S4.finally Switched capacitor converter it have S1,S2,S3,and S4 switches, by using these four switches the three converter are combined in a single MPC.

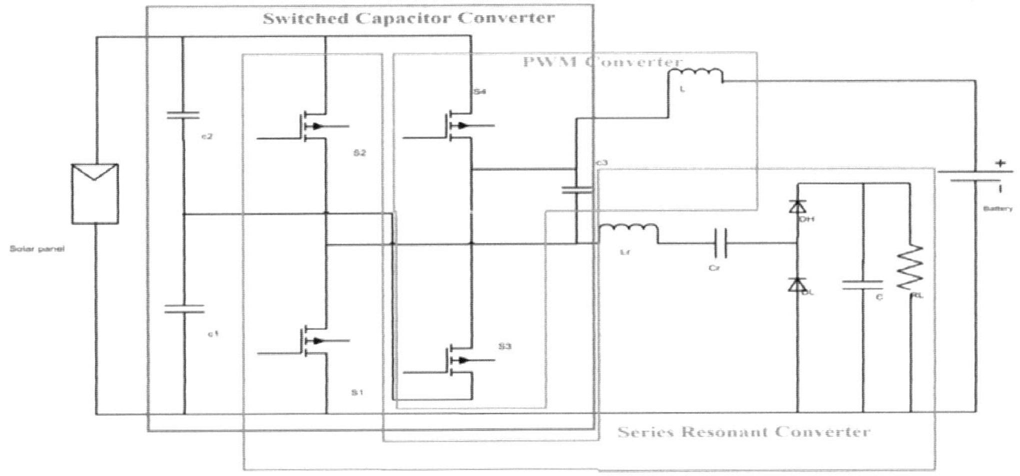

Figure 1 Block Diagram of Multi-Port Converter.

2.1. Switched Capacitor Converter Based Multiport Converter Major Benefits

In examination with the traditional PV framework, the framework is drastically streamlined by incorporating 3 converters into a 1 unit. Moreover, the complete switch tally can likewise be decreased, subsequently accomplishing the circuit rearrangements; the proposed Multi-port requires four mosfet switches altogether, though the all-out mosfet switch tally is the traditional framework is ten. Notwithstanding the framework and circuit-level rearrangements, scaled down plan is additionally achievable gratitude to the Switched capacitor converter. The proposed Multi-port is fundamentally a crossover Switched capacitor converter that can extraordinarily cut back inductors, as announced.

2.2. PWM- Pulse Width Modulation

Pulse width modulation (PWM) is a strategy, which the exchanged voltages are created for various frequencies and voltages. PWM gives an approach to diminish the All-out Symphonies Twisting (THD) of burden current. The THD necessity can be met when the yield of a PWM inverter is sifted since the unfiltered yield of a PWM based reversed will have a moderately high contortion. There are numerous PWM plans, the most mainstream ones being sinusoidal PWM (SPWM). It very well may be actualized utilizing simple systems, the remaining PWM require the utilization of a chip. The PWM innovation redresses the yield voltage as indicated by the changing the Width of the exchanging recurrence in the oscillator side.

2.3. Circuit of the System with Mosfet

This MPC presents a recently structured, non-segregated single-input double yield 3 level of dc-dc converter. With a support of control procedure, the converter profits by the 3 levels and multiport system. Inferable from its 3 level structure, the given converter has the upsides of decreased voltage weight on diodes and switches, improved effectiveness and diminished aloof parts size.

IOP Conf. Series: Materials Science and Engineering **906** (2020) 012008 doi:10.1088/1757-899X/906/1/012008

3. Principle Operation

There are three modes of operation in this multi-port converter each operation are used with different types of switches. The red line in figure 2,3and4 denotes the active switching elements in the converter. This converter consists of four mosfet switches S1, S2, S3 and S4.

3.1. Mode 1

In mode 1, the mosfet switches S1 and S3 turns on and the current passes through the capacitor C2 to the S2 switch, from the switch S2 to the capacitor C3,from the capacitor C3 to the switch S4. Voltages at the Vx and Vw, are in their elevated state. Capacitors are associated in parallel in the switched capacitor converter charge and release one another so their voltages are brought together and the current flow in SRC.

Mode 1: (0.5 < d2 & d1 < 1) & (d1 > d2).

3.2. Mode 2

In mode 2,the MOSFET switches S4 and S2 turns on and the current passes through the capacitor C2 to the S2 switch, from the switch S2 to the capacitor C3,from the capacitor C3 to the switch S4. Voltages at the Vx and Vw, are at their elevated level. Meanwhile, there is no current flow in SRC.

Mode 2: (0.5< d2 & d1 < 1) & (d1 < d2).

3.3. Mode 3

In Mode 3,the multi-port converter is similar to mode 2 operation here the current*iLr1* and *iLr2* become zero. The SRC is inactive in this mode.

Mode 3: (d2+0.5 < d1 < 1) & (0 < d2 < 0.5).

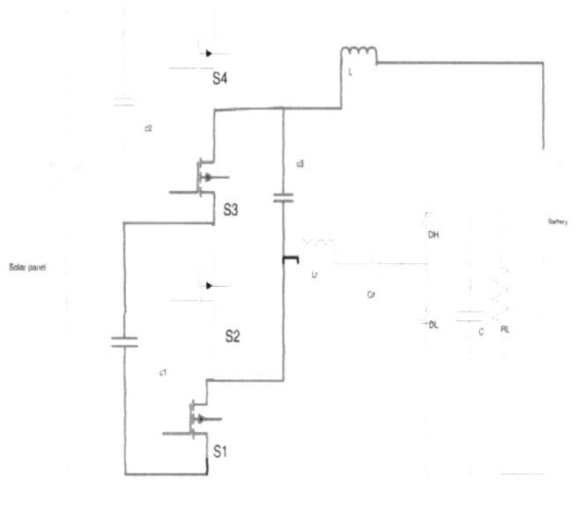

Figure 2. mode 1

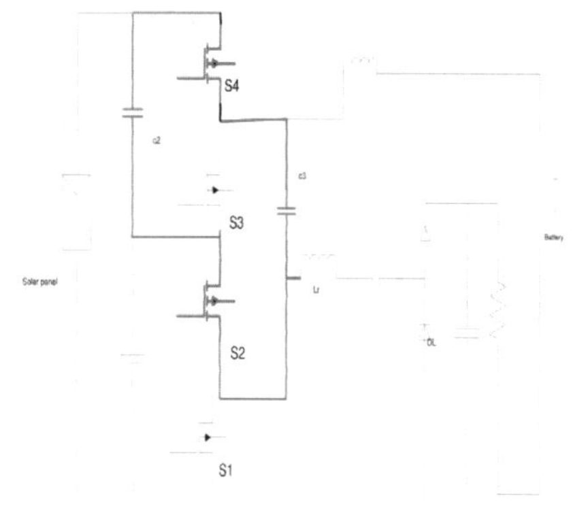

Figure 3. mode2

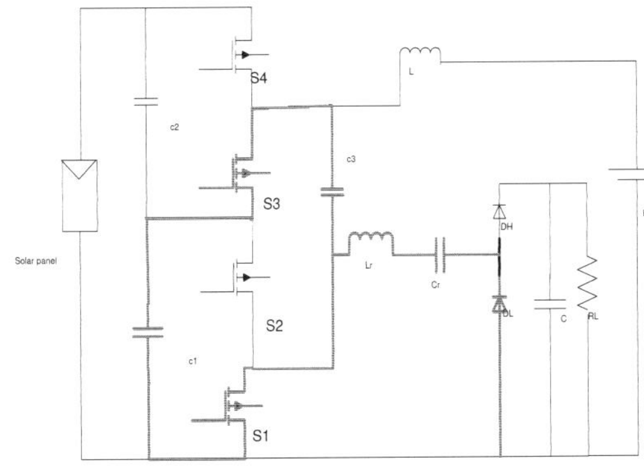

Figure 4. mode 3

From Table I and the switching sequences S1, S2, S3 and S4 are given, the output voltages conversion ratio can be gained for mode 1 as follows:

For the L1 inductor,

$$Vin\{d2-0.5\} + \{Vin-Vo1/2\}(d1-d2)+(Vin-Vo1)(1-d1)=0$$

Hence,

$$Vo1/Vin=1/(2-d1-d2) \quad (1)$$

Table 1 Operating Range of The Converter

Table I

Mode	Duty-cycle	Voltage limits
1	0.5<d2&d1<1 d1>d2	Vin/2<Vo
2	0.5<d2&d1<1 d1<d2	0<Vo<Vin/2
3	d2+0.5<d1<1	Vo<2Vin

$$(-Vo2)(d2-0.5)+(Vo1/2-Vo2)(d1-d2)+(Vo1/2-Vo2)(1-d1)=0$$

Hence,

$$Vo2/Vo1=1-d2 \quad (2)$$

Thus,

$$Vo2/Vo1=(1-d2)/(2-d1-d2) \quad (3)$$

The voltage gains in mode 2and 3 can also be obtained in the same way as the above procedure.

Voltage gains for all three cases become

$$Vo1/Vin= 1/(2-d1-d2),\ \text{mode } 1\&2 \quad (4)$$

$$1/(1-d2),\ \text{mode3}$$

$$Vo1/Vin=(1-d2)/(2-d1-d2), \text{ mode } 1\&2 \quad (5)$$
$$(d1-d2)/(1-d2), \quad \text{mode3}$$

From (4) and (5), it can be seen that d1 and d2 are the control values for output voltages v1 and v2. In mode 1 and 2, the step-up output voltage is related to both duty cycle d1 and d2, while in the mode 3, it is only related to d2. On the other hand, the step-down output voltage in all 3 cases is similar to both duty-cycle d2 and d1. More detailed analysis of the control strategy will be tested in the following system.

4. **Simulations and Hardware Model**

The figure 5 shows the Simulink model of the proposed system. The power generated from the dc source as PV systems is passed over the subsystem. The output of the converter is used to link the single phase light load and if excess power is generated from PV more than load, the surplus power passes to the multi-port through link. Simulation input is highly given so comparatively the output is higher than hardware output.

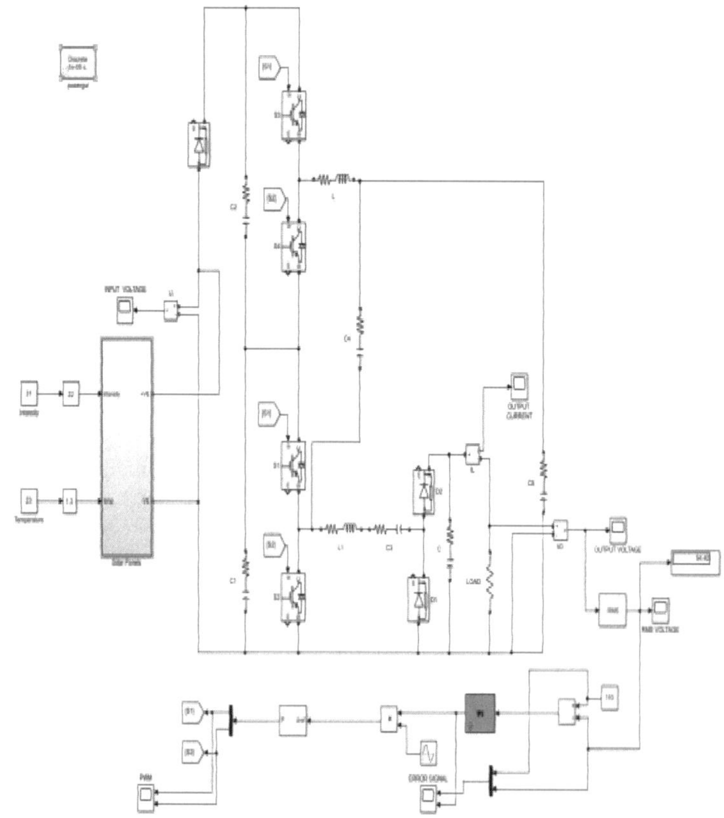

Figure 5. Simulation In Matlab.

Based on the power generated (low or high voltage) there will be a variation in the graph (In other words, the power generated is directly proportional to the variation in the graph).Figure 6 shows the current and voltage waveforms at the load while the connection of Load.

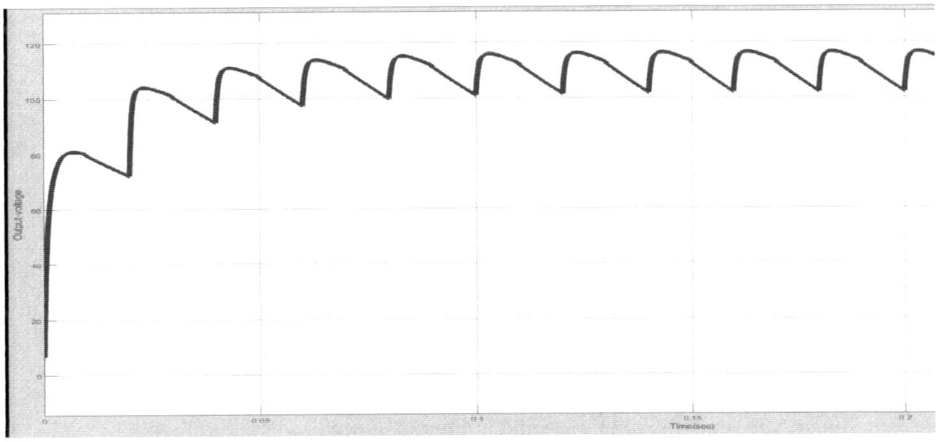

Figure 6. Simulation Output.

The waveform shows that the voltage and the current with the time in X-axis and amplitude in Y-axis.

Table2 Experimental Results

Parameter	Value
Total output power	300W
Input voltage	15.4V
Output voltage V1	36V
Output voltage V2	30.3V
resistive load	20 ohm
Switching frequency	20 kHz

4.1 Hardware analysis

A prototype of 300W converter has been designed for the proposed system the DC source is taken instead of Solar-panel. The maximum power obtained from the panel is when the panel is 15-28 degree inclined. Accordingly the maximum power is set to 30W.In the converter we use MPPT based on perturb and observation method. The input voltage is taken as 15.4V given in figure 8. Every single converter are verified and tested separately by giving different pulse to the multi-port converter(figure 8 to figure 12).the converter is tested in two stage, pulse s1&s2 and pulse s3&s4 in figure 9,10.

From the multi-port converter output waveform 15.4V stepped-up to 36V.the output voltage V2 is step-down to30.3V.

From the bidirectional PWM converter the pulse output waveform is tested. The PID controller is used for the control operation of the system.

Figure 7. Hardware Output

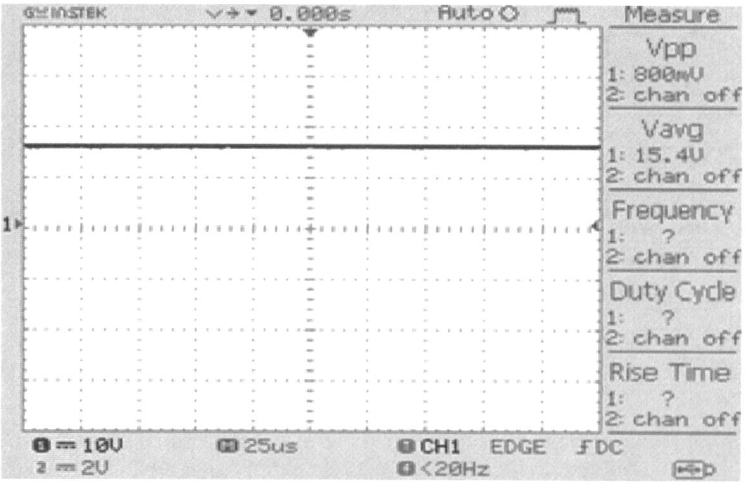

Figure 8. Input voltage in DSO for converter

IOP Conf. Series: Materials Science and Engineering **906** (2020) 012008 doi:10.1088/1757-899X/906/1/012008

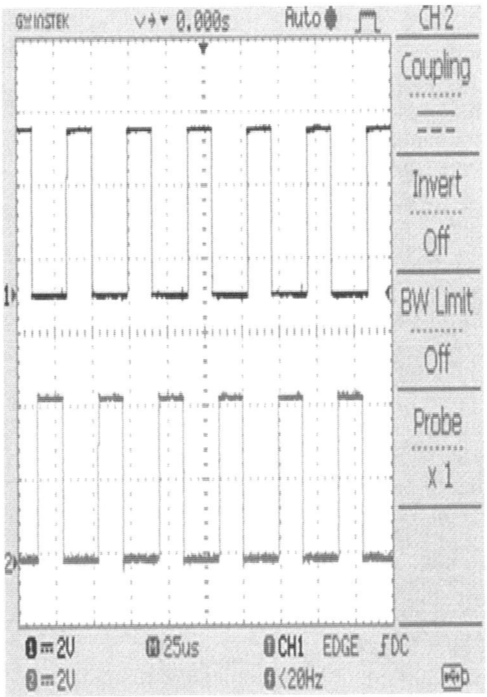

Figure 9. Pulse Wave Form for converter S1&S2.

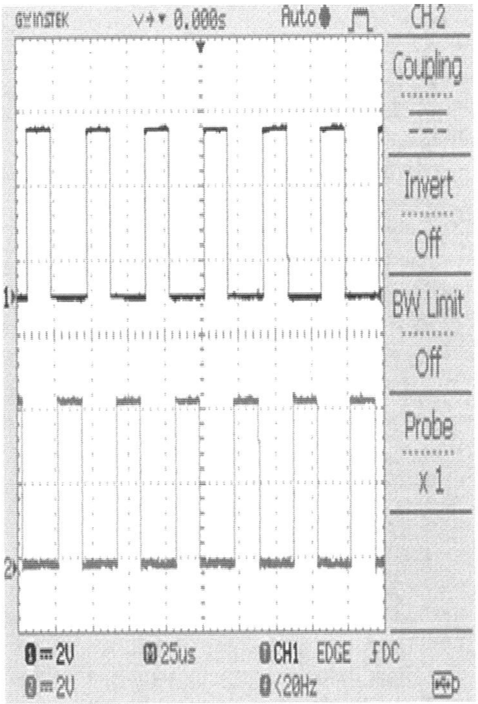

Figure 10. Pulse Wave For S3&S4 .

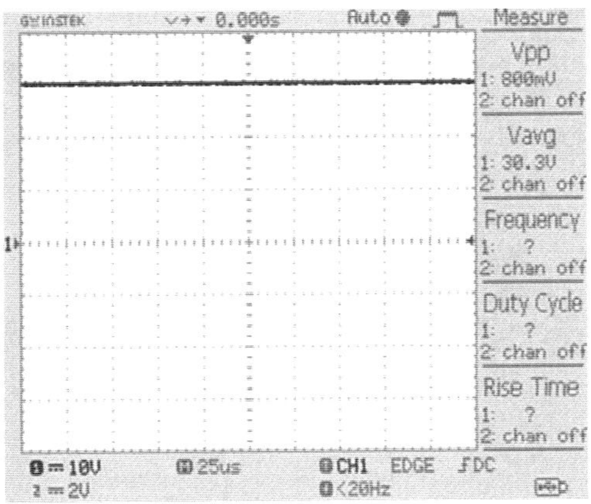

Figure 11. Output voltage V1.

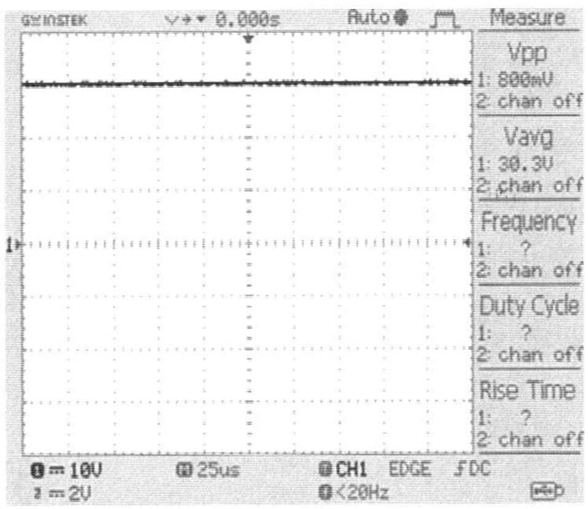

Figure 12. Output voltage V2.

5. Conclusions

The switched capacitor –multiport converter for solar system has been said in this paper. The MPC are determined by incorporating a stepping tool such as series resonant converter(src), switched capacitor converter(scc) and bidirectional PWM. and these are integrated in an single system by sharing the mosfet switches the Square wave voltages made at trading center points in the SCC are utilized to drive course of action full tanks in the Series Resonant Converter and a divert inductor in the PWM-converter. The proposed system are given in mat lab and the simulation for the multiport system is taken in an pwm waveform. The prototype of the system is verified with the matlab .The characteristics of the MPC matched with simulation results model.

6. References

[1] Atrash.H.A, Tian.F, and Batarseh.I, "Tri-modal half-bridge converter topology for three-port interface," IEEE Trans. Power Electron., vol. 22, no. 1, pp. 341–345, Jan. 2007.

[2] Debnath.D and Chatterjee.K, "Two-stage solar photovoltaic-based standalone scheme having battery as energy storage element for rural deployment," IEEE Trans. Ind. Electron., vol. 62, no. 7, pp. 4148–4157, Jul. 2015.

[3] Indragandhi V, Keerthi Vardhan , Arunkumar. G ; Raja Singh R Banumathi. S, "Implementation of a DC Micro-grid for House Hold Applications", 8th International Conference on Power and Energy Systems (ICPES), 2018.

[4] Krishnaswami.H and Mohan.N, "Three-port series-resonant dc–dc converter to interface renewable energy sources with bidirectional load and energy storage ports," IEEE Trans. Power Electron., vol. 24, no. 10, pp. 2289–2297, Oct. 2009.

[5] Li.W, Xu.C, H. Luo, Hu.Y, He.X, and Xia.C, "Decoupling-controlled triport composited dc/dc converter for multiple energy interface," IEEE Trans. Ind. Electron., vol. 62, no. 7, pp. 4504–4513, Jul. 2015.

[6] Qian.Z, Rahman.O.A, Atrach.H.A, and Batarseh.I, "Modeling and control of three-port dc/dc converter interface for satellite applications," IEEE Trans. Power Electron., vol. 25, no. 3, pp. 637–649, Mar. 2010.

[7] Sundararaju.K, Banumathi.S,Saranya.S,Sowmiya.R ,Susithra.S, Yuvarani.P, "Smart Power Management in Microgrid", International journal of pure and applied mathematics Academic Publications, ISSN ; 1314-3395, Vol.118, No.20, pp 2239-2246, 2018.

[8] Kumar. K, Ramji Tiwari, Ramesh Babu. N and Prabhu. K.R, "Analysis of MISO Super Lift Negative Output Luo Converter with MPPT for DC Grid Connected Hybrid PV/Wind System" Energy Procedia Vol.145 pp.345 – 350 July 2018

[9] Su.G.J and Tang.L, "A multiphase, modular, bidirectional, triple-voltage dc–dc converter for hybrid and fuel cell vehicle power systems," IEEE Trans. Power Electron., vol. 23, no. 6, pp. 3035–3046, Nov. 2008.

[10] Sun.X, Shen.Y, Li.W, and Wu.H, "A PWM and PFM hybrid modulated three-port converter for a standalone PV/battery power system," IEEE J. Emerging Sel. Topics Power Electron., vol. 3, no. 4, pp. 984–1000, Dec. 2015.

[11] Tao.H, Kotsopoulos.A, Duarte.J.L, andHendrix. M.A.M., "Transformercoupled multiport ZVS bidirectional dc–dc converter with wide input range," IEEE Trans. Power Electron., vol. 23, no. 2, pp. 771–781, Mar. 2008.

Design of a Microgrid Architecture for Rental E-Bike Charging Stations

Ghanishtha Bhatti, R Raja Singh
Department of Energy and Power Electronics,
School of Electrical Engineering, VIT Vellore.

Correspondence email: rrajasingh@gmail.com

Abstract. This paper focuses on developing a sustainable charging station for public motorbike rental services. Electric motorbikes or electric bicycles (both referred to as E-bike) are compact electric vehicles which are primarily battery powered and driven solely by electric motors. The work proposes a microgrid architecture to develop a novel model for the rental E-bike charging station that acts as a charger for multiple nodes simultaneously as well as acting as a docking/parking station for these vehicles. The model has been built to propose a system which increases the energy savings to rental companies while reducing reliance on public grid during peak hours. The use of distributed generation makes the system more resistant to failure while moving towards a future of eco-friendly and power efficient technology.

1. Introduction

The decline in non-renewable resources such as crude oil has become so significant, that the transition to sustainable energy has become a priority for law and policy makers. This has led to the research and development of renewable technology such as distributed generation (DG). However direct transition to fully DG supplied PEVs is not yet viable due to cost and power regulation constraints to prevent mismatches with the load. To bridge this gap, the concept of microgrids combines the conventional grid based supply, renewable DG energy and storage technology to actively regulate and maintain balance in the system [1]. Microgrids provide a reliable and efficient alternative architecture to increase the small scale renewable energy penetration while minimizing the energy cost [1-3], predominantly in case of urban areas.

Plug-in electric vehicles are predicted to become an integral part of the vehicle market in future years, and represent mankind's leap towards achieving low-carbon automobiles. Currently a notable amount of automobile manufacturers produce plug–in electric vehicles (PEV), either battery based electric vehicles or hybrid electric vehicles. It is also forecasted that during the next few years almost all major automotive manufacturers will introduce PEV models [1]. While the average consumer is skeptical about shifting to electric vehicles, the shift provides many consumer benefits such as lower fuel costs, and lower need of maintenance, because the average electric vehicle has around one tenth the moving parts that a conventional vehicle does. Policy has also affected the consumer outlook towards electric vehicles, and many public communities provide added perks such free parking, discounted insurance and other purchasing incentives to promote lowering vehicular carbon footprint. Although PEVs have various merits, they do place a substantial load on the real-time power supply. PEVs draw large amount of current when charging, which may have a detrimental effect on the public grid.

Microgrid architectures pose potential for innovation, therefore a multitude of academic study has delved into utilizing them for optimizing PEV infrastructure. A charging station model is presented in [13] that allows fast charging (25 kWh per hour at peak) using a bidirectional AC/DC converter. One of the more popular methodology is using solar energy for PEV charging stations [4-6], however most of these models fail to propose a feasible design for the market. Another aspect of increasing energy savings and efficiency is selling energy back to the grid with models such as vehicle to grid strategy (V2G) [7]. This model combined with energy management algorithms such as day ahead scheduling, ensures long term profits. A similar technique that benefits from using the excess energy during vehicle charging is shown in [8]. Coming to more integrated approaches, [9] proposes a novel architecture for grid connected charging stations that includes PV supply as well. This model has also taken the advantage real-time energy management algorithm that continuously uses feedback to reduce errors. For more complex loads, such as entire buildings, a different layered approach which requires multiple agents is needed, such as the one proposed in [10]. The research in [13] showcases a building integrated DC microgrid which utilizes several different renewable energies and present a visionary construction for residences of the future. Many studies focus on controlling the charging parameters of PEVs at peak times to reduce stress on the grid such as [11], though these works lack in description of the system when the grid power is disconnected. The study brings light upon the DC/DC boost converter which allows compatibility with the DC link. The most popular battery technology in PEVs is Lithium-Ion batteries. However, proposals for unconventional media such as superconducting magnetic energy [12] present scope for greater optimization. A DC bus has been used to integrate all these components and the interface is again made possible by power electronic converters.

While a lot of the focus with PEV is towards electric cars, another growing area of popularity is electric two wheel vehicles such as E-bikes and E-scooters. Two wheelers gain popularity as the traffic congestion in metropolitan cities increases, as well as with the advent of narrower road and lesser parking space. In India amongst the 26.27 million vehicles sold in 2019, 80% were two wheelers [14]. Furthermore in recent years there has been increasing popularity amongst app based bike rentals for commuting rather than purchase of a personal vehicle. There were also notably 1.56 lakh new units of EVs (excluding E-rickshaws) purchased in 2019, the majority of which were electric two wheelers. A study into academia shows some novel configuration for charging electric two wheelers. Wireless energy transfer using field effect transistors, proposed in [15], is shown to be a viable charging strategy for motorcycles in remote locations. Other contemporary work such as [16] show potential for reimagining electric two wheelers as a new class of hybrid electric vehicles allowing a varied supply of energy from both conventional and renewable sources. The logic controller design for an E-bike charging station proposed in [17] provides potential for more secure usage of public charging stations, even though the RFID framework suggested is less reliable for security purposes.

This work aims to contribute to this developing field by proposing a microgrid based electric motorcycle/bike charging station. This model aims at utilization by rental services to offer a chargeable commute service via electric bikes with solar energy based charging and docking station simulation. The utility of charging an array of rental vehicles simultaneously as they park offers a substantial cost saved on fuel as well as a significant reduction in greenhouse gas emissions. These features make the contribution of this paper novel. The remainder of this paper is organized as follows. Section 2 presents the microgrid architecture of the proposed charging station, considering the real-world constraints and trends. Section 3 provides a mathematical conceptual understanding of the electrical systems involved. It also details the use of photovoltaic arrays to harness energy. Section 4 introduces the input data of the simulation and evaluates the parameters observed when modelling a microgrid for this application. Section 5 illustrates the obtained simulation results when performed with MATLAB Simulink and evaluates the model with these results. Section 6 concludes and summarizes the novelty and contributions of the proposed framework.

IOP Conf. Series: Materials Science and Engineering **906** (2020) 012009 doi:10.1088/1757-899X/906/1/012009

2. System Description

The proposed system is based on a DC microgrid. As show in Figure 1, the microgrid is energized from a 6 kW local solar array, 30 kW solar farm as well as connection to the public grid. All these components are centrally DC converted and linked to the charging station which facilitates battery charging for EVs. Rather than using a grid reliant model for the charging station which is the more commercially utilized, a microgrid approach is suggested. The microgrid approach helps to create a low cost, efficient and locally resilient system that reduces the load on the public grid, an imminent concern in cosmopolitan cities at peak hours. Furthermore it ensures that the system has contingency in case of power failure or unexpected electrical faults to the main grid. The reliability boosts the operational capability for the rental agency as well as improved and uninterrupted service to the end user. The demotion of the public grid as a backup to this system rather than the primary source also invokes the possibility to sell excess energy. This kind of architecture is known as Vehicle to Grid model (V2G), however this work has not extended to V2G for the purposes of this paper. Microgrids operate in various modes which make their operation flexible. There are two major modes considered for modelling the microgrid architecture [18].

- **Hybrid Mode**
  The breaking mechanism is closed to connect to the utility grid. The utility grid equips all the feeders and the distributed generation based microgrid provides some surplus power to the main grid.
- **Islanding Mode**
  The grid breaker is opened in case of electric faults thus preventing the utility grid from providing power. Microgrid can operate independently preventing power outage to load when the utility grid is not usable.

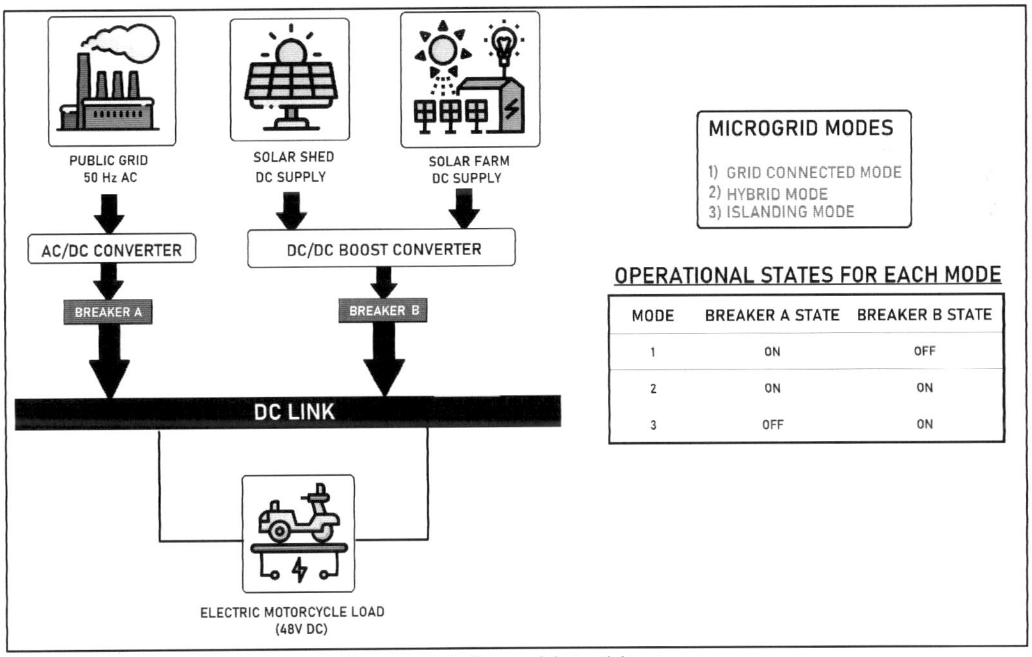

Figure 1. Microgrid Architecture

As part of this study both these modes have been modelled in this work, in order to ensure the feasibility of the model when prototyped. Furthermore for the purpose of comparative analysis the model has also been run in gird connected mode (Public Grid only) as shown in Figure 1. Solar Energy is considered as the distributed resource for this model as it represents the following features which are optimal for this application and the environment:

1. Solar energy is theoretically unlimited
2. Higher power density than most conventional renewable resources
3. It is a non-polluting resource and does not aid greenhouse effect.
4. It is suitable for both high and low power devices
5. Solar energy is free (excluding the harvesting equipment). Over the years the energy and cost savings exceed fuel or grid power for vehicles.
6. PV systems are low maintenance and have long life.
7. Do not occupy much ground space as they may be pole-mounted
8. Can be used in residential areas as they do not produce noise or hindrance for civilians.
9. Can harvest energy in most open areas at a significant rate
10. Largest market in India among renewable resources

The model of commercial usage of the system prototype based on this architecture is illustrated in Figure 2. The user simply rents the electric motor bike for a period using a digital authorization such as a smartphone app. Once they have used the bike or want to charge it they may simply return it to the dockings station which detects completion of the users purchase and also charges the motorbike for next use. Overall this model incurs great benefits to the rental service as costs of fuelling the electric vehicles is greatly reduced, and the system requires minimum intervention by human employees, which reduces labour cost as well. The produced PV electricity is intended as the primary source for charging of Electric bikes. The two separate PV arrays are located in different geographical areas for maximizing reliability and availability of solar energy to the system. The first array is placed on a shed mounted over the electric motorbike docking area which acts as a shade to prevent vehicle heating while also harvesting energy with low transmission loss. To prevent the possibility of system failure in case of low local irradiance, another solar array is used as a power source which is routed from a solar farm.

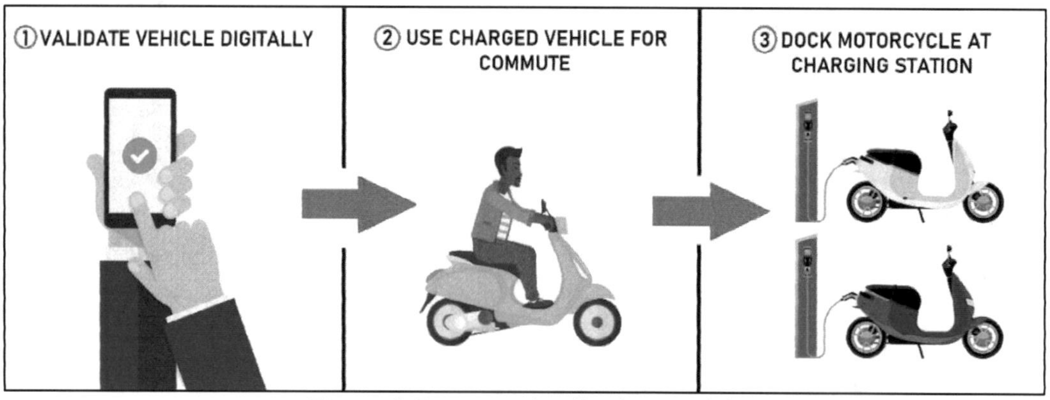

Figure 2. Prototype Application Model.

3. Mathematical Modelling of PV powered E-bike charging station

To be able to successfully build this architecture, the mathematical models that govern the constituent input component, such as the photovoltaic array and power electronic converters must be determined. On the output side, to simulate the charging of the EVs (Electric bikes) they are modelled as batteries whose parameters must priorly computed. Their charging equations such as voltage behaviours and battery state of charge (SOC) need also be determined to evaluate the model performance.

3.1. Photovoltaic Array

Solar power has been determined in the model using input of irradiance (beam and diffuse radiation included) as well as ambient temperature. In order to calculate the solar radiation absorbed on the cell surface, S, it is mandatory to measure incident radiation, air mass, and incidence angle. With these parameter it is possible calculate solar power (S) using Equation 1.

$$S = M(G_b R_b(\tau\alpha)_b + G_d(\tau\alpha)_d \frac{1 + \cos\beta}{2} + G\rho_g(\tau\alpha)_g \frac{1 + \cos\beta}{2} \quad (1)$$

Here M is the air mass modifier, τα the fraction absorbed by an absorber plate. G gives the radiation (the subscript indicates whether the radiation is beam (b), diffuse (d) or ground (ρg). β is the Angstrom turbidity coefficient and R_b the ratio of beam radiation on the plane to that on a horizontal surface at any time. The solar power S is measured in W/m^2.

The other factor affecting the modelling and output of PV arrays is temperature. While it might be thought that heat increases the amount of solar energy received by solar cell, the solar energy that can be harvested is dependent on the irradiance. However, this does not imply that heat or temperature does not affect temperature. Efficiency ηs of the solar cell given in Equation 2, will highlight this relationship. Here I_{max} represents maximum current and V_{max} indicates the maximum voltage.

$$\eta s = \frac{(I_{max} \times V_{max})}{P_{in}} \quad (2)$$

When the temperature increases, the current will increase, while the voltage will decrease. As the voltage decreases faster than the current increases, the overall efficiency is decreased. Therefore, the power output can be presented as a function shown in Equation 3.

$$P = P(S, T_{mod}) \quad (3)$$

Here T_{mod} represents the temperature of module given by the relation in Equation 4. The parameter k_T defines the clearness index of the location. T_{air} represents the temperature of air and G represents total incident radiation.

$$T_{mod} = T_{air} + k_T G \quad (4)$$

The power output of the cell (P) can be calculated using Equation 5. P_{nom} is the peak power, in kWh.

$$P = \frac{P_{nom} \times S}{1000} \quad (5)$$

3.2. Power Electronic Converters for the Microgrid

DC-DC converters are essential to photovoltaic array based systems when connecting them to loads or grid. Converters allow manipulation of the output voltage to meet the required the levels of the load or the grid. In consideration with function of this model a boost converter is required in conjunction to each PV array in order to step up the voltage to the required level of the EV load. The parameters required to compute the power stages and component specifications are as follows [19]:

1. V_{OUT}: nominal output voltage
2. $V_{IN(min-max)}$: the input voltage range
3. $I_{OUT(max)}$: maximum output current
4. F_s : minimum switching frequency of the converter

In order to determine the value of the inductance, the duty cycle (D) of the converter must be calculated using Equation 6 where ηc represents the efficiency of the converter.

(6)

$$D = 1 - \frac{V_{IN(\min)} \times \eta c}{V_{OUT}}$$

The switching current, which is determined by the duty cycle, drives the operation of the switch which is commonly an IGBT or MOSFET. This switching current calculation is detailed in Equation 7.

$$I_{SWmax} = \frac{\Delta I_L}{2} + \frac{I_{OUTmax}}{1 - D} \tag{7}$$

ΔI_L represents inductor ripple current. With these parameters it is possible to calculate the inductance (L) using Equation 8.

$$L = \frac{V_{IN} \times (V_{OUT} - V_{IN})}{\Delta I_L \times \text{Fs} \times V_{OUT}} \tag{8}$$

Similarly, the input capacitor value is determined from the data sheet. The value is kept at minimum so that the input voltage is stabilized for the switching power supply. In the case of the output capacitor, it is used to achieve the required level of output ripple voltage. Equation 9 allows computation of this capacitance [19].

$$C_{OUT} = \frac{I_{OUTmax} \times D}{\text{Fs} \times \Delta V_{OUT}} \tag{9}$$

3.3. Electrical Vehicle Modelling

For the purposes of this work, the primary interest lies in the charging behaviors of the electrical vehicle or more specifically the electric motorbike. This is why we may model solely the battery subsystem of the automobile. While considering the battery the most important aspect is to select the required battery chemistry. In the case of most EVs, lithium ion batteries are used and the same battery chemistry will be utilized for the purpose of this paper.

Lithium ion batteries are selected because they have the highest coulombic efficiency (CE) values compared to alternate rechargeable batteries (> 99%). CE refers to the efficiency of charge at which electrons are transported in batteries and refers to the ratio of the total charge obtained from the battery to the total charge given to the battery during a full cycle [20]. The usable battery capacity is characterized by SOC of battery defined as the ratio of remaining capacity to fully charged capacity shown in Equation 10.

$$SOC = \frac{Q - Q_{min}}{Q_{max} - Q_{min}} \tag{10}$$

In the above equation Q is the charge currently stored, Q_{min} is the charge that remains in the battery when minimum voltage (open circuit) is reached and Q_{max} is the charge at maximum open circuit voltage. For aiding modelling the SOC must be defined, as given in Equation 11.

$$\text{SOC} = \text{SOC}_0 - \frac{1}{\text{SOC}_0} \int \text{i dt}. \tag{11}$$

SOC_0 is the initial SOC level of battery and i is the battery current. To get a more comprehensive estimation of SOC, a few additional parameters must be considered. Firstly the voltages based SOC (SOC_V) given by Equation 12.

$$\text{SOC}_v = \frac{1}{\text{a1}} (V_{OC} - a_0) \tag{12}$$

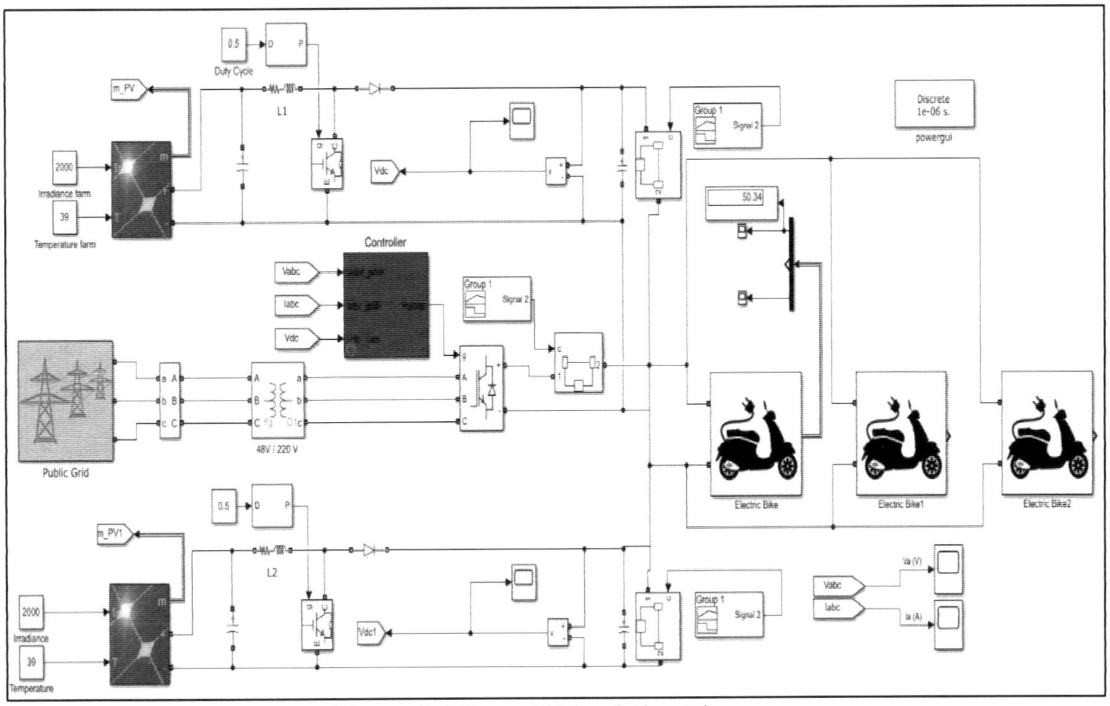

Figure 3. Microgrid Simulation Diagram

V_{OC} defines temperature dependent open circuit voltage, a_0 is terminal voltage at 0% SOC, and a1 represents the terminal voltage at 100% SOC. On the other hand current based SOC (SOC_i) which is determined by the Coulomb Counting method (Equation 13). Here C_p represents the battery capacity.

$$SOC_i = \frac{1}{C_p}(C_p - \int i\, dt) \tag{13}$$

A systematic correction factor (η) is also required and is a function of initial SOC level. The equation that corrects the discharging profile values is given in Equation 14.

$$\eta = (1 - \frac{SOC_0}{100}) \tag{14}$$

Finally, using these parameters the modified equation for SOC can be obtained as shown in Equation 15. Here w is an additional weighting factor.

$$SOC = w \times SOC_v + (1 - w)(SOC_i - \eta) \tag{15}$$

4. Implementation of the Proposed E-Bike Charging System

The proposed microgrid framework has been modelled in MATLAB Simulink with the advanced and customizable blocksets. The complete simulation model, shown in Figure 3, consists of dual PV arrays, power electronic converters, the public grid and electric motorbike loads to simulate the charging behavior. These subsystems are categorized by their parameters which customize the function of this architecture towards the application of a multiport charging station.

4.1. PV System Parameters

Physical considerations that entail the sizing of the PV system are the irradiance and temperature models. The temperature and irradiance datapoints have been modelled based on conditions in Vellore, Tamil Nadu during the month of April. The average global horizontal irradiance is 2000 W/m2 as per data from Meteonorm [21] and the mean temperature is considered as 39°C. Indian market models have been considered while modelling the PV array to ensure compatibility with the Indian Grid System. The dual PV cell is modelled closely in specification with SunPower SPR-305E-WHT-D. The array to be utilized locally (at the charging station) has 10 parallel strings of solar cells with each string consisting of 2 series connected cells. The second array is modelled to resemble a solar farm for larger fault tolerance to the system. This array consists of 50 parallel strings with 2 series connected solar cells in each string.

4.2. Public Grid Parameters

A standard three-phase source models the public grid at 220V and 230,000 MVA [22] operating at 50Hz. The public grid connection is carried out by a three-phase AC/DC converter. A simple voltage source controller mechanism accompanies the grid converter for feedback based regulation of output voltage [23].

4.3. Power Electronic Converter Parameters

Two major circuits have been used as part of this model for power electronic conversion.

11. **Three-phase AC/DC converter:** The role of a converter in the microgrid is to operate as an interface between energy generation and consumption points. An IGBT circuit is used with this purpose. The driving gate pulse of the converter is provided by a VSC controller which acts as a regulator by taking feedback from the grid voltage and current as well as the DC output voltage of PV array.

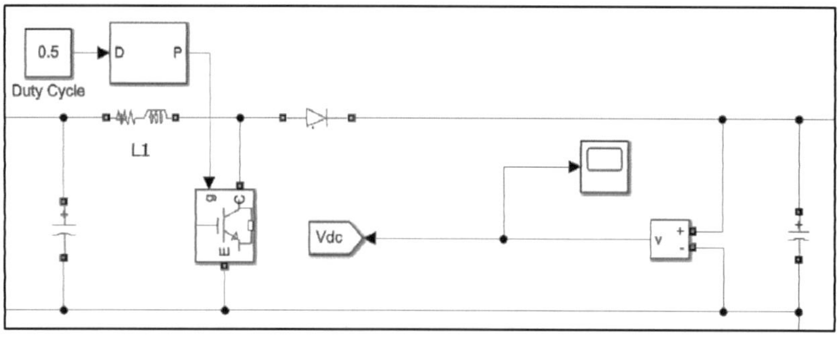

Figure 4. DC/DC Boost Converter Circuit

12. **DC/DC Boost Converter (Figure 4):** This has been used to regulate the voltage level from the PV source to the required output level of the Electric bike load. The switching frequency (duty cycle) has fixed at the minimal achievable value of 0.5. Without any tracking or MPPT this is the standard testing value. The specifications of the converter are detailed in Table 1.

Table 1. DC/DC Boost Converter Parameters

Sl. No	Parameter	Value
1	Input Capacitance	0.1 mF
2	Inductance	5 mH
3	Duty Cycle of Switching	0.5
4	Output Capacitance	12 mF

4.4. Model Modes and Autonomous Switching

In order to evaluate the feasibility of the microgrid architecture the model switches between different modes during simulation which entail various energy source combinations. The modes are as follows:

- **Grid Connected Mode:** Standard architecture for PEV charging, through direct connection to the public grid. This mode is the control for testing the performance of the microgrid.
- **Hybrid Mode:** The ideal functioning mode for the microgrid which utilizes power from both the AC public grid as well as two separate DG sources (in this case PV arrays).
- **Islanding Mode:** In case a fault or failure of the public grid occurs, the microgrid islands itself and supplies power independently from the DG source to the load.

In order to switch the model between various modes in case of a fault, the grid circuit is built with breaker mechanisms that islands the charging station from the grid. The circuit breakers are controlled by the proposed algorithm.

4.5. E-bike Battery system parameters

The load is considered to be conventional electric bike of economic class that are most commonly utilized in the Indian market such as the Vespa Elettrica™. The battery technology employed is Lithium Ion batteries which are commonly used in the market. For simulation the Simulink battery blockset [24] has been utilized along with experimental and commercial customization. The aging and temperature effects on the battery have not been developed as part of this study. The specifications have also been modelled from commercial vehicles as shown in Table 2.

Table 2. Electric Bike Model Parameters

Sl.No	Parameter	Value
1	Nominal Battery Voltage	48 V
2	Battery Capacity	86 Ah
3	Battery Energy	4.2 kW/h
4	Forecasted Battery Life	1,000 cycles
5	Optimal Charging Time	4 hours
6	Initial Battery SOC	50 %

5. Results and Discussion

To validate the DC microgrid for PEVs charging station modeling approach, the system simulation is performed with MATLAB Simulink. The simulation is developed to model the E-bike charging station under as a standalone microgrid in islanding mode and hybrid mode.

The simulation has a run time of 3s and the battery response time has been lowered to produce quantifiable changes in SOC without expending limited computational resources for the purposes of proof of concept. Furthermore, to optimize the simulation with regards to processing power, it has been run on the Simulink Accelerator mode using the Just-in-time algorithm. In this workflow model Simulink generates for the top-level model only an execution engine in memory excluding the referenced models. Therefore, the inherent C compiler is not used during run-time, increasing execution speed and reducing processing required. The charging profile of the implemented system is shown in Figure 5.

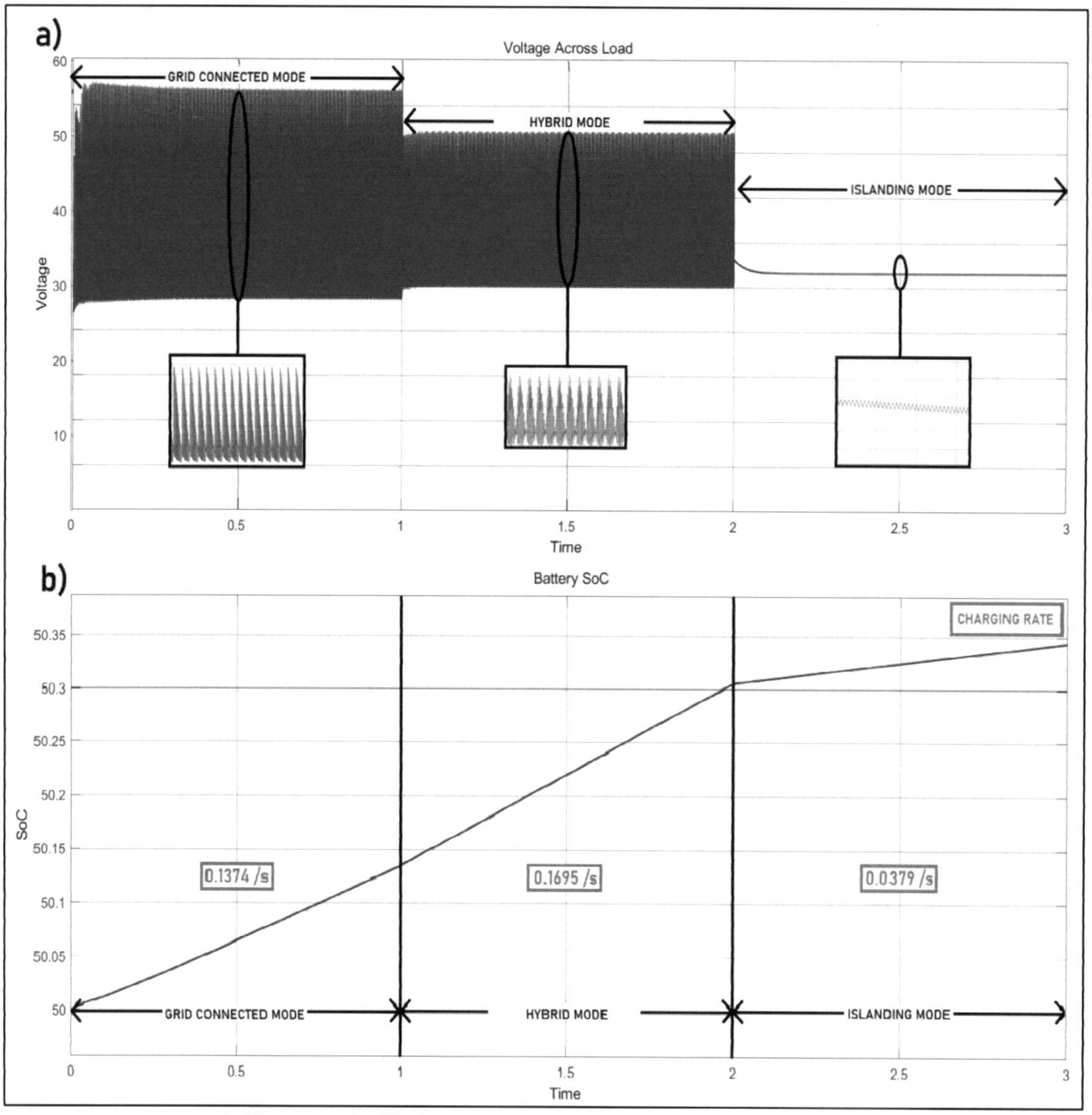

Figure 5. Charging Profile in various modes a) Load Voltage, b) Battery SOC

5.1. Grid Connected Mode

This mode shows the standard model of operation for PEV charging. It has a calculated charging rate of 0.1374% per second in the response accelerated model. Furthermore a ripple in the output voltage at the load end is observed fluctuating between the limits of 27.721 to 56.224 V. This ripple is likely present due to incomplete suppression of the AC waveform after rectification. While the value of ripple in the circuit is minimal, it represents lost power and decreased efficiency. It may be amended in the future by introducing electronic filters or improved regulator at the output level to supress the ripple. This regulation may also be achieved using ripple suppressing devices such as capacitors.

5.2. Hybrid Mode

This mode introduces the microgrid architecture in place. Predictably, the rate of SOC growth increases to 0.1695 % per second with the injection of two DC PV arrays as sources. This injection also considerably lowers the output ripple between 30.156 and 52.589.

5.3. Islanding Mode
The microgrid islanding that comes into utilization during faults is presented in this mode. The rate of charging is observed to be 0.0379 % per second. While this represents a drop of 0.1316 % per second, we can observe that the renewable distributed generation framework is able to supply the necessary energy to support the system without requiring the system to be taken offline. This supports the feasibility of this model as a prototype as well as demonstrates the efficiency and reliability of the microgrid architecture for the electric motorbike charging stations. It can also be observed that there is a reduction in output ripple voltage which is due to disconnecting the major AC source from the microgrid. The output ripple is between the limits 31.144 and 32.045 V.

6. Conclusion
Shifting from conventional, polluting and depleting resources has become crucial as we progress into an age where these resources have become increasingly scarce. Electric Vehicles play a major role in this shift. The consumer market grows towards adopting electric vehicles especially more pocket-friendly alternatives such as electric motorcycles and scooters because of the savings they offer. While most countries have a rapidly growing market for electric vehicles, the infrastructure for maintaining them is still limited. Charging stations, especially in India, are extremely rare, which poses a barrier to expanding the electric motorcycle market. In order to propose a charging infrastructure which is both feasible and easy to implement this study proposes a microgrid model to support the increasingly popular bike rental business. The use of distributed generation allows renewable energy to be integrated into the charging system and the microgrid concept that facilitates islanding allows for reducing load on the conventional grid.

The developed simulation model in this study was evaluated with respect to the ideal load of a Vehicle Lithium Ion Battery and its power, voltage and current parameters were verified to be sufficient to charge such a load. The model has been developed as a proof of concept and does not venture into intricacies of vehicle dynamics, battery capacities or cost benefit analysis. Introducing these elements form the future scope of this elements and hold the possibility of optimization of the power flow and conversion.

7. References
[1] Locment F and Sechilariu M 2015 Modeling and simulation of dc microgrids for electric vehicle charging stations. *Energies* vol 8 pp 4335-4356

[2] Manna D and Goswami S 2017 Micro-grid operation and control *National Conference on Recent Trends in Engineering and Technology (NCRTET-2017) Tripura (W), India* Vol: ISBN: 978-93-87380-61-5

[3] Patterson B T 2012 Dc, come home: dc microgrids and the birth of the "enernet" *IEEE Power and Energy Magazine* vol 10 no. 6 pp 60-69

[4] Corchero C, Cruz-Zambrano M, Heredia F.-J, Cairo J.-I, Igualada-Gonzalez L and Romero-Ortega A 2012 Optimal sizing of microgrids: a fast charging station case *2012 9th International Conf. on the European Energy Market* pp 1-6

[5] Guo Y, Xiong J, Xu S and Su W 2016 Two-stage economic operation of microgrid-like electric vehicle parking deck *IEEE Transactions on Smart Grid* vol 7 no. 3 pp 1703-1712

[6] Mouli C, Ram G, Bauer P and Zeman M 2016 System design for a solar powered electric vehicle charging station for workplaces *Applied Energy* vol 168 pp 434-443

[7] Armstrong M, Moussa C, Adnot J, Galli A and Rivière P 2012 Optimal recharging strategy for battery-switch stations for electric vehicles in france *Energy Policy* vol 60

[8] Van Roy J, Leemput N, Geth F, Büscher J, Salenbien R and Driesen J 2014 Electric vehicle charging in an office building microgrid with distributed energy resources *IEEE Transactions on Sustainable Energy* vol 5 no. 4 pp 1389-1396

[9] Mohamed A, Salehi V, Ma T and Mohammed O 2014 Real-time energy management algorithm for plug-in hybrid electric vehicle charging parks involving sustainable energy *IEEE Transactions on Sustainable Energy* vol 5 no. 2 pp 577-586
[10] Wang Z, Wang L, Dounis A and Yang R 2012 Integration of plug-in hybrid electric vehicles into energy and comfort management for smart building. *Energy and Buildings* vol 47 pp 260–266
[11] Goli P and Shireen W 2014 Pv integrated smart charging of phevs based on dc link voltage sensing *IEEE Transactions on Smart Grid* vol 5 no. 3 pp 1421-1428
[12] Liu Y, Tang Y, Shi J, Shi X, Deng J and Gong K 2015 Application of small-sized smes in an ev charging station with dc bus and pv system *IEEE Transactions on Applied Superconductivity* vol 25 no. 3 pp 1-6 Art no. 5700406
[13] Strunz K, Abbasi E and Huu D N 2014 Dc microgrid for wind and solar power integration *IEEE Journal of Emerging and Selected Topics in Power Electronics* vol 2 no. 1 pp 115-126
[14] Indian Automobile Industry Report 2020 *Copyright © 2010-2020 India Brand Equity Foundation*
[15] Widjanarka W and Sutiyatno S 2020 Experiment electrical motorcycle used solar energy with wireless transfer energy method for transportation in isolated regions *Journal of Physics: Conference Series* vol 1517 pp 012098.
[16] Wu and Xu 2019 The optimal control of fuel consumption for a heavy-duty motorcycle with three power sources using hardware-in-the-loop simulation *Energies* vol 13 pp 22
[17] Inácio C, Silva I, Faria J, Pombo J, Calado M and Mariano S 2020 Controller design and experimental validation of a solar powered e-bike charging station *KnE Engineering International Congress on Engineering — Engineering for Evolution* pp 846-856
[18] Prakash M M and Paul J 2016 Control of microgrid for different modes of operation *International Journal of Engineering Research & Technology (IJERT)* vol 05 issue 05
[19] Hauke B 2014 Basic calculation of a boost converter's power stage *Texas Instruments Inc* SLVA372C
[20] Miao Y, Hynan P, Jouanne A V and Yokochi A 2019 Current li-ion battery technologies in electric vehicles and opportunities for advancements *Energies* vol 12
[21] TANGEDCO 2020 Solar irradiance data in tamil nadu *Retrieved from https://www.tangedco.gov.in/linkpdf/solar%20irradiance%20data%20in%20Tamil%20Nadu.pdf,*
[22] Chowdhury T 2015 Design of a micro-grid system in matlab/simulink *International Journal of Innovative Research in Science, Engineering and Technology* vol 04 pp 5262-5269
[23] Abootorabi M 2020 power_PVarray_grid_det *MATLAB Central File Exchange* retrieved May 2020 (https://www.mathworks.com/matlabcentral/fileexchange/52434-power_pvarray_grid_det)
[24] Kuperman A, Levy U, Goren J, Zafransky A, and Savernin A 2013 Battery charger for electric vehicle traction battery switch station *IEEE Transactions on Industrial Electronics* vol. 60, no. 12, pp. 5391-5399.
[25] Piaggio & C 2018 Vespa Elettrica *Retrieved from https://elettrica.vespa.com/_common/pdf/EN-DIGITAL-VespaElettrica-10-2018.pdf*

Acknowledgments
We would extend our gratitude to the faculty members and staff members at Vellore Institute of Technology with whose support we were able to procure the necessary software resources to develop our model and run analysis on it.

IOP Conf. Series: Materials Science and Engineering **906** (2020) 012010 doi:10.1088/1757-899X/906/1/012010

DUAL INPUT FULL BRIDGE ISOLATED DC TO DC CONVERTER

Himani Shimpi [1], G Arunkumar [2], Santoshkumar M. Hunachal [3], Ajay Bhosale [4], Rajan Kumar Jaysawal [5].
[1,4,5] Student at Vellore Institute of Technology, Vellore, India.
[2] Associate professor at Vellore Institute of Technology, Vellore, India.
[3] Reserch scholar at Vellore Institute of Technology, Vellore, India.

shimpihimani@gmail.com

Abstract. This paper presents isolated dc to dc converter. It has dual input followed by two cascaded multilevel H bridge inverter and rectifier. This topology reduces the no of switches and components. Hence the cost of inverter is reduced and complexity is decreased. This is a major advantage over conventional topologies. This topology uses two DC sources which are asymmetrical, because of that seven level output voltage is obtain on the primary side of the transformer and step-up voltage obtain across the load. Switching techniques are used for proper working on this isolated converter and hence fulfil the objective of reducing overvoltage and stress in semiconductor devices. Simulation results are obtained in PSIM software.

Keywords— Multilevel inverter (MLI), Cascaded H-Bridge (CHB)

1. Introduction

For high power and high voltage application, bridge converter gained popularity and increased more attention in industry applications. As a full-bridge converter is an isolated DC-DC converter, it rectifies the disadvantage of boost converter. In various applications like switched-mode power supply, military application, battery charging system of a vehicle, medical application, renewable energy and distributed power system isolated converters are required. In many applications there is a need for electricity security demand. Hence to fulfil this demand isolated DC-DC converter should be cascaded in such a manner that consists of inverter high-frequency transformer and then rectifier. Hence scheme is like DC-AC-DC. The use of high-recurrence transformer can understand voltage proportion modification in fluctuates application events and diminish the volume of the transformer. In proposed topology, semiconductor devices like MOSFET are preferred, as there is a need for high-frequency application and MOSFET has high switching capability. As inverter is less efficient and it has demerits like high cost and high switching loss, hence multilevel inverter is preferred instead of it. The term called Multilevel originated from the three-level converter. In recent days for high voltage high power application MLI used. It produces a staircase pattern waveform; it seems to be like a sinusoidal wave, as many levels increased harmonic reduce and approximately smooth waveform obtained which is like a sinusoidal waveform. Compared to conventional inverter MLI output voltage has fewer harmonics. To reduce the total harmonic distortion, number of levels of MLI increase to N level. In MLI on reduction in step voltage between levels, the switching stress on the device is also reduced.
The multilevel inverter is classified as Neutral point diode clamped, Flying capacitor inverter and H bridge type inverter. In the proposed topology H-Bridge MLI used, is connected at the primary side of the transformer. H-bridge in cascaded fashion is the most attractive topology in Multilevel Inverter as its construction is easy and control techniques are simple. In proposed topology Two H-Bridge are connected in series and each having a separate DC source. Generally, both dc sources have the same

value but in proposed topology, both dc sources are in 1:2 ratios so that number of levels is increased than normal H-Bridge converter. Because of new topology by using fewer semiconductor devices the number of steps increases the cost of the overall system is reduced. If the ratio of two voltage sources increases then the number of levels again increases. Hence as per our requirement, we can boost the voltage. As per requirement transformer winding ratio also help to boost the voltage.

2. Different topology of multilevel inverter

2.1. Diode clamped multilevel inverter

To restrain the voltage pressure on power devices, diode is used in topology for clamping the voltage. A voltage (Vdc) is applied across each capacitor so that voltage stress over each clamping device is bounded to Vdc by using a clamping diode. This is the first topology which is practically implemented. In this topology configuration is as follow:

Required input voltage source= $K-1$ … (1)

Required number of switches for K-level= $2(K-1)$ … (2)

Required diodes for clamping voltage= $(K-2)(K-2)$ … (3)

As this topology has some disadvantage like number of levels are related to no of clamping diode hence system gets complex for higher output voltage applications [1]. The system is not flexible. Hence proposed H-bridge DC-AC converter topology is efficient for DC-DC full-bridge converter.

2.2. Flying capacitor multilevel inverter

It is similar to diode clamped inverter only the dissimilarity is instead of a diode; capacitor is used for clamping purposes. At the dc side it has a tree-like structure of the capacitor where the voltage of each capacitor is different than the next capacitor. The size of the voltage levels in the output waveform is obtained by the increase of voltage between two nearby capacitor legs. Configuration for this topology is followed:

The required number auxiliary capacitors= $(K-1)(K-2)$ … (4)

The required number of switches for K -level = $2(K-1)$ … (5)

The required number of DC-link capacitors= $(K-1)$ … (6)

As this topology have some disadvantages like purchasing capacitors for the same level of voltage. Also efficiency of this topology is very poor and it has switching realization problem. Circuits become more costly, heavily bulky and complicated than diode clamped circuit. Hence for higher voltage applications use of flying capacitor-based inverter topology is not efficient.

2.3. The cascaded H-bridge inverter

The proposed topology of Dual input isolated Full-Bridge DC-DC converter, two sources H-Bridge inverter is connected in a cascaded fashion. In H-Bridge inverter which has more than one DC source is used and each inverter produces different voltage levels. And the total produce voltage of the MLI is a sum of voltage level which is produced by each cell.

The no of level in output voltage is $2K = 1$ … (7)

Where k= number of input sources. The main feature of this H-Bridge MLI is less number of semiconductor devices like switches and diode. As in H-Bridge inverter topology, sources are different so isolation is provided to each source, and also a rating of each source is low hence voltage stress on switches is reduced and at last voltage levels are added so the output voltage is boosted. H- Bridge inverter is more flexible and easy to reconstruct in case of fault. Fig.1. Shows cascaded H –Bridge inverter in which both sources have the same value it is nothing but symmetrical Cascaded H –Bridge inverter [2].

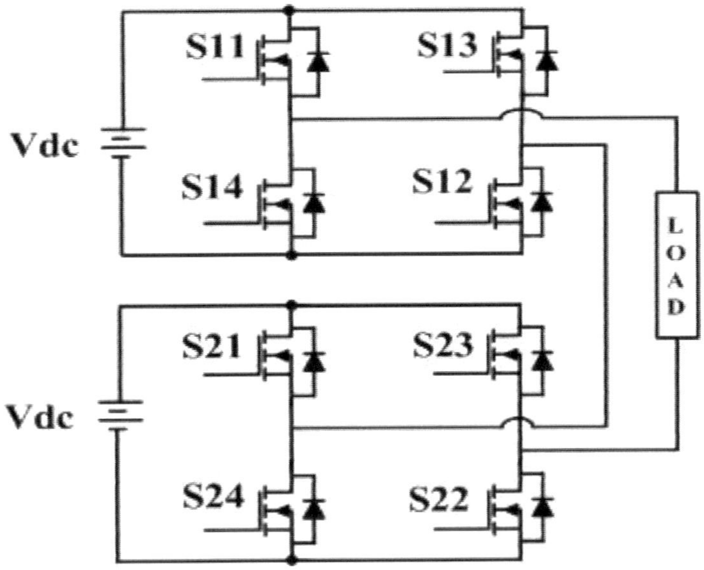

Fig.1 cascaded type H-bridge inverter

3. Proposed topology

In this paper dual input are used which are asymmetrical input voltage source. For high-frequency application eight MOSFET switches are used. Generally MOSFET is a semiconductor device and is used for an application that needs high frequency, load variation and long duty cycle. MOSFET has high voltage stress withstand capacity. Mosfet switches are used, as they have fast switching capability. In proposed topology according to switching sequence, we get 7 level output voltage at primary of the transformer which is an output of a multilevel inverter. To get this seven-level output voltage using eight switches, Two H-Bridge inverter is connected in a cascaded fashion. At the secondary of the transformer, output voltage is boosted according to the transformer turns ratio and as high-frequency transformer is used hence a size of the transformer is reduced though voltage is increased. Further to convert AC voltage in constant DC full bridge rectifier is used. Inverter output voltage is not proper sinusoidal it is a staircase wave but because of the increasing number of level total harmonic distortion (THD) reduced and get approximately sinusoidal output.

In proposed topology two H-Bridge inverter are used which have an asymmetric voltage source. As shown in fig.2 this cascaded H-Bridge inverter is different than conventional topology.

The number of voltage level = $\mathbf{2P + 1}$, where P signifies the number of sources ... (8)

Number of switches required= $\mathbf{2(K - 1)}$, where K denotes no of level ... (9)

Hence in conventional H-bridge topology 5 levels can be obtained using 2 sources and 8 switches. But in proposed topology as an asymmetric voltage source (i.e. voltage sources are in 2:1 ratio) are used we can get seven levels of output voltage using the same two sources and 8 switches. Also if voltage source in the ratio of 3:1 then using the same two voltage source and 8 switches nine levels of output voltage is obtained. Hence this proposed topology is more advantageous than conventional one as more no of output voltage level obtained using fewer devices so cost and complexity of system get reduced. Comparison between conventional topologies and proposed topology is showing in table 2.

Using proper switching sequence voltage levels can be increased. Table.1 shows the switching sequence. According to table 1, sources are 12 Volt and 6 Volt respectively. When both sources are acting then S11, S12, S21, S22 conducting and obtain 18 voltage output levels. For 12 volt output voltage only one source is acting that is 12 volt at that time S11, S12, S22 conduct and mosfet internal diode conduct. For 6 volt S11, S12 conduct also internal diode of mosfet is forced to conduct. At time of 0 voltages no one switch conduct. Similarly for obtaining negative voltage level proper switching sequence is necessary. For -6V S13, S14 conducts. For -12 volt S13, S14, and s24 conducts. Similarly for -18 voltage output S13, S14, S23, S24 conducts. Hence as per switching sequence 7 level of output voltage is obtained as 18 V, 12V , 6V, 0V, -6V, -12V, -18 V.

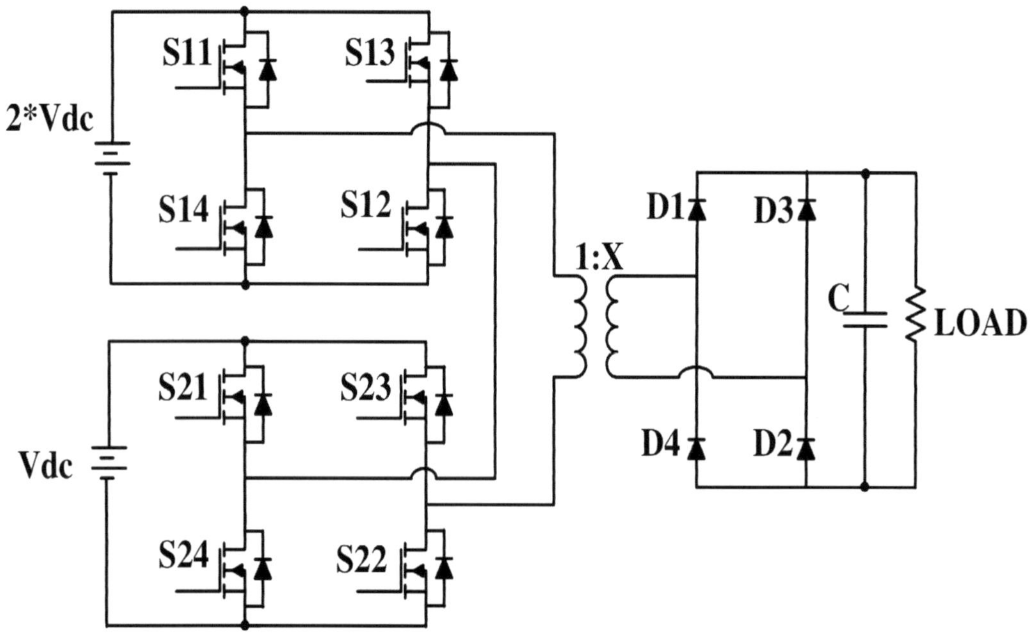

Fig2: Dual input full-bridge dc-dc isolated converter

Voltage level	S 11	S 12	S 13	S 14	S 21	S 22	S 23	S 24
18	1	1	0	0	1	1	0	0
12	1	1	0	0	0	1	0	0
6	1	1	0	0	0	0	0	0
0	0	0	0	0	0	0	0	0
-6	0	0	1	1	0	0	0	0
-12	0	0	1	1	0	0	0	1
-18	0	0	1	1	0	0	1	1

Table 1. Switching sequence of switches

Sr. no	Characteristics	Multilevel Inverter Types			
		Diode Clamped	Flying Capacitor	Cascaded H-bridge	Proposed cascaded H – Bridge
1	No of level (K)	5	5	5	7
2	Switches required	2(K-1)	2(K-1)	2(K-1)	2(K-3)
3	Diode	2(K-1)	2(K-1)	2(K-1)	0
4	Clamping diode	(K-1)(K-2)	0	0	0
5	Balancing Capacitor	0	[(K-1)(K-2)] / 2	0	0
6	DC- bus capacitor	(K-1)	(K-1)	0	0
7	Auxiliary capacitor	0	(K-1)(K-2)	0	0
8	DC source	1	1	2 symmetric fashion	2 asymmetric fashion

Table 2. Comparison of topologies with proposed topology

According to comparison given in table number 2. Cascaded H- bridge inverter topology is better than conventional h- bridge topology and also other topologies. As it required less number of devices, cost is reduced.

4. Calculations and Simulation results

As in proposed topology staircase modulation technique is used. Using equal partition method, switch conduction period is calculated as follows.

$$\propto_i = (i \times 180) \div K \qquad \dots (10)$$

Where, i = 1, 2… [(K-1)/2] and K= no of level. Conduction period according to firing angle is given in table3. Design inductor value and capacitor value for proposed topology is given in table 3. Inductor and Capacitor are act as dc side backend filter.

Switches	Conduction period	
S 11	$\propto_1 = 25.71^\circ$	$\propto_6 = 154.29^\circ$
S 12	$\propto_1 = 25.71^\circ$	$\propto_6 = 154.29^\circ$
S 13	$\propto_7 = 205.71^\circ$	$\propto_{12} = 334.25^\circ$
S 14	$\propto_7 = 205.71^\circ$	$\propto_{12} = 334.25^\circ$
S 21	$\propto_3 = 77.14^\circ$	$\propto_4 = 102.6^\circ$
S22	$\propto_2 = 51.42^\circ$	$\propto_5 = 128.58^\circ$
S23	$\propto_9 = 257.14^\circ$	$\propto_{10} = 282.84^\circ$
S24	$\propto_8 = 231.42^\circ$	$\propto_{11} = 308.55^\circ$

Table 3. Switches conduction angle

For full bridge converter Output voltage equation is given as follows

$$V_0 = \frac{V_{in}\, n_2}{2n_1\,(1-D)} \quad \dots (11)$$

Inductor critical value calculated by following equation no. 12

$$L_{CRITICAL} = 2\left(\frac{n_1}{n_2}\right)^2 R(1-D)^2\left(D-\frac{1}{2}\right)T \quad \dots (12)$$

4.1. Waveform of seven level output voltage

The input voltage of topology is 12 volt and 6 volt. Hence 7 level output voltage is obtained as 18 V, 12V, 6V, 0V, -6V, -12V, -18V. Seven level output voltage of MLI on the primary side of the transformer is shown in fig.3

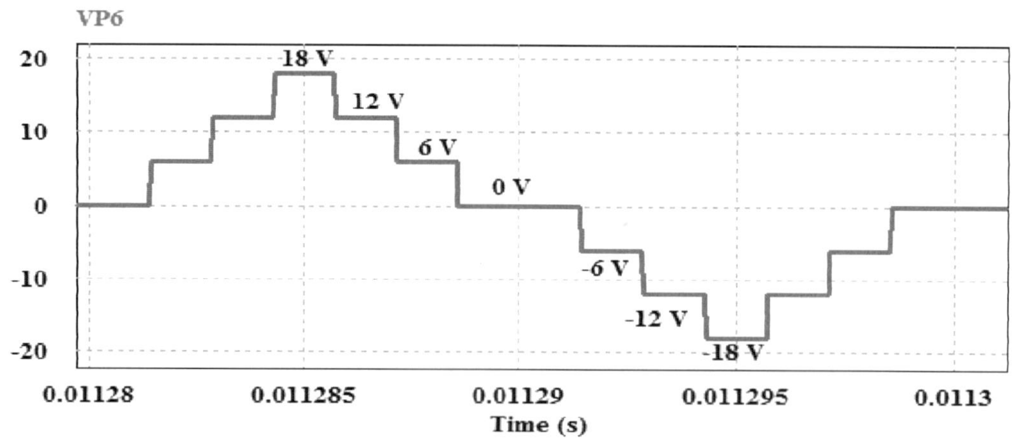

Fig.3 Waveform of seven-level output voltage

4.2. Waveform of input current from two sources

Average value of input current of each source is shown in fig. 4

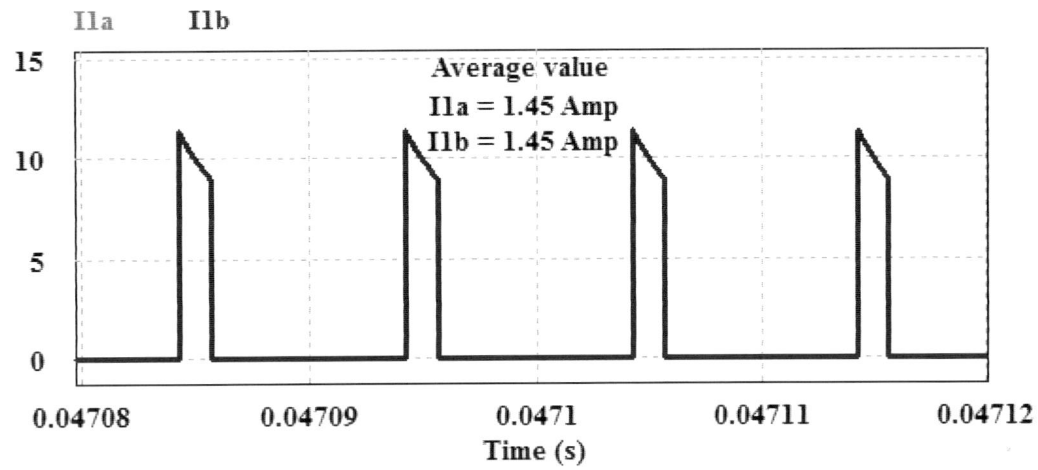

Fig.4 waveform of the average value of current

4.3. Waveform of output voltage

Output voltage waveform of dual input full bridge dc-dc converter when the transformer ratio is 1:2. As input sources voltage is 12 and 6 hence maximum voltage is 18 volt obtaining at primary side of transformer.

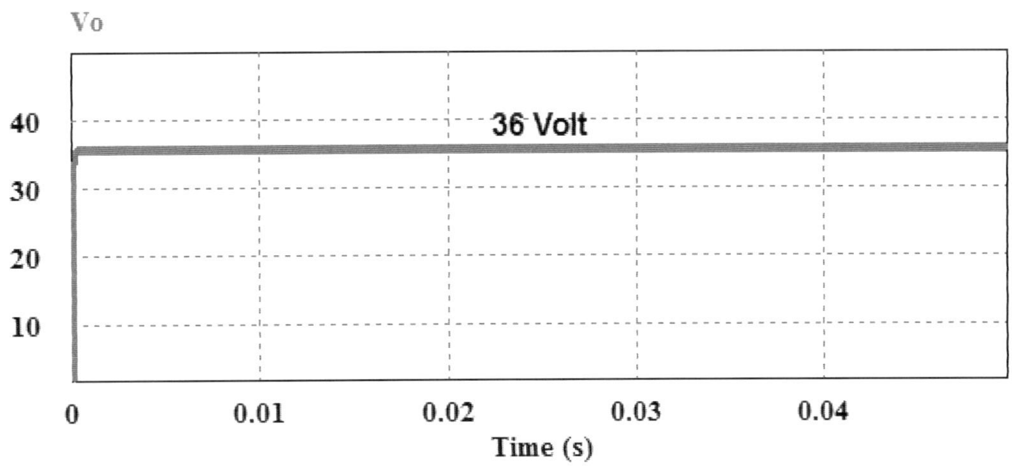

Fig.5 waveform of output voltage

4.4. Waveform of output power

Output power 25 watt is shown in Fig. 6

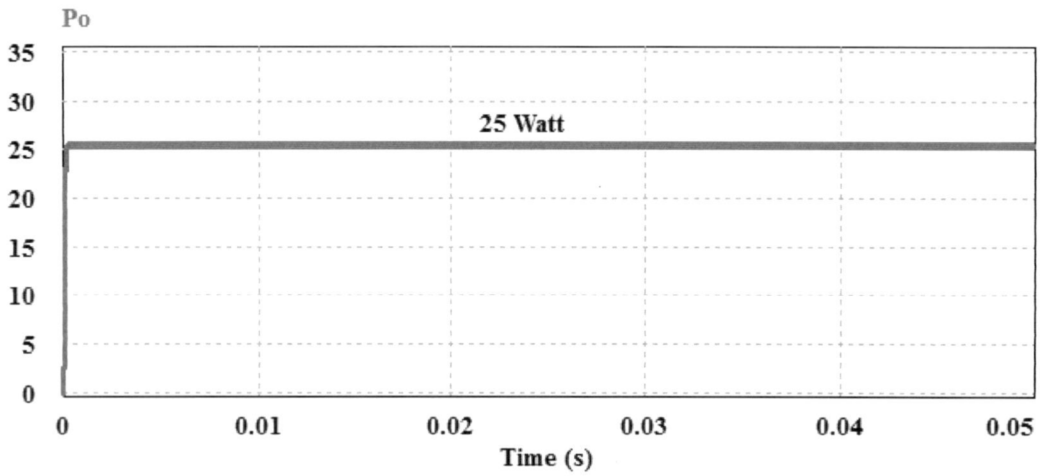

Fig.6 The output power waveform

From simulation waveform V out = 36 volt, Ia average = 1.45 Amp Ib average = 1.45 Amp and output power = 25.52 watt, Input power = (12*1.45) + (6*1.45) = 26.1 Watt, Hence the efficiency of dual input full-bridge isolated DC to DC converter is 97%

4.5. The output voltage waveform

When the transformer ratio is 1: 20 then using same 12V and 6 V input source, 351V output voltage obtained. Output voltage is shown in fig.7

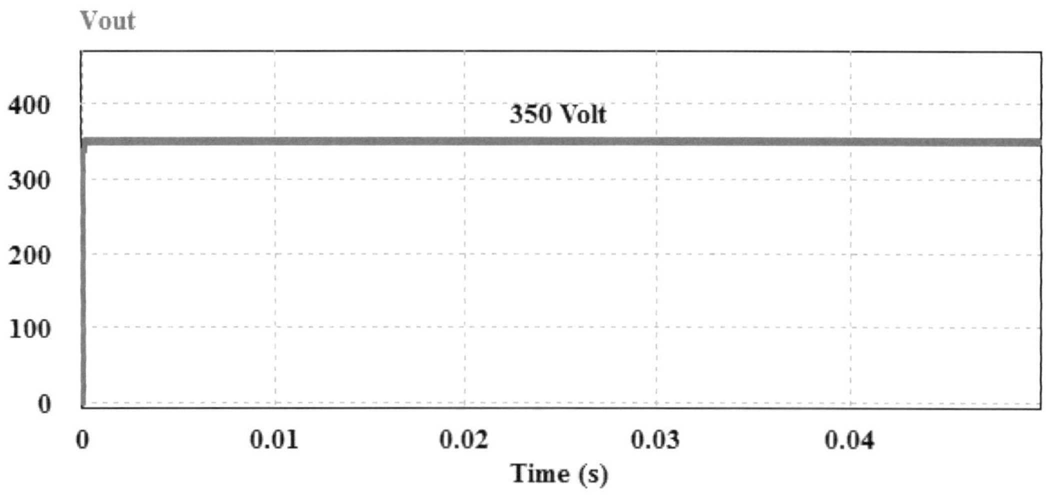

Fig.7 waveform of output voltage

5. Conclusion

In this paper, using two asymmetrical input source, numbers of voltage level increased on the transformer's primary side. The boosted output voltage is obtained using a high-frequency transformer which has a high transformer turns ratio. Hence the efficiency of the Full bridge converter is increased using two cascaded multilevel inverter. Also cost is reduced because number of devices gets reduced. According to simulation result, this topology has low power losses, high reliability, high power density and a wide range of handling input and output voltage.

6. Reference

[1] M. Pahlevani, S. Eren, H. Pahlevaninezhad, I. Askarian, and S. Bagawade, *"Digital current sensorless control of current-driven fullbridge DC/DC converters," IEEE Trans. Power Electron.*, vol. 33, no. 2, pp. 1797–1815, Feb. 2018.

[2] N. Sandeep, and U. R. Yaragatti, *"A Switched-Capacitor-Based Multilevel Inverter Topology With Reduced Components,"* IEEE Trans. Power Electron., vol. 33, no. 7, pp. 5538-5542, July 2018.

[3] E. Zamiri, N. Vosoughi, S. H. Hosseini, R. Barzegarkhoo, and M. Sabahi, *"A New Cascaded Switched-Capacitor Multilevel Inverter Based on Improved Series–Parallel Conversion With Less Number of Components," IEEE Trans. Ind. Electron.*, vol. 63, no. 6, pp. 3582-3594, June 2016.

[4] W. J. Cha, J. M. Kwon, and B. H. Kwon, *"Highly efficient asymmetrical PWM full-bridge converter for renewable energy sources," IEEE Trans. Ind. Electron.*, vol. 63, no. 5, pp. 2945–2953, May 2016.

[5] R. Kathiresan, P. Das, T. Reindl, and S. K. Panda, *"A novel ZVS DC–DC full-bridge converter with hold-up time operation," IEEE Trans. Ind. Electron.*, vol. 64, no. 6, pp. 4491–4500, Jun. 2017.

[6] K.K. Gupta, A. Ranjan, P. Bhatnagar, L.K. Sahu, S. Jain, "Multilevel Inverter Topologies with Reduced Device Count: A Review," IEEE Trans. on Power Electron., vol.31, no.1, pp.135–151, Jan. 2016.

[7] M. Malinowski, K. Gopakumar, J. Rodriguez, M.A. Perez, *A survey on cascaded multilevel inverters,* IEEE Trans. Industr. Electron. 57 (7) (2010) 2197–2206.

[8] Y. Ko, M. Andresen, G. Buticchi, and M. Liserre, *"Power Routing for Cascaded H-Bridge Converters," IEEE Trans. Power Electron.*, vol. 32, no. 12, pp. 9435-9446, Dec. 2017.

[9] International Journal Of Engineering And Computer Science ISSN: 2319-7242 Volume 3 Issue 6 June, 2014 Page No. 6341-6346 *"Analysis of Three Level Diode Clamped Multilevel Fed Induction Motor Drive"* Sadhana pandey, Sulochana Wadhwani, Praveen Bansal

[10] *"Asymmetric cascaded half-bridge multilevel inverter without polarity changer",* Ahmed, A Sheir, M Orabi - Alexandria engineering journal, 2018 - Elsevie

IOP Conf. Series: Materials Science and Engineering **906** (2020) 012011 doi:10.1088/1757-899X/906/1/012011

Luminous power improvement in InGaN V-Shaped Quantum Well LED using CSG on SiC Substrate

Manikandan M[1], Nirmal D[2], Prajoon P[3], Dhivyasri G[4], Chandran V[5]

[1,4,5] Department of Electronics and Communication Engineering, KPR Institute of Engineering and Technology, Coimbatore, Tamil Nadu, India
[2]Department of Electronics and Communication Engineering, Karunya Institute of Technology and Sciences, Coimbatore, Tamil Nadu, India
[3]Department of Electronics and Communication Engineering, Jyothi Engineering College, Thrissur, Kerala, India

Corresponding author: m.manikandan@kpriet.ac.in

Abstract. This paper presents the design and simulation of Silicon Carbide (SiC) based technology, Indium Gallium Nitride (InGaN) Multiple Quantum Well (MQW) Light-Emitting Diode (LED) with a Compositionally Step Graded (CSG) InGaN barrier and V-Shaped well in the active region. The simulations are obtained in Silvaco Computer Aided Design simulator and parameters such as Internal Quantum Efficiency (IQE) with respect to input current, spontaneous emission in regard to wavelength and power versus current in the device are theoretically studied. The CSG InGaN barrier LED with V-shaped quantum well shows substantial growth in output power when compared to the CSG GaN barrier structure with conventional MQW. The high carrier confinement in the V-shape well causes, transportation/injection of hole and change in band bending due to polarization effect. Moreover, lattice-matched SiC substrate over GaN material increases the InGaN V-shaped MQW LEDs radiative recombination rate which in turn leads to high output power. The optical luminous power of 160mW and 82% of peak IQE, emitting wavelength at 460 nm and 200mA of injection current is obtained for the proposed LED. The enactment of the V shape MQW CSG-InGaN device technology is a good alternative choice for commercial and industrial lighting applications.

1. Introduction

Among various semiconductor materials, III compound group Nitride materials plays a vital role in Nano scale lighting and display due to their tunable and direct energy bandgap from 0.7eV to 6.2eV [1].GaN-based LEDs are mainly used in visible light communication, television display and mobile platforms. There are several factors which degrades the performance (in terms of emission efficiency) of the device mainly electron leakage [2]-[7], current crowding [8], auger recombination [9]–[11], poor injection of hole carrier [12], [13], polarization effect [14], [15] and defects [16]. Xuna Li et al. proposed a structure using P-type aluminium gallium nitride Electron Blocking Layer (EBL) to reduce the electron leakage which intern leads to less hole injection efficiency [17].Yen et al. proposed n-type aluminium gallium nitride EBL structure which increases the output power, hole injection efficiency & photoelectric performance [18]-[20].Due to strong electric field the position of electron and hole wave function in MQW may get change which leads to more time consumption for carrier recombination. Due to strong electric field the carriers, may spill out from the active region. To overcome this MQW technique is proposed [17].This paper focuses comparison

IOP Conf. Series: Materials Science and Engineering **906** (2020) 012011 doi:10.1088/1757-899X/906/1/012011

of InGaN/GaN based LEDs with V-shape MQW and normal MQW with CSG barrier for the improvement of output power.

2. LED device structure parameters

2.1Device Structure

The parameters of compositionally step graded InGaN MQW LED device in cross-sectional representation is given in Figure 1. The LED device is designed using undoped GaN buffer layer with 3 μm thickness and n-type GaN with 6 μm thickness. Then silicon concentration of about $5x10^{18}cm^{-3}$ is also used for the device design. Consecutively six period v- shaped MQW active region with InGaN materials are placed along SiC substrate. The quantum well comprises of 3 nm-thick (6 in number) $In_{0.16}Ga_{0.84}N$ layer and quantum barrier consist of compositionally 18 nm-thick (6 in number) graded InGaN layers. A 3 nm thick p-$Al_{0.2}Ga_{0.8}N$ Electron Blocking Layer (EBL) is present on the top of active region and a p-GaN capping layer with a Magnesium (Mg) concentration of $1x10^{19}cm^{-3}$. The entire device is in rectangular shape with the chip size dimension of 300 μm x 300 μm.

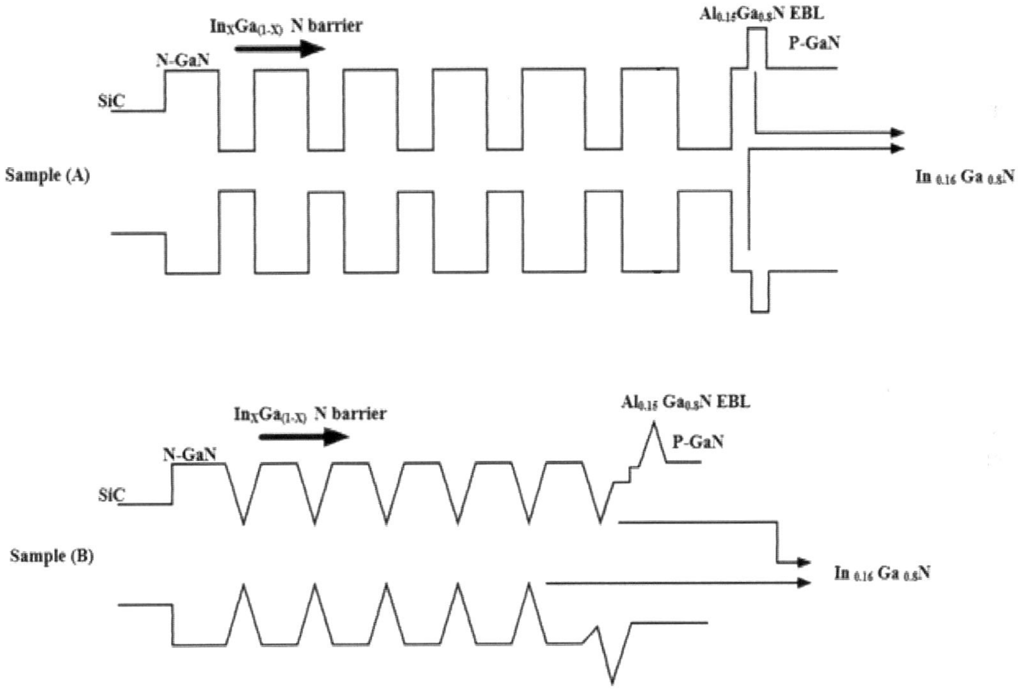

Figure 1. CSG-InGaN based LED Structure with V-shape MQW

2.2 Model expression

The ternary alloy of InGaN is expressed as

$$E_g(In_yGa_{(1-y)}N)=yE_g(InN)+(1-y)E_g(GaN)-1.4y(1-y) \quad (1)$$

The AlGaN energy band gap is expressed by

$$E_g(Al_yGa_{(1-y)}N)=yE_g(AlN)+(1-y)E_g(GaN)-0.7y(1-y) \quad (2)$$

Recombinaion of electrons and holes takes places in active region of the device, which emits in the form of photon.

The integral of emitted photon at each point gives the spontaneous emission rate.

$$R^{sp}{}_{active} = \int_0^{\infty} r^{sp}(\hbar\tilde{\omega})\, \rho^{opt}(\hbar\tilde{\omega}) d(\hbar\tilde{\omega}) \quad (3)$$

The total output power of LED device is determined by the following expression

$$P^{P}_{total} = \int_{active} \int_0^{\infty} R^{sp}_{active}(\hbar\tilde{\omega})\, d(\hbar\tilde{\omega}) dV \quad (4)$$

Where, $\hbar = h/2\Pi$, h is the planks constant and V is the total volume of MQW active region.

3. V-shape multiple quantum well

In recent times, optical and electrical properties are improved by observing the energy, and also the disruptions of interfacial strain facilitates the blue InGaN-based LED performance [21, 22]. Tomita et al. used an elevation angle annula dark field SEM transmission and atom sample tomography to confirm the presence of thin quantum wells in the slope of V-pit region, and Indium (In) concentrations and thickness were much lower than those of the flat region. The dislocations in the V-shaped pits serve as energy barriers for lateral charge transport [23]. In 2014, Kim et al. addressed the power of V-pit and their energy barrier from their facets orientation and revealed that higher V-pit energy barrier heights in InGaN quantum wells effectively overcomes non-radiative recombination at active region, thereby increasing the IQE [24]. Chan et al. used super lattice layers to control the V-pits nanoscale to attain the optimum V-pits size and high-efficiency InGaN/ GaN LEDs of blue wavelength [25].

4. Results and discussion

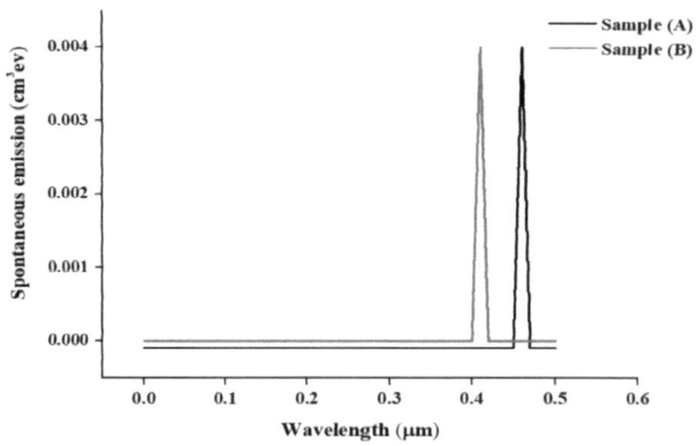

Figure 2. Wavelength vs Spontaneous emission

In figure 2, two sample devices with different wavelengths spontaneous emission rate is illustrated. From the obtained results, it is found that a maximum emission is occurring at 460 nm wavelength using V-shape MQW rather than normal MQW, which indicates that the LED emits lights in blue spectrum.

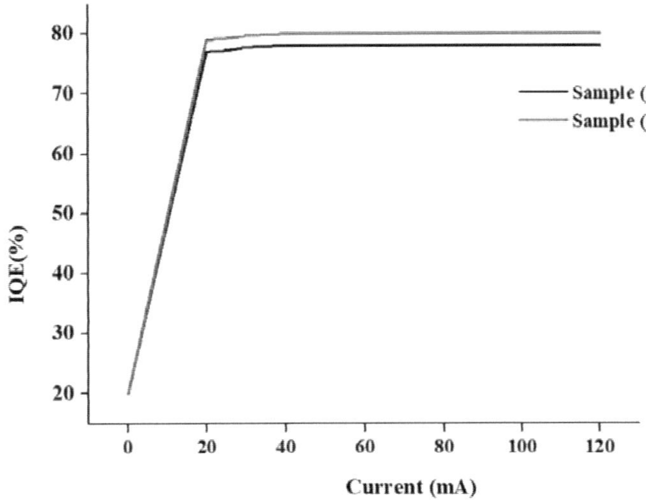

Figure 3. Current (mA) vs IQE (%)

In figure 3, IQE versus current characteristics of the two sample are illustrated. At low injection current, all the samples show similar characteristics and the IQE peaks up to 80 %. It is observed that IQE gradually decreases in a high injection current due to the efficient droop effect.

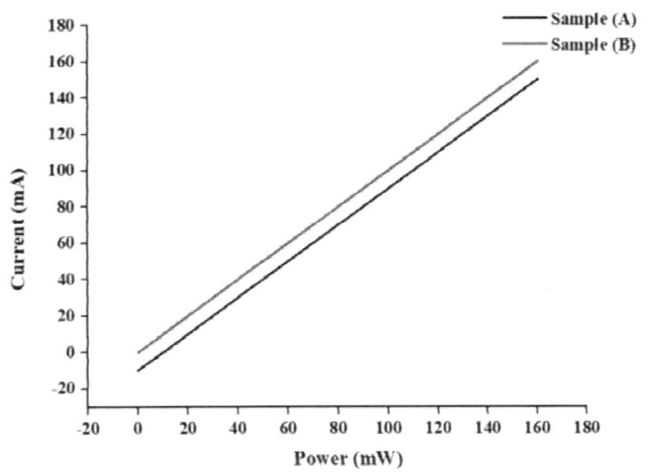

Figure 4. Power (mW) vs Current (mA)

The experimental IQE is valued from the wall plug efficiency (ηp) and is given by the expression

$$\eta P = \frac{P_{out}}{IV} = \eta_v \eta_{ex} \quad (5)$$

$$\eta_{ex} = \frac{P_{out}}{IV} = \eta_{int} \eta_{extr} \quad (6)$$

Where - η_{extr} - optical extraction efficiency
η_v - voltage efficiency
η_{int} -internal quantum efficiency
V_g bandgap voltage of the active layer

In figure 4, shows the optical luminous power and peak IQE obtained is 160 mW using V-shaped MQW since high electron reconfinement takes place in v-shape MQW rather than normal shape MQW which leads to high output power. Near to the p-contact concentration of the hole in the last quantum well is found to be maximum.

5. Conclusions

The simulated result of V-shape MQW based CSG InGaN barrier LED (sample B) is compared with conventional MQW based CSG InGaN barrier LED (sample A). From the results it can be concluded that sample A requires the light output power of 155 mW and 78% efficiency droop at 200 mA injection current. Where else sample B LED renders higher light output power of 160 mW, 82% of efficiency with at 200 mA injection current. Thus the proposed LED is more advantageous in lowering the overall light fixture cost which in turn reduces the consumption of energy in solid state lighting and visible light communication systems.

Acknowledgements

The authors would like to extend their thanks to VLSI Laboratory at Karunya University and KPR Institute of Engineering and Technology for providing the support to perform this research work.

References

[1] Davydov, V. Y., Klochikhin, A. A., Seisyan, R. P., Emtsev, V. V., Ivanov, S. V., Bechstedt, F., ... & Semchinova, O. (2002). Absorption and emission of hexagonal InN. Evidence of narrow fundamental band gap. *physica status solidi (b)*, *229*(3), r1-r3.

[2] Ni, X., Li, X., Lee, J., Liu, S., Avrutin, V., Özgür, Ü... & Evans, K. R. (2010). InGaN staircase electron injector for reduction of electron overflow in InGaN light emitting diodes. *Applied Physics Letters*, *97*(3), 031110.

[3] Schubert, M. F., Xu, J., Kim, J. K., Schubert, E. F., Kim, M. H., Yoon, S., & Park, Y. (2008). Polarization-matched GaInN/AlGaIn N multi-quantum-well light-emitting diodes with reduced efficiency droop. *Applied Physics Letters*, *93*(4), 041102.

[4] Kim, M. H., Schubert, M. F., Dai, Q., Kim, J. K., Schubert, E. F., Piprek, J., & Park, Y. (2007). Origin of efficiency droop in GaN-based light-emitting diodes. *Applied Physics Letters*, *91*(18), 183507.

[5] Choi, S., Ji, M. H., Kim, J., Jin Kim, H., Satter, M. M., Yoder, P. D., & Ponce, F. A. (2012). Efficiency droop due to electron spill-over and limited hole injection in III-nitride visible light-emitting diodes employing lattice-matched InAlN electron blocking layers. *Applied Physics Letters*, *101*(16), 161110.

[6] Piprek, J., & Li, S. (2010). Electron leakage effects on GaN-based light-emitting diodes. *Optical and quantum electronics*, *42*(2), 89-95.

[7] Zhao, H., Liu, G., Zhang, J., Arif, R. A., & Tansu, N. (2013). Analysis of internal quantum efficiency and current injection efficiency in III-nitride light-emitting diodes. *Journal of Display Technology*, *9*(4), 212-225.

[8] Li, C. K., & Wu, Y. R. (2011). Study on the current spreading effect and light extraction enhancement of vertical GaN/InGaN LEDs. *IEEE transactions on electron devices*, *59*(2), 400-407.

[9] Delaney, K. T., Rinke, P., & Van de Walle, C. G. (2016). Erratum: "Auger recombination rates in nitrides from first principles"[Appl. Phys. Lett. 94, 191109 (2009)]. *Applied Physics Letters*, *108*(25), 259901.

[10] Bertazzi, F., Goano, M., & Bellotti, E. (2010). A numerical study of Auger recombination in bulk InGaN. *Applied Physics Letters*, *97*(23), 231118.

[11] Tan, C. K., Zhang, J., Li, X. H., Liu, G., Tayo, B. O., & Tansu, N. (2013). First-principle electronic properties of dilute-As GaNAs alloy for visible light emitters. *Journal of Display Technology*, *9*(4), 272-279.

[12] Zhao, H., Liu, G., Arif, R. A., & Tansu, N. (2010). Current injection efficiency induced efficiency-droop in InGaN quantum well light-emitting diodes. *Solid-State Electronics*, *54*(10), 1119-1124.

[13] Liu, J. P., Ryou, J. H., Dupuis, R. D., Han, J., Shen, G. D., & Wang, H. B. (2008). Barrier effect on hole transport and carrier distribution in In Ga N/ Ga N multiple quantum well visible light-emitting diodes. *Applied Physics Letters*, *93*(2), 021102.

[14] Schubert, M. F., Xu, J., Kim, J. K., Schubert, E. F., Kim, M. H., Yoon, S., & Park, Y. (2008). Polarization-matched Ga InN/ Al Ga In N multi-quantum-well light-emitting diodes with reduced efficiency droop. *Applied Physics Letters*, *93*(4), 041102.

[15] Arif, R. A., Ee, Y. K., & Tansu, N. (2007). Polarization engineering via staggered InGaN quantum wells for radiative efficiency enhancement of light emitting diodes. *Applied Physics Letters*, *91*(9), 091110.

[16] Monemar, B., & Sernelius, B. E. (2007). Defect related issues in the "current roll-off" in InGaN based light emitting diodes. *Applied Physics Letters*, *91*(18), 181103.

[17] Li, X., Sun, H., Cai, J., Yang, M., Zheng, H., Sun, H., & Guo, Z. (2015). Study of blue InGaN/GaN light-emitting diodes with n-AlGaN layer as space layer and without the P-AlGaN electron blocking layer. *Journal of Display Technology*, *11*(2), 170-174.

[18] Chen, S. W., Li, H., Chang, C. J., & Lu, T. C. (2017). Effects of nanoscale V-shaped pits on GaN-based light emitting diodes. *Materials*, *10*(2), 113.1.

[19] Peng, W., Qu, S., Cong, G., & Wang, Z. (2006). Synthesis and structures of morphology-controlled ZnO nano-and microcrystals. *Crystal growth & design*, *6*(6), 1518-1522.

[20] Ivill, M., Pearton, S. J., Norton, D. P., Kelly, J., & Hebard, A. F. (2005). Magnetization dependence on electron density in epitaxial ZnO thin films codoped with Mn and Sn. *Journal of applied physics*, *97*(5), 053904.

[21] Kaneta, A., Funato, M., & Kawakami, Y. (2008). Nanoscopic recombination processes in InGaN/GaN quantum wells emitting violet, blue, and green spectra. *Physical Review B*, *78*(12), 125317.

IOP Conf. Series: Materials Science and Engineering **906** (2020) 012011 doi:10.1088/1757-899X/906/1/012011

[22] Tsai, C. L., & Wu, W. C. (2014). Effects of asymmetric quantum wells on the structural and optical properties of InGaN-based light-emitting diodes. *Materials*, *7*(5), 3758-3771.

[23] Tomiya, S., Kanitani, Y., Tanaka, S., Ohkubo, T., & Hono, K. (2011). Atomic scale characterization of GaInN/GaN multiple quantum wells in V-shaped pits. *Applied Physics Letters*, *98*(18), 181904.

[24] Kim, J., Cho, Y. H., Ko, D. S., Li, X. S., Won, J. Y., Lee, E., ... & Kim, S. (2014). Influence of V-pits on the efficiency droop in InGaN/GaN quantum wells. *Optics express*, *22*(103), A857-A866.

[25] Chang, C. Y., Li, H., Shih, Y. T., & Lu, T. C. (2015). Manipulation of nanoscale V-pits to optimize internal quantum efficiency of InGaN multiple quantum wells. *Applied Physics Letters*, *106*(9), 091104.

Performance of statistical control tools in identifying the defect on plunger- A review

N.ARAVINDHAN , CHIDAMBARANATHAN BIBIN , D.ROKESH,D.JAGADESH, S.PRAVIN, N.RAMKUMAR

Department of Mechanical Engineering,RMK College of Engineering and Technology

Email ID: aravindhn25389@gmail.com

Abstract. In this paper, we carried out a study to upsurge the brilliance of the plungers used in the braking system. The main focus has been taken on various defects that occur in the plungers during manufacturing. In this work, we analyzed the types of defects that transpire in the plunger. The foremost issue is acknowledged using tools like Pareto graphic representation, root and effect diagram, Failure Mode and Effect Analysis, Trend Analysis and Control chart. We identified more frequently occurring defects and major causes and effects of these defects in plungers. Based on the result of the analysis, interpretation measures are taken to overcome the defects of the plungers that will help to improve the overall efficiency of the brake system in vehicles.

Keywords: Plungers, Braking system, Pareto, Failure mode, Trend Analysis.

1. Introduction

Plunger plays a vital role in the braking system and it is a very crucial part of the hydraulic braking system [1]. So, the plungers are required to have a high accuracy in its quality. Very small defects in the Plunger may also lead to a major accident so it is very important to improve the quality of the plungers [2]. In general, a brake system of a vehicle is operated to slow down or to stop the vehicle when the vehicle is in running condition. The brake pedal is associated with a plunger, as it enforces hydraulic oil through a sequence of tubes along with hoses to the braking component [3]. The piston forces the brake pads adjacent to the brake are bound to the wheel may freeze down or stop. In this research study, an attempt was made by the researchers to identify and analyze various types of defects that occur in plunger during manufacturing and the root causes for the occurrence of defects. The investigation of defects in plungers is made based on unremitting enhancement [4]. It requires coordination within the organization members and effectively using the quality tools in their improvement activities and decision-making process [5]. In the current scenario, many different tools are available in the market and each tool is used for specific purposes and differentiates among them. In further there are different types of tools format is required like statistical, analytical, and clerical [6]. The use of statistical quality control tools had increased on both the practical and theoretical levels. This analytical data is further used to measure the causes of defects using DMAIC, Cause and Effect analyses, Pareto chart [7]. In today's scenario, the business firm is at the speed of the iterative development and this development records the feedback from the iteration for further development which implements the software-based quality tools like Pareto and six sigma and it also gathers feedback from the customer end to further rectification [8].

1.1 Plunger

It is a component inside the wheel cylinder of the braking system. The brake pedal is forced towards the plunger within the wheel cylinder [9]. This leads to the transporting of fluids to all portions of the wheel [10]. Many researchers indicated that the performance is better for a component at low pressure and heat [11]. So inturn the parameters need to be optimized [12].

2. Defects Identified in Plungers

These are the defects identified on the plunger while manufacturing is classified as Barrel diameter, damage, or wrong size, Collar diameter damage or wrong size, Damage, Drill depth over or undersize, Drill diameter oversize, Overall diameter wrong size, Seal Groove diameter oversize, Seal diameter poor finish, Seal face chatter mark, Short feed, Stem diameter oversize, Line marks, Nick marks, Wrong indexing, Runout [13],[14]. The ion implantation was advanced to avoid the deterioration in the plunger [1].

2.1 Causes of Defects

According to our research and data collected from the literature, there are some major causes of defects that will affect the quality of the plunger while manufacturing and may affect the performance of the plunger [15], [16]. These are the ways of rejection identified while the first operation and it is caused by primary faults that may occur on the machine or the operator. They are Material rejection, End bit rejection, Power cut, Process, Tryout, Operator, Setting, Tool is broken [8]. The defects due to heat are most common on engines and it affects the performance of engines too [17], [18].

3.Methodology

3.1 Research Design

In this study, the researcher has used the analytical type of research [19]. The researcher intends to collect secondary data on defect variance in plunger at brake system, such as collar diameter oversize, barrel diameter oversize, collar diameter undersize, drill depth oversize, drill depth undersize, conveyor damage, etc [20]. The root cause of the defects is analyzed and effective measures are taken to eliminate the occurrence of defects in plunger and that helps to advance the braking system used in automobiles [5], [19]. The research design needs a novel approach for improvising the performance of the wear rate on the plunger [4], [21]. The data are collected from various literature such as Quality control, Production, Manufacturing Engineering, Materials departments in industries that are taken into account for sampling the process [21], [22]. The researcher collected sample data on defects of brakes plunger. The sample data was collected for one month [23].

3.2 Limitations

On bearing in mind about time constraints for carrying out the research, only "Post- machining process defects" are considered to in this research study [24]. The defects which occur in the "Machining process" will be considered and analyzed in a future study by using statistical control tools [5], [25].

4. Results and Discussion

4.1 Fault investigation through trend chart

Trend analysis is the extensive tradition of gathering in sequence to spot a pattern [26]. Even though trend analysis is frequently used to forecast upcoming events, it also shows the path of future results and uncertain events in the past [27]. The above trend analysis shows that the rejection rate varies continuously concerning time [28].

4.2 Pareto Analysis

A Pareto chart is an systematic technique for segregating the issues for erdicating the factors affecting the quality . The pareto chart gives a transparency of the parent defect which intiates other sub defects. It prioritize the defects using its relative consequence. From the above analysis, it is shown that vital few and trivial many defects of Plunger manufacturing.

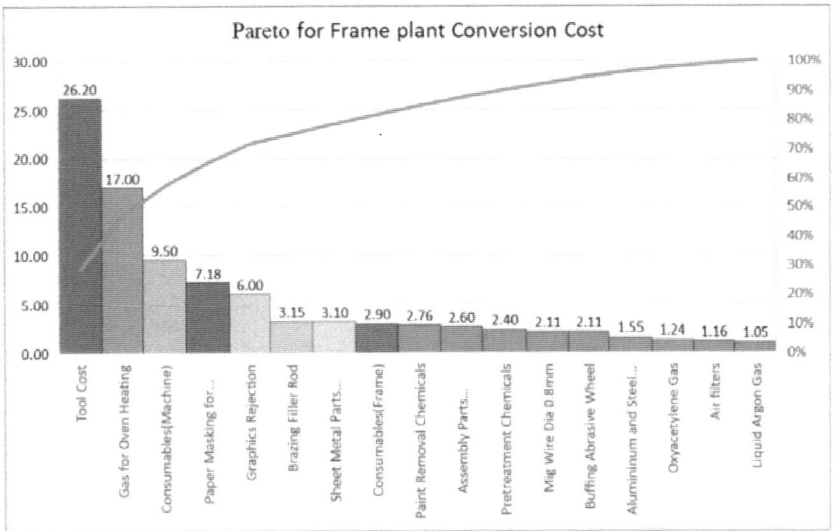

Figure 1. Pareto Analysis [5]

4.3 Cause and Effect Analysis

The cause and effect diagram is used to discover all the impending or real cause. The causes are characterized by their effects. It is also known as the "Fishbone diagram or Ishikawa diagram". We categorize the foremost reimbursement, and the Causes are in order according to their altitude of substance. From on top of the chart, it explains the assortment of causes of the incidence of damages in plunger the foremost causes of harm are done by the man mistakes.

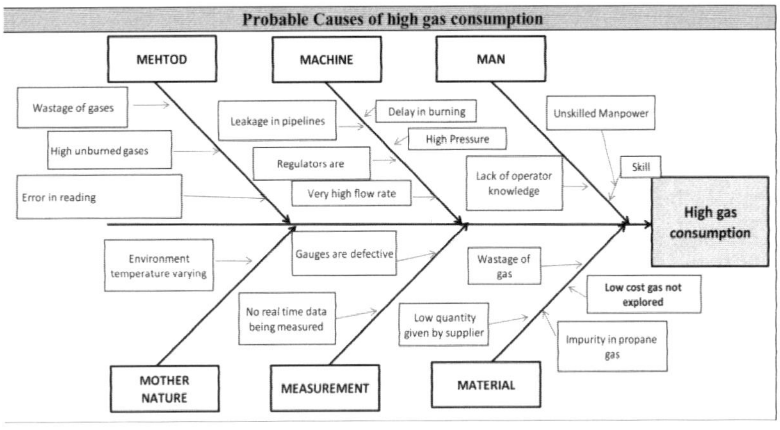

Figure 2: Cause and Effect Analysis for sample high gas consumption plot [5]

4.4 Failure Modes & Effects Analysis (FMEA)

Failure Mode and Effect Analysis is an approach for performing risk analysis, it a sequential process to categorize all possible letdown [29]. This mode of approach is an everlasting process for identical creativeness. It used to describe, recognize, and eradicate existing problems and probable prospective failures ahead of them to accomplish the customer. Every risk factor has been calculated from every process and notified from 1- Lowest risk and 10-Highest risk. From the above FMEA – Failure Mode Effective Analysis table it was originated that the probable possessions of the failures and their cause and their impending significance. By prioritizing the foremost RPN of all prospective failure to decide upon behavior to shrink risk.

4.5 Control Chart

The control chart is a graph that shows the variations in each process over time. In a control chart, the midline is called average, a high procession is for better control limit and the minor line is for subordinate control limit. By correlating the active data to these outlines, we can terminate whether the process is in control or not. Statistical process control is used to follow good quality control practices in the manufacturing process. And it is used to detect the unnatural patterns and deviations in the process.

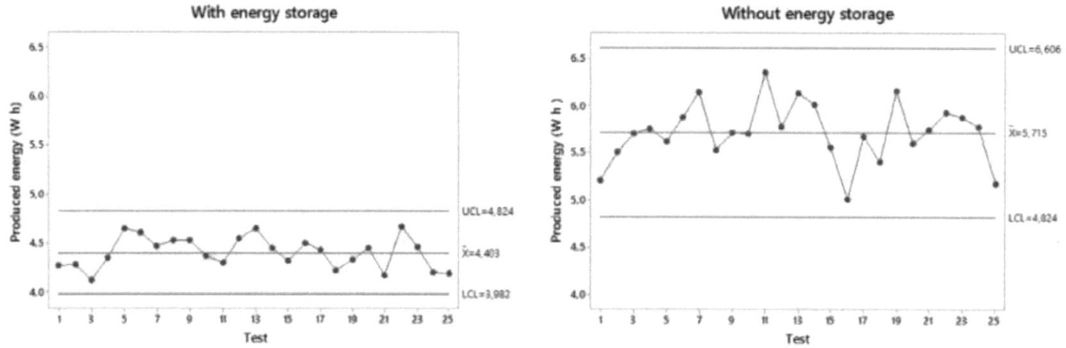

Figure 3. Control chart for Energy production [20]

Analysis of various defects in plungers using a statistical Quality control tool will help to identify the frequency of occurrence of defects, the causes, and its impacts. The interpretation from the statistical analysis will help to take effective measures to control and eliminate the causes of various defects in plungers during manufacturing This in turn help the companies to reduce the rejection rates of plungers and the cost of Quality. By reducing the rejection rates of defects, rework time, and rework costs can be minimized [30].

Analysis of various defects in plungers using a statistical Quality control tool will help to identify the frequency of occurrence of defects, the causes, and its impacts. The interpretation from the statistical analysis will help to take effective measures to control and eliminate the causes of various defects in plungers during manufacturing [31]. This in turn helps the companies to reduce the rejection rates of plungers and the cost of Quality. By reducing the rejection rates of defects, rework time, and rework costs can be minimized. The plunger stress analysis during the time is already shown in figure 9 he overcame the design based on ion implantation [32].

5. Conclusion

From the above research of defects in plunger manufacturing, the researchers have identified the various defects occurring more frequently and the causes of these defects. In this Analytical type of research Pareto chart is an ideology to identify and split the imperative few from trivial many defects. Trend analysis identifies that the rejection rate continuously varies concerning time. Then the cause and effect diagram is used to recognize the mixture of cause for the incidence of the major defect and the diagram highlights the errors produced while manufacturing of plunger. Finally, the FMEA tool is used to analyze the various effects that are caused by the defects and this tool also shows the various prevention and detection for the failure.

References

[1] C. Yin, W. Dai, and Y. Lian, "Optimization of the Manufacturing Defects of Plunger Unit based on Ion Implantation," *Procedia CIRP*, vol. 56, pp. 46–49, 2016, doi: 10.1016/j.procir.2016.10.014.

[2] P. E. Lang, D. C. Carslaw, and S. J. Moller, "A trend analysis approach for air quality network data," *Atmos. Environ. X*, vol. 2, no. March, p. 100030, 2019, doi: 10.1016/j.aeaoa.2019.100030.

[3] K. H. Singh, A. Kumar, and R. Kumar, "Optimization of Quality and Performance of Brake Pads Using Taguchi's Approach," vol. 5, no. 7, pp. 632–639, 2014.

[4] R. A. Kumar, A. Devaraju, and S. Arunkumar, "Experimental Investigation on Mechanical Behaviour and Wear Parameters of TiC and Graphite Reinforced Aluminium Hybrid Composites," *Mater. Today Proc.*, vol. 5, no. 6, pp. 14244–14251, 2018, doi: 10.1016/j.matpr.2018.03.005.

[5] V. Shivajee, R. K. Singh, and S. Rastogi, "Manufacturing conversion cost reduction using quality control tools and digitization of real-time data," *J. Clean. Prod.*, vol. 237, p. 117678, 2019, doi: 10.1016/j.jclepro.2019.117678.

[6] AN. Aravindhan, K. M. J. Vasanth, R. V. Kumar, M. Jayasurya, S. S. Prakash, and V. Sabareeshwaran, "Materials Today : Proceedings A novel approach for improving the performance of air engine to achieve zero-emission for a pollution-free environment," *Mater. Today Proc.*, no. xxxx, pp. 1–5, 2020, doi: 10.1016/j.matpr.2020.02.930.

[7] S. Gopinath, P. K. Devan, C. Mohan, L. R. Krishna rao, P. Lokesh kumar, and S. Vara Prasad, "A review on influence of injection timing and injection pressure on DI diesel engine fuelled with low viscous fuel," *Mater. Today Proc.*, 2020, doi: https://doi.org/10.1016/j.matpr.2020.04.070.

[8] P. Devan, S. Gopinath, K. Rajesh, and S. Madhu, "Improving the characteristics of engine oil using nanofluid as coolant in combat vehicles," *Mater. Today Proc.*, no. xxxx, pp. 2–6, 2019, doi: 10.1016/j.matpr.2019.12.005.

[9] S. Gopinath and P. K. Devan, *Optimization and Prediction of Reaction Parameters of Plastic Pyrolysis oil Production Using Taguchi Method*, vol. 39,Issue2. Iranian Journal of Chemistry and Chemical Engineering (IJCCE), 2019.

[10] T. Ajiskrishnan, C. Bibin, S. Gopinath, N. Aravindhan, and M. Krishna Kumar 5 1 Engineer, "Vision-Based Position and Orientation Estimation of an Aerial Vehicle," *IJSRD-International J. Sci. Res. Dev.*, vol. 3, no. 01, pp. 2321–0613, 2015.

[11] C. Bibin, P. Seenikannan, and N. Kanthavelkumaran, "Study On Waste Heat Recovery In An Internal Combustion Engine," vol. 1, no. 10, pp. 1–4, 2012.

[12] C. Bibin and P. Seenikannan, "Experimental analysis of energy recovery from an internal combustion engine exhaust using rankine cycle," *Adv. Mater. Res.*, vol. 768, no. x, pp. 158–163, 2013, doi: 10.4028/www.scientific.net/AMR.768.158.

[13] N. Sadasivan, M. Balasubramanian, R. Venkatesh, S. Vigneshram, and T. Sunil, "Influence of equal channel angular pressing in an acute angle die with a back pressure notch on grain refinement, torsion and mechanical properties of aluminium," *Materwiss. Werksttech.*, vol. 50, no. 2, pp. 155–164, 2019, doi: 10.1002/mawe.201700102.

[14] N. Sadasivan and M. Balasubramanian, "Equal channel angular pressing of gas tungsten arc welded AA6061 alloy," *Weld. World*, 2020, doi: 10.1007/s40194-020-00897-0.

[15] S. Gopinath, V. B. Murugan, and N. Dineshbabu, "Optimizing the Various Parameters for Biodiesel Production by Transesterification," vol. 4, no. 02, pp. 716–718, 2015.

[16] Bibin.C, Vinoth Kumar.K Muruga Pandi.D, "Fabrication of an Offroad Hybrid Vehicle," vol. 1, no. 1, pp. 2–4, 2017.

[17] Gopinath S, Devan P K, and Bibin C, "Experimental Study and Optimization of Injection Parameters for a Biodiesel Using Taguchi Method," *Int. J. Appl. Eng. Res.*, vol. 9, no. 26, pp. 9292–9296, 2014.

[18] C. Bibin, M. J. Selvaraj, and S. Sanju, "Flutter analysing over an aircraft wing during cruise speed," *Procedia Eng.*, vol. 38, pp. 1950–1961, 2012, doi: 10.1016/j.proeng.2012.06.238.

[19] N. Xiao, H. Z. Huang, Y. Li, L. He, and T. Jin, "Multiple failure modes analysis and weighted risk priority number evaluation in FMEA," *Eng. Fail. Anal.*, vol. 18, no. 4, pp. 1162–1170, 2011, doi: 10.1016/j.engfailanal.2011.02.004.

[20] M. G. de Andrade *et al.*, "Statistical quality control for the evaluation of the uniformity of microsprinkler irrigation with photovoltaic solar energy," *Renew. Sustain. Energy Rev.*, vol. 78, no. May, pp. 743–753, 2017, doi: 10.1016/j.rser.2017.05.012.

[21] L. Gorjan *et al.*, "Ceramic protection plates brazed to aluminum brake discs," *Ceram. Int.*, vol. 42, no. 14, pp. 15739–15746, 2016, doi: 10.1016/j.ceramint.2016.07.035.

[22] S. Sri Karthikeyan, E. Balakrishnan, S. Meganathan, M. Balachander, and A. Ponshanmugakumar, "Elemental analysis of brake pad using natural fibres," *Mater. Today Proc.*, vol. 16, pp. 1067–1074, 2019, doi: 10.1016/j.matpr.2019.05.197.

[23] K. O. Davis, R. P. Brooker, H. P. Seigneur, M. Rodgers, A. C. Rudack, and W. V. Schoenfeld, "Pareto analysis of critical challenges for emerging manufacturing technologies in silicon photovoltaics," *Sol. Energy*, vol. 107, pp. 681–691, 2014, doi: 10.1016/j.solener.2014.06.017.

[24] W. Y. Mengesha, Yonatan; Singh, Ajit Pal; Amedie, "Quality Improvement Using Statistical Process Control Tools in Glass Bottles Manufacturing Company," *Int. J. Qual. Res.*, vol. 7, no. 1, pp. 107–126, 2013.

[25] A. Joshi and L. M. Jugulkar, "Investigation and Analysis of Metal Casting Defects and Defect Reduction By Using Quality Control Tools," 2014.

[26] C. Bibin, P. Seeni Kannan, and P. K. Devan, "Performance, emission and combustion characteristics of a Di diesel engine using blends of punnai oil biodiesel and diesel as fuel," *Therm. Sci.*, vol. 2018, pp. 1–18, 2018, doi: 10.2298/TSCI180325233B.

[27] T. M. Yeh, F. Y. Pai, and C. C. Yang, "Performance improvement in new product development with effective tools and techniques adoption for high-tech industries," *Qual. Quant.*, vol. 44, no. 1, pp. 131–152, 2010, doi: 10.1007/s11135-008-9186-7.

[28] E. Arteaga-López, C. Ángeles-Camacho, and F. Bañuelos-Ruedas, "Advanced methodology for feasibility studies on building-mounted wind turbines installation in urban environment: Applying CFD analysis," *Energy*, vol. 167, pp. 181–188, 2019, doi: 10.1016/j.energy.2018.10.191.

[29] T. H. Setiawan, B. Adryfan, and C. A. Putra, "Risk Analysis and Priority Determination of Risk Prevention using Failure Mode and Effect Analysis Method in the Manufacturing Process of Hollow Core Slab," *Procedia Eng.*, vol. 171, pp. 874–881, 2017, doi: 10.1016/j.proeng.2017.01.382.

[30] S.Arunkumar, R.Ashokkumar,M.SubramaniSundaram,K.M.SukethKanna,S.Vigneshwara, "Materials Today : Proceedings Optimization of wear behaviour of Al7075 hybrid metal matrix composites using Taguchi approach," Mater. Today Proc., no. xxxx, pp. 1–5, 2020, doi.org/10.1016/j.matpr.2020.05.453

[31] S.Arunkumar, M.Subramani Sundaram, K.M.SukethKanna ,S.Vigneshwara, "Materials Today : Proceedings Optimization of wear behaviour of Al7075 hybrid metal matrix composites using Taguchi approach," Mater. Today Proc., no. xxxx, pp. 1–5, 2020, doi.org/10.1016/j.matpr.2020.05.453

[32] RAJESH K; DEVAN P K; BharthSai Kumar G K. "Parametric Optimization and Biodiesel Production from Coconut Fatty Acid Distillate". *Iranian Journal of Chemistry and Chemical Engineering (IJCCE)* , 2020 (Online), http://www.ijcce.ac.ir/article_39119.html

Optimized energy management control over price-responsive demands

N. Loganathan[1] **M. Shanmugapriya** [2] **Dr. T. Logeswaran** [3] **M. Dhanapal** [4]

[1]Assistant Professor, Department of Electrical and Electronics Engineering, Sri Krishna College of Engineering and Technology, Coimbatore, Tamil Nadu.

[2]Assistant Professor, Department of Electrical and Electronics Engineering, KCG College of Technology, Chennai, Tamil Nadu.

[3]Assistant Professor (Senior Grade), Department of Electrical and Electronics Engineering, Kongu Engineering College, Erode, Tamil Nadu

[4]Assistant Professor, Department of Electrical and Electronics Engineering, K.S.Rangasamy College of Technology, Tiruchengode, Tamil Nadu.

E-mail: logukirsh@gmail.com, shanmugapriya.muthusamy@gmail.com, logeskongu@gmail.com, dhanapal2007@gmail.com

Abstract. The fair and efficient allocation for all energy management systems is one of the fundamental principles. In order to achieve optimum efficiency, classic allocation strategies aim to optimize the total surplus of customers. This paper proposes the procurement, storage and selling of energy at appropriate times with a new energy management algorithm. Uncertainty is modeled using PSO optimization techniques in relation to both DER production level and the price of energy obtained from / sold to the main grid. The test results show the advantages of the model in energy management performance for a cluster of cost-effective demands.

Keywords. Energy Management System, Renewable Energy Systems, Smart Grid.

1. Introduction

In the last decades, the growth of electricity demand and pollution from the generators has led to an increase in its price. The energy management at demand level has drawn considerable attention under these circumstances. The key goal of energy management is to improve energy efficiency, thereby reducing energy costs and indirectly reducing emission levels [1, 2]. The bidirectional communication infrastructure and decision-making tools provide basic energy management requirements in electrical networks. The development of smart grid technologies [3] will fulfill the first requirement. The creation of the methods for decision-making on requests is a problem under consideration, and this paper is also the principal concern. One of the main concepts in every problem of economic allocation based on the basic principles of the economy is the simultaneous achievement of equity and efficiency. These two criteria must therefore be considered at the same time when allocating energy to requirements. Economics defines equity and efficiency as: Efficiency: Maximum use of social resources is referred to as efficiency [4]. That is to say, whether the allocation of resources for a company's members maximizes its profit or its total surplus, the allocation will be efficient. Equity: Equity is regarded as equity as a fair share of income generated from the capital of a business by its members. In other words, an allocation is fair and equitable if no member of a company prefers other people's advantages or surplus to himself. If a society's members gain unequal advantages or surpluses from equal wealth (due to their different attributes and characteristics), an attempt to achieve equality decreases the productivity of the society. And it costs the company equally and equitably.

Consequently, absolute equity and productivity in the allocation cannot be accomplished at the same time and, in effect, a trade-off between these two goals should be considered [6]. This challenge is being taken into consideration when describing a community as a collection of requirements under an energy management system and energy as the resource allocated. A new approach to address this problem is proposed in this paper. The proposed energy control system (EMS) is therefore designed in the following way. Demands provide the EMS responsible for its energy supply with consumption information (levels and marginal utilities). Based on energy price data and demand ranges, the EMS decides on the hourly energy consumption for each requirement and transfers the total energy consumption to energy suppliers in a positive way. Here, two energy sources are therefore considered, namely the main grid and a wind farm. All requests and vendors can be conveyed by EMS.

To address the EMS issue, the aim should be to maximize the utility for the demand clusters (cost reduction), subject to a set of restrictions that shape demand requirements, i.e. minimum day energy consumption, maximum and minimum hourly load levels, load collection and drop ramp limits, energy storage limits and power availability from the main grid and DER.. SEES, which is represented using a dc model without loss, covers both the demand and DER.

Linear programming (LP) or nonlinear programming (NLP) can be used to formulate DR optimization problems. Binary decision variables are often used to determine the status (On or OFF) of the different consumers and devices, in which case mixed-integer linear (MILP) or mixed-integer nonlinear (MINLP) programming may be used. In reference (7) authors use MILP to optimize DR and schedule the use of renewable energies, batteries and electric vehicles for the residential community grid. A reduction issue of the energy costs purchased from the residential community was overcome in this optimization. Reference [8] uses MILP to achieve minimization of costs in smart micro grid constructions with regard to DR optimization and day-to-day operation. Two different intelligent buildings with 30 and 90 houses are composed of this case study. The optimal schedule of household appliances is identified during the optimization process. Reference [9] is also used to illustrate how strategies such as DR can be suitable in any region, given that the renewable energy is highly penetration able. Reference [10] provides a case in point of NLP for DR optimization, in which the unit engagement problem for a computer grid is resolved. The problem of optimization is the reduction of load and the incentives paid for each interval. Reference [11] gives another example of MINLP, which involves the reduction of gas and grid power purchases by taking various loads into account over time. It decides the optimal day-to-day preparation for energy centers. For an aggregator, the DR specification has been extended to final customers over time. It operates as a service provider and this service provider must be paid for by the DR services. A thermostatically controlled loads aggregation for DR are presented in Reference [12]. In this case, the consumption of air conditioning is regarded as charge. Aggregation services are not limited to applying DR programs, a generalized energy storage aggregation can be found in Reference [13]. The storage of aggregators is used for energy and regulatory activities. It is also possible to incorporate DR services targeting individual users without contracts or service providers [14, 15]. The applications are treated as autonomous since no aggregator is linked to the user. Normally, the user has in his house a device to control the loads if the application is independent. A PV inverter is used in reference [16] while a home energy management system is used in reference [17]. Passive loads (i.e. air conditioning, fridges, laundry maker) and active loads (e.g. DG, ESS, vehicle-to-grid, pv) can be used for the controllable loads. The DR applies to discrete loads in References [19] with only two states: on or off. In recent years, its application in electricity systems has grown, with a focus on artificial intelligence (AI). Metaheuristics is an important part of AI to solve problems with optimization. These techniques have reasonable performance to solve problems with engineering by seeking an almost optimal solution with a reduced calculation load.

In problems with a large range of decision variables, metaheuristics can be used and adapted easily to a problem which has many limitations [20]. A photovoltaic, storage, and power network shifting strategy for the optimal functioning of the priced demand response is used in the comparison [18]. The Home Energy Management System implements the optimization algorithm. The PSO algorithm is also used in reference [21]. Other algorithms such as genetic algorithms[22], simulated ringing and differential evolution are often used to solve DR optimization problems[23], taking into account the variation in the price of electricity imposed on us by DSO to reduce consumption. This paper proposes the optimization of the system using the PSO approach to solve the optimization problem and compares the results to a RO topology.

2. Problem Description:

In that case, both the main grid and a stochastic DER (for example a small wind producer), can provide energy demand from the demand cluster. The cluster also has an energy storage system which can be used as necessary. Furthermore, the demand cluster is able to sell electricity in hours with high prices to the main grid. The arrangement between requests and suppliers requires the cluster's EMS to collect suppliers' hourly details several minutes before power consumption This data and the technical / economic data of the market are used by the EMS to change its hourly energy usage to optimize the maximum utility of the cluster. Following is the specific role of the EMS:

1) In an hour ten minutes before the supply of energy, the main grid submits its price. While the EMS operator has noticed the prices for the initial hours, they are uncertain for the next few hours.

2) The wind producer shall produce the wind power within 10 minutes of delivery. The rates of wind power are estimated for 24 hours a day.

(3) EMS specifies the daily electricity and energy usage and quantities supplied by the main grid, the wind turbine and storage unit for each application. The EMS determines the energy fraction that is stored and the energy fraction that is sold to the main grid.

4) Decision-makers and demands (e.g. 5 min. before hour) are aware of the hourly decisions.

3. Problem Description:

The following are the demand response models for the hourly cluster of requests:

$$min[\lambda_t^S e_t^S + \lambda_t^W \breve{e}_t^{AW} - \sum_{i=1}^{N_c} u_{i,t}(e_{i,t})] + \sum_{h=1}^{24-t}[\{\lambda_{t+h}^S\}e_{t+h}^S + \lambda_{t+h}^W\{e_{t+h}^{AW}\} - \sum_{i=1}^{N_c} u_{i,t+h}(e_{i,t+h})]$$

The objective function is the less complete utility of all requirements for 24 hours. The energy price of the main grid in (λ_T^S)hour t is known, but the prices for the next 24-t hours of the day, $y_{t+h}^S \geq 0,\ h = 1, \ldots, 24-t$,are unknown, and are represented by brackets The 24-hour wind energy prices are recognized. Notice that the EMS has to purchase all the available wind energy output by taking-or-pay contract. The term referring to the cost of wind power is also an objective term and can be excluded.

3.1. Proposed Algorithm

Given the energy management algorithm in the previous sub-section of the PSO, the following is described:

1) Start t=1.

(2) The energy cost from the central grid shall be sent to the EMS operator for the current hour, e.g. 10 minutes before the hour.

(3) The wind generator shall give the wind power to the EMS manager, e.g. 10 minutes before the hour to be generated. Price sensitive demands must submit their present and remaining hours to the EMS dispatcher, e.g. 10 minutes before an hour, their load rates and usefulness.

4) The EMS operator calculates the limit for energy prices and wind production for the 24-t-hour remaining 24-t-days on the basis of historical energy price series from DER 's leading grid and wind power productions. Note that the EMS operator is responsible for the risk associated with the simulation of uncertainty, so that the correct projection techniques are utilized to accurately assess these trust limits.

5) The EMS operator solves the PSO and obtains energy fractions from the main grid, from the wind generator and the storage unit, as well as energy for each individual order, the energy quantity to be stored and the energy quantity to be sold to the principal grid in an hour.

6) The EMS operator informs energy suppliers and demands, e.g. 5 minutes before hour of the optimal decisions made in step 6 above.

7) Update the hour counter t $\leftarrow$t+1. If<25, go to Step 2. When t=25 finishes the algorithm for the day and goes on to the next day. Note that in step 5 the EMS incorporates the new hourly information it gets, which updates the confidence limits forecast for the uncertain figures for the rest of the day.

4. Case Study:

4.1. Data

The energy management of 7 criteria is shown in Fig in the 5-bus system. For analysis, 1 is taken into account. Bus 5 includes a wind farm and power storage. Bus 1 connects the main grid. The average power injection from / to the principal grid is 400 MW. Tables I and II respectively contain network data and technical demand data. In Table II, the last column gives the initial burden for every request. The storage capacity is 200 MWh and the two conversion efficiencies are 90%. The power transferred to the warehouse and the amount of energy stored in the warehouse at the beginning of hour 1 are zero. The total power supplies currently contain 292 Megabytes (MS), 22 Megabytes (MWs), and 0 Megabytes (MWs). Table III provides the data on the minimum hourly demand load. While requirements 1 to 4 are zero in most hourly consumption, requests 5 to 7 are less flexible. The elasticity of demand represents an all-day multi-block utilities feature. Table IV shows the base multi-block utility function. The usefulness for the specifications of 1 to 4 is obtained by increasing the base utility by 0.8, 1.0 and 1.2 respectively in 1-8, 9-16 and 17-24 hours. The basic utilities are multiplied by 1.1, 0.9 and 1.0 in hours 1–8, 9–16, and 17–24, for hourly demand utility 5 to 7.

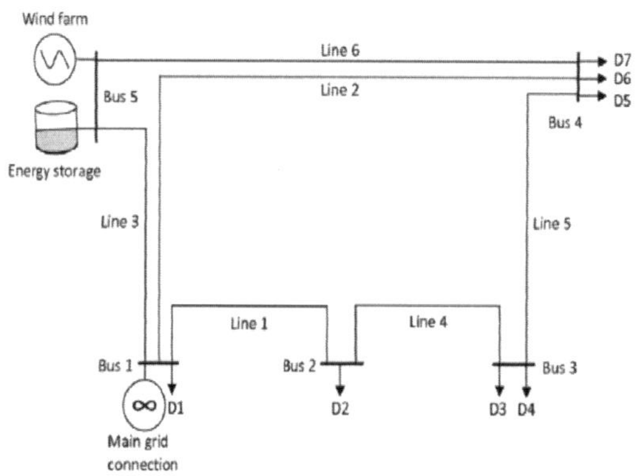

Figure 1. 5 – Bus Network

Table 1 : Network Information

Line	From	To	Capacity[MW]	Reactance[Ohm]
1	1	2	250	0.0341
2	1	4	150	0.0264
3	1	5	400	0.0564
4	2	3	350	0.0249
5	3	4	240	0.0342
6	4	5	240	0.0582

Table 2 : Least Hourly Demand Level

Demand	E^{day}	D^{max}[MW]	r^U[MW/h]	r^D[MW/h]	$d_{i,1}$[MW]
1	100	10	6	6	6
2	100	10	6	6	6
3	500	70	40	40	30
4	600	70	45	45	26
5	1200	150	100	100	78
6	1000	150	90	90	79
7	1500	200	110	110	89

Table 3 : Technical Data for Demands

Time h	d_1^{min} [MW]	d_2^{min} [MW]	d_3^{min} [MW]	d_4^{min} [MW]	d_5^{min} [MW]	d_6^{min} [MW]	d_7^{min} [MW]
2	0	0	0	0	0	0	0
3	0	0	0	0	0	0	0
4	0	0	0	0	0	0	0
5	0	0	0	0	0	0	0
6	0	0	0	0	0	0	0
7	0	0	0	0	0	0	0
8	0	0	0	0	10	20	20

9	0	0	0	0	10	20	20
10	0	0	0	0	10	20	20
11	0	0	0	0	10	20	20
12	0	0	0	0	10	20	20
13	0	0	0	0	10	20	20
14	0	0	0	0	10	20	20
15	0	0	0	0	20	40	40
16	0	0	0	0	20	40	40
17	1	1	2	2	20	40	40
18	1	1	2	2	20	40	40
19	1	1	2	2	20	40	40
20	1	1	2	2	20	40	40
21	1	1	2	2	20	40	40
22	1	1	2	2	20	40	40
23	1	1	2	2	10	20	20
24	1	1	2	2	10	20	20
25	0	0	0	0	0	0	0

5. Results and Discussion

The demand response problem of the cluster of demands in the following cases is investigated in order to examine the effectiveness of the PSO based EMS:

Without SG: The EMS receives the prices and the wind output in real-time and thus 24 unknown prices / wind outputs without the SG technology. The objective function of this state is the following:

$$\sum_{h=1}^{24}\left[\lambda_h^{S,min} e_h^S - \sum_{i=1}^{N_c} u_{i,h}\left(e_{i,h}\right)\right] + \beta^S T^S + \sum_{h=1}^{24} \varepsilon_h^S$$

In SG: real-time prices and wind production via SG technology are available. The proposed algorithm is considered for the use of this information. The results are shown in numbers. 2 and 3, representing, respectively, for the case and for non-SG the fractions of energy supplied by / from the main grid and supplied by the producer of wind. Figure 4, on the other hand, shows the energy transferred to the storage unit and the energy supplied from the storage unit.

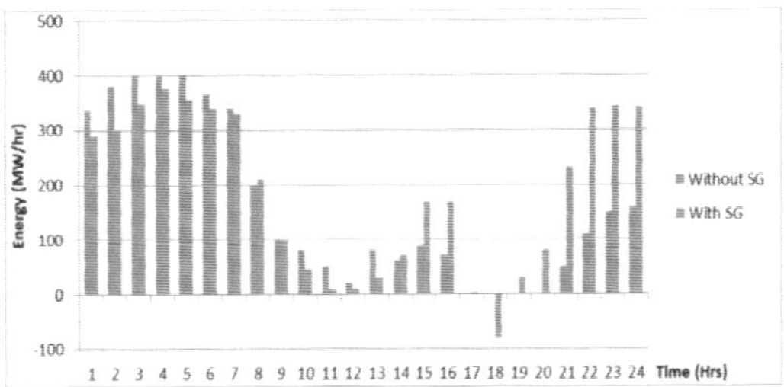

Figure 2. Total Energy Supplied By/Sold to the Main Grid

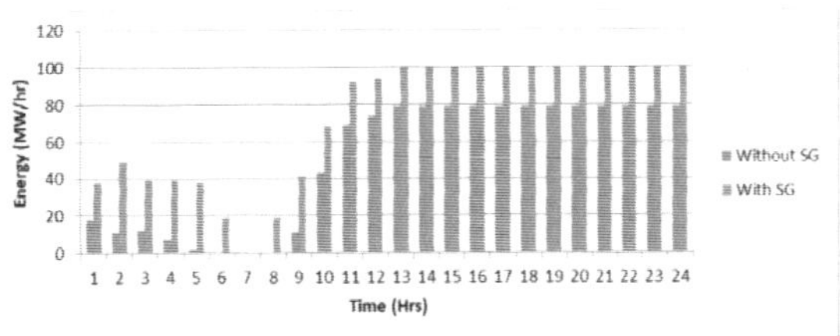

Figure 3. Whole Power Provided by the Wind

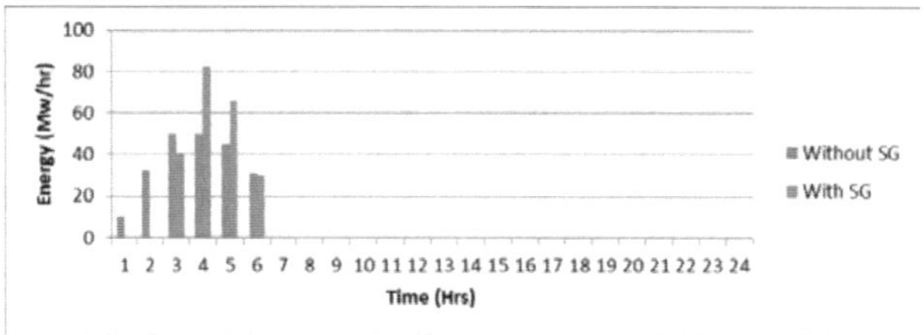

Figure 4. (a) Power Transported to the Storing Part

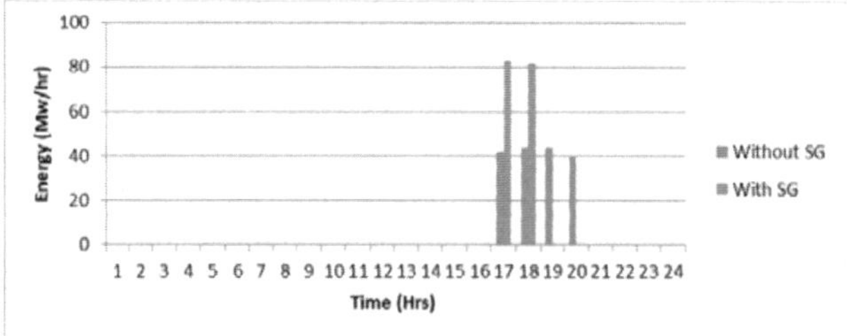

Figure 4. (b) Power Transported from the Storing Part

Table 4 : Real Information and Prediction Boundaries of Energy Cost

Time (h)	Actual [$/MWh]	Lower bound [$/MWh] [RO method]	Upper bound [$/MWh] [RO method]	Lower bound [$/MWh] [PSO method]	Upper bound [$/MWh] [PSO method]
1	47.45	42.93	61.52	45.62	60.94
2	48.39	37.31	61.75	43.46	60.11
3	45.28	31.63	56.45	39.63	55.43
4	46.38	30.17	53.28	42.15	50.12
5	48.72	30.12	54.63	46.15	53.72
6	50.71	31.6	62.9	45.45	60.72
7	60.01	37.71	65.46	40.71	63.62
8	60.52	34.09	70.66	37.26	68.23
9	60.56	37.15	73.51	49.23	69.51
10	60.48	34.53	72.27	52.63	70. 32
11	60	37.25	74.93	55.23	71.23
12	60	37.44	75.51	54.23	72.23
13	59.46	37.46	75.67	51.23	75.68
14	59.18	35.64	74.94	54.16	70.23
15	57.05	35.05	74.15	49.63	73.15
16	57.93	36.31	75.46	48.63	72.31
17	73.76	53.82	92.86	65.89	91.00
18	87.55	69.96	107.98	75.62	108.23
19	79.77	64.2	105.68	65.29	103.23
20	70.3	56.89	99.07	61.23	98.05
21	61.16	46.16	86.16	51.23	95.63
22	59.87	40.85	80.73	52.36	79.23
23	54.36	32.29	71.65	42.56	69.23
24	51.97	32.12	70.49	40.25	68.53

Note that the main grid energy prices, as stated in Table 4, range from hrs 17 to 20 above the demands of the services. Therefore, in all cases, the strong model leads to a small amount of energy being acquired from the main grid and even to energy being sold within several hours to the main grid. The demand cluster utilizes the energy in the storage facility during these hours. The storage facility makes it possible for the cluster of demands to store energy in hours when the main grid prices are small and to use energy in hours when the major grid prices are high. On the other hand, the energy prices of the major grid are lower in hours 1 to 6 than those of demand. The robust model thus leads to energy acquisition from the main grid as much as possible. The main difference between the amount of energy consumption in the two cases, with and without SG, lies on the fact that in the case without SG, the EMS schedules the energy consumption at the beginning of the day, while in the case with SG, the EMS takes advantages of bidirectional communication to reschedule the energy consumption and to adapt it to the actual energy price and the actual wind power production hour by hour. With the SG technology, wind energy is used in its entirety and only about 75% of the total usable wind energy is used without SG.

6. Conclusion

Detailed digital simulations predict a flexibles management mechanism even under uncertainty with PSO-based optimization techniques. Similarly, contact in two ways often offers an efficient response to requests. The study concludes that the topology proposed focuses on the efficient use of energy from resource sources. The micro grid with hybrid optimization technology can be expanded to boost the EMS for the future.

References

[1] Energy efficiency: Reduce energy bills, protect the environment. The U. S. Department of Energy (DOE) and the U.S. Environmental Protection Agency (EPA).
[2] Wang F, Chen A. Energy management handbook: BSR publication; 2012.
[3] Qian LP, Zhang Y, Huang J, Wu Y. Demand response management via real-time
[4] Mankiw N Gregory. Principles of economics. South-Western Educational Publishing; 2003.
[5] Varian HR. Intermediate microeconomics: a modern approach. 8th ed. W.W.Norton & Company; 2009.
[6] Nicola P. Efficiency and equity in welfare economics. Berlin Heidelberg: Springer-Verlag; 2013.
[7] Nan, S.; Zhou, M.; Li, G. Optimal residential community demand response scheduling in smart grid. Appl. Energy, vol. 210, pp. 1280 – 1288, 2018.
[8] Zhang, D.; Shah, N.; Papageorgiou, L.G. Efficient energy consumption and operation management in a smart building with micro grid. Energy Convers. Manag, vol. 74, pp. 209–222, 2013.
[9] Pina, A.; Silva, C.; Ferrão, P. The impact of demand side management strategies in the penetration of renewable electricity. Energy, vol. 41, pp. 128–137, 2012.
[10] Nwulu, N.I.; Xia, X. Optimal dispatch for a microgrid incorporating renewables and demand response. Renew. Energy, vol. 101, pp. 16–28, 2017.
[11] Alipour, M.; Zare, K.; Abapour, M. MINLP Probabilistic Scheduling Model for Demand Response Programs Integrated Energy Hubs. IEEE Trans. Ind. Inform. 2018,
[12] Zhou, X.; Shi, J.; Tang, Y.; Li, Y.; Li, S.; Gong, K. Aggregate Control Strategy for Thermostatically Controlled Loads with Demand Response. Energies, 12, pp. 79 – 88, 2019.
[13] Yao, Y.; Zhang, P.; Chen, S. Aggregating Large-Scale Generalized Energy Storages to Participate in the Energy and Regulation Market. Energies, vol. 12, Pp. 1024 – 1030, 2019.
[14] Spknola, J.; Faria, P.; Vale, Z. Photovoltaic inverter scheduler with the support of storage unit to minimize electricity bill. Adv. Intell. Syst. Comput., vol. 619, pp. 63–71, 2017.
[15] Hussain, B.; Khan, A.; Javaid, N.; Hasan, Q.; Malik, S.A.; Ahmad, O.; Dar, A.; Kazmi, A. A Weighted-Sum PSO Algorithm for HEMS: A New Approach for the Design and Diversified Performance Analysis. Electronics, vol. 8, pp. 180 – 190, 2019.
[16] Shen, J.; Jiang, C.; Li, B. Controllable Load Management Approaches in Smart Grids. Energies, vol. 8, pp. 11187–11202, 2015.
[17] Kong, D.-Y.; Bao, Y.-Q.; Hong, Y.-Y.; Wang, B.-B.; Huang, H.-B.; Liu, L.; Jiang, H.-H. Distributed Control Strategy for Smart Home Appliances Considering the Discrete Response Characteristics of the On/Off Loads. Appl. Sci, vol. 9, pp. 457 – 465, 2019.
[18] Jordehi, A.R. A review on constraint handling strategies in particle swarm optimization. Neural Comput. Appl., vol. 26, pp. 1265–1275, 2015.
[19] Faria, P.; Vale, Z.; Soares, J.; Ferreira, J. Demand Response Management in Power Systems Using

Particle Swarm Optimization. IEEE Intell. Syst vol. 28, pp. 43–51, 2013.
[20] Shehzad Hassan, M.A.; Chen, M.; Lin, H.; Ahmed, M.H.; Khan, M.Z.; Chughtai, G.R. Optimization Modeling for Dynamic Price Based Demand Response in Microgrids. J. Clean. Prod., vol. 222, pp. 231 - 241, 2019.
[21] Aghajani, G.R.; Shayanfar, H.A.; Shayeghi, H. Demand side management in a smart micro-grid in the presence of renewable generation and demand response. Energy, vol. 126, pp. 622–637, 2017.
[22] Hu, M.; Xiao, F. Price-responsive model-based optimal demand response control of inverter air conditioners using genetic algorithm. Appl. Energy, vol. 219, pp. 151–164, 2018.
[23] Qian, L.P.; Wu, Y.; Zhang, Y.J.A.; Huang, J. Demand response management via real-time electricity price control in smart grids. Smart Grid Netw. Data Manag. Bus. Model. pp. 169–192, 2017.

IOP Conf. Series: Materials Science and Engineering **906** (2020) 012014 doi:10.1088/1757-899X/906/1/012014

Solar powered highly efficient Seven-level inverter with switched capacitors

Dr.A.Nazar Ali [1] R. Jai Ganesh[2] D. Sivamani[3] D. Shyam[4]

[1] Associate Professor, Department of Electrical and Electronics Engineering, Rajalakshmi Engineering College, Rajalakshmi Nagar, Thandalam, Tamil nadu

[2] Assistant Professor, Department of Electrical and Electronics Engineering, K. Ramakrishnan College of Technology, Trichy, Tamilnadu.

[3] Assistant Professor, Department of Electrical and Electronics Engineering, Rajalakshmi Engineering College, Rajalakshmi Nagar, Thandalam, Tamil nadu

[4] Assistant Professor, Department of Electrical and Electronics Engineering, Rajalakshmi Engineering College, Rajalakshmi Nagar, Thandalam, Tamil nadu

E-mail: nazarali.a@rajalashmi.edu.in

Abstract. In this article, a seven-level inverter powered by solar has been proposed to achieve a sinusoidal output voltage with high efficiency and enhanced power quality. This system consists of active inverter and flipped condenser clamping. It gives output voltage level of 2/3. By connecting the switched condenser branch in the front or back end series to the inverter output a seven-level output voltage wave shape can be obtained with ample non-work switching states to match the voltage of switching condensers. The consequence of the proposed method is validated using MATLAB SIMULINK for simulation. Comparing with conventional seven-level inverter topology, the proposed design used a minimal number of switches using a basic modulation process.

1. Introduction

Given the low capacity, high performance and high strength quality, staggered inverters have been widely used in renewable power source transition applications. There's a generating resources for phased inverter exams and their topologies and regulatory techniques [1,2]. In a way of saying, the five-level inverter has been condensed and can usually be divided into three get-togethers: a diode-cinched inverter, a moving capacitor inverter, and an H-connect inverter that drops. The higher yield voltage is a big stress that affects staggered inverter execution. As these topologies of five levels are difficult to loosen to topologies of up to seven levels confined by length, size, effectiveness and strength of action. In this way, discovering an modified topological configuration to reach seven-level yield voltage is justifiable and alluring.

Relevant tests were carried out on the seven stage inverter. A seven-level inverter topology containing two unsymmetrical DC power sources has been shown which in actual and dynamic conditions cannot guarantee clasped condenser voltage stability. A felling configuration of the 7L inverter topology with a lone source that involves specific force switches regardless of everything. Another type of seven-level dynamic unbiased point clasped inverter topology that uses half-connected PEBB is seen, and it encounters colossal number of DC sources. Seven-level cross breed inverter topology with a single DC is in, whereas topology includes complex capacitor voltage balance calculation. If the topology of the staggered inverter is shown, a fitting tweak technique is crucial, which selects yield execution, such as consistency and efficacy of the waveform. Regulation strategies were planned, for example, for staggered inverters for space vector beat width tweak (PWM), transporter attitude balance, and bearer stage step balance. More precisely, for space vector PWM invention, as yield voltage levels increase, the sum of fundamental vector and the dreary states' decision are usually multi-faceted, and are not ideal for seven-level inverter[3].

In the control technique for stage resistance aura is used to boost the complex, neutral point cinched inverter of seven stages. Despite the fact that the problem of multistep bounce in yield voltage is known, additionally raises the yield voltage symphony material. In the sine PWM technique with a sole bearer is used to accomplish modification of the seven-level inverter blended drop[4,5].

2. Seven Level Inverter Topology and Operating Principles:

The altered topology comprises of a working cinching inverter and swapped capacitor component and the past is liable for the staggered voltage, whereas the present is liable for redundancy growth. The shared part of the capacitor contains three switches S9, S10, and S11 as well as two clipped C3 and C4 condensers seen in the Fig. 1, capable of providing two voltages ranges (E and 2E) with the C3 and C4 voltages E tested. As the states of the switches S9, S10, and S11, C3 and C4 are in game plan or identical association, the shared portion of the condenser is configured to make the voltage levels E and 2E[6].

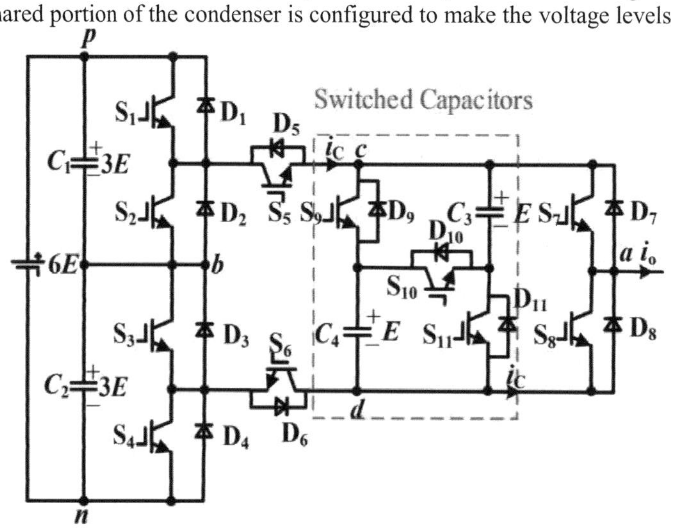

Figure 1: 7L inverter with switched capacitors

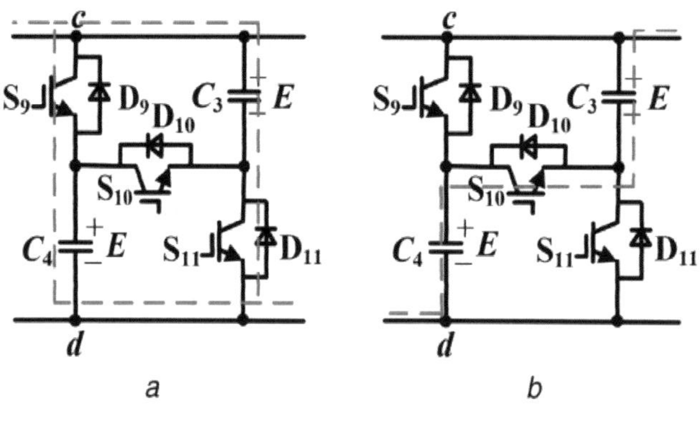

Figure 2: Switching States of switched capacitor part (a) Parallel state, (b) Series state

Just where the existing iC of the braced condenser component flows through dynamic responsiveness with respect to c and flows out point d, the traded condenser portion is going along with the dynamic clipping inverter, and literally, The traded condenser component would be in reverse association with the dynamic clasping inverter if iC runs out point c and streams to point d. The seven-level voltage from −3E to 3E can be formed with the interim E by associating exchanged condenser component in forward or backward arrangement to the dynamic bracing inverter. The DC controller comprises two condensers in scheme collaboration with the appraised C1 and C2 voltages3E. It is likely to agree that the unbiased point b is zero potential and yield current io when streaming out of point a.

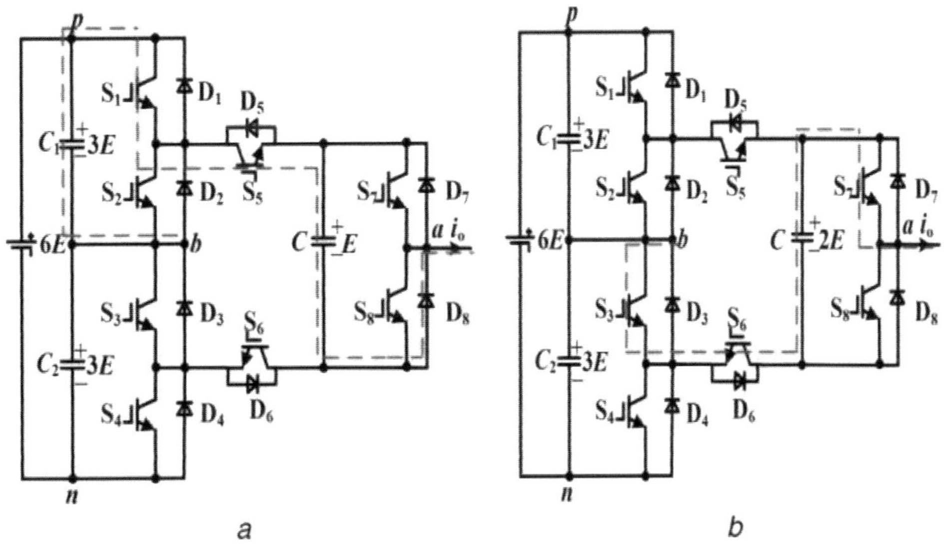

Figure 3: Switching states of active clamping inverter at 2E level of output voltage (a) Switching state a, (b) Switching state b

Working standards of exchanged capacitor part

When switching conditions of switches S9, S10, and S11, C3 and C4 are in arrangement or equivalent association, the exchanged capacitor part produces voltage levels E and 2E. While, there is an rise in unregulated exchange states at E and 2E levels, which is beneficial to balance the voltage of the C3 and C4 condensers. Operation requirements for the exchanged condenser portion appear in Fig. 2. By Fig. 2a, during the switches S9 and S11 the ON and S10 are blocked, the C3 and C4 condensers are attached in equal proportions, within hence the shared capacitor component yield voltage is E. Out of Fig. 2b, the S9 and S11 switches are blocked and the S10 is ON, the C3 and C4 braced capacitors are attached in arrangement and the shared capacitor component yield voltage is 2E. In all cases, the dynamic cinching inverter is connected forward in spite of the fact that the current stream between the clipped capacitor C3 and C4 and out of them is aligned with the dynamic bracing inverter while they are in backwards[7,8].

Working standards of dynamic cinching inverter

The component of the exchanged condenser can be considered as a clasped condenser C all around, where the voltage is both E and 2E levels. Fig. 3 shows the current way and the exchange rates of seven level inverter yield voltages for the 2E level. By Fig. 3a, when the voltage of the reciprocal clasped capacitor C is E and the S1, S3, S5, and S8 switches are ON. If the current io yield is positive, the exchanged condenser portion is in forward association with the circuit, and the present way is S1 to S5 to D8 to a b to C1; if the current io yield is negative, the exchanged condenser portion is in rearward association with the circuit and the present way is a to S8 to C to D5 to D1 to C1 to b.
By Fig. 3b, while the reciprocal clasped capacitor voltage C is 2E and the S1, S3, S6, and S7 switches are ON[9]. Whenever the current io yield is positive, the exchanged part of the condenser is in rear association with the circuit and the present way is S3 to D6 to C to S7 to a b; whenever the current io yield is negative, the exchanged part of the condenser is in forward association with the circuit and the present way is D7 to C to S6 to D3 to b.

Working standards of novel seven level inverter with exchanged capacitors

Considering the course of action of voltages obtained from the DC link capacitors and shared capacitors, the stronger yield waveforms of seven stages produce. In Table 1 the courses of voltage operation were listed based on their state. In table 1 the word uab upn, ucd means seven-level inverter yield voltage, DC link yield voltage, capacitor voltage exchanged. The gift"/" is circumvented to the clipped condenser; io > 0 reveals that the yield current streams point an out, and io < 0 reveals that the yielding current streams point a. '+' speaks to charging the clasped condenser, while '–' speaks to free the clipped condenser[10]. At E and 2E stages of yield voltage, there are two excess exchange states, separately, so the cinched condensers are in charge and release states accordingly. Choosing exchange states under various current headings is important, resulting in a waveform of seven stages of yield voltage. For the planned 7L inverter, there are absolute 12 moving states from 1 to 12. Exchanging states at level 2E are explored in depth, for example.

Switching state 2: The switches S1, S3, S5, S8, S9 and S11 are followed, and the positive cathodes of the DC-link capacitor C1 and the clamped capacitor C3 and C4 are paired in similar association with C3 and C4. Also, the voltage over C1 is 3E, C3 and C4 are all E, so 2E degree of output voltage is gained after 3E short E. As shown by this condition of trade, the present manner appears in Fig. 4a. A. The present ways are represented as S1 to S5 to S9 to C4 to D8 to a to b to C1 and S1 to S5 to S11 to D8 to a to b to C1, C3 and C4 both charge during this process, whenever the performance current io is certain. Where the current io production is negative, the present forms are shown a to S8 to C4 to D9 to D5 to D1 to C1 to b and a to S8 to D11 to D3 to D5 to D1 to C1 to b, C3 and C4 all release along those lines[11,12].

Switching state 3: Switches S1, S3, S6, S7, and S10 are guided, so in the scheme connection negative terminals of the clamped condenser C4 and zero potential point b are aligned with C3 and C4. Furthermore, as the measured voltages of the DC relation are zero, C3 and C4 are all E, after 0 contains 2E; 2E is then obtained. As shown by this condition of trade, the present manner appears in Fig. 4b. whenever the performance current io is definite, the current paths are defined as S3to D6 to C4 to S10 to S7 to a to b, subsequently both C3 and C4 are in release state. Once again, if the output current io is negative, the present forms are as a to D7 to C3 to D10 to C4 to S6 to D3 to b, in C3 and C4 both charge.

Table 1 : Seven level output voltage and states of clamped capacitors

S. no.	uab	upn	ucd	C3/C4 states	
				io>0	io< 0
1	3E	3E	/	/	/
2	2E	3E	E	+	−
3	2E	0	2E	−	+
4	E	0	E	−	+
5	E	3E	2E	+	−
6	0	0	/	/	/
7	0	0	/	/	/
8	−E	0	E	+	−
9	−E	−3E	2E	−	+
10	−2E	−3E	E	−	+
11	−2E	0	2E	+	−
12	−3E	−3E	/	/	/

This is equivalent to researching the other ten moving states types. Table 2 displays seven current voltage ranges for different exchange states. In Table 2, '1' and '0' speak similarly to on and off switch conditions; '/' speaks to switch state may be either 1 or 0. There are four sets of correlative switches (S1, S2), (S3, S4), (S5, S6), (S7, S8), and (S9, S10) and three sets of switches (S1, S3), (S2, S4), and (S9, S11) providing a common exchange signal for each pair, independently, such that S1, S5, S7, and S9 will work autonomously to dispose of illicit exchange states[13,14].

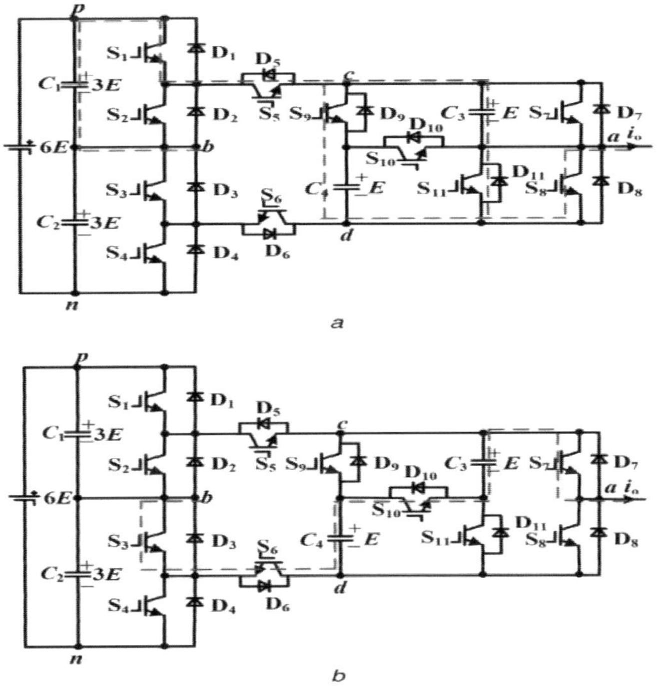

Figure 4: Switching states of 7L inverter with switched capacitors at 2E level (a) Switching state 2, (b) Switching state 3

3. Results and Discussion

The specifications for the suggested structures are listed in Table 2, where bearer recurrence fs is 10 kHz, L and C5 speak separately to inductance and capacitance of the channel inducer and condenser, and R speaks to stack interference[15]. Fig. 5 shows the effects of seven-level output voltage, inductor current and independent output voltage on leisure. By Fig. 5a, a waveform of 7L yield voltage is acquired and its pinnacle estimate is 300 V, whereas the valley estimate is −300 V and each yield voltage stage is 200 V. Fig. 5b shows the present of the inductor with an excess of ~10 A. At Crucial recurrence of 50 Hz in Fig. 5c, the isolated output voltage is nearly sinusoidal. Figs. 7a and b display sun-oriented symphonic range with seven-level voltages and distinct output voltage, respectively. THD is 24.06, while THD is 0.53 channel yield voltage. The main abundance of output voltage is 300 V, which means that the strong value is 220 V. Fig. 6, shows the Matlab Simulink model of solar fed seven level inverter [16-18].

Table 2: Values of elements in simulations

Parameter	Value
upn	750 V
uab	300 V/50 Hz
fs	10 kHz
L	300 μH
R	32.26 Ω

IOP Conf. Series: Materials Science and Engineering **906** (2020) 012014 doi:10.1088/1757-899X/906/1/012014

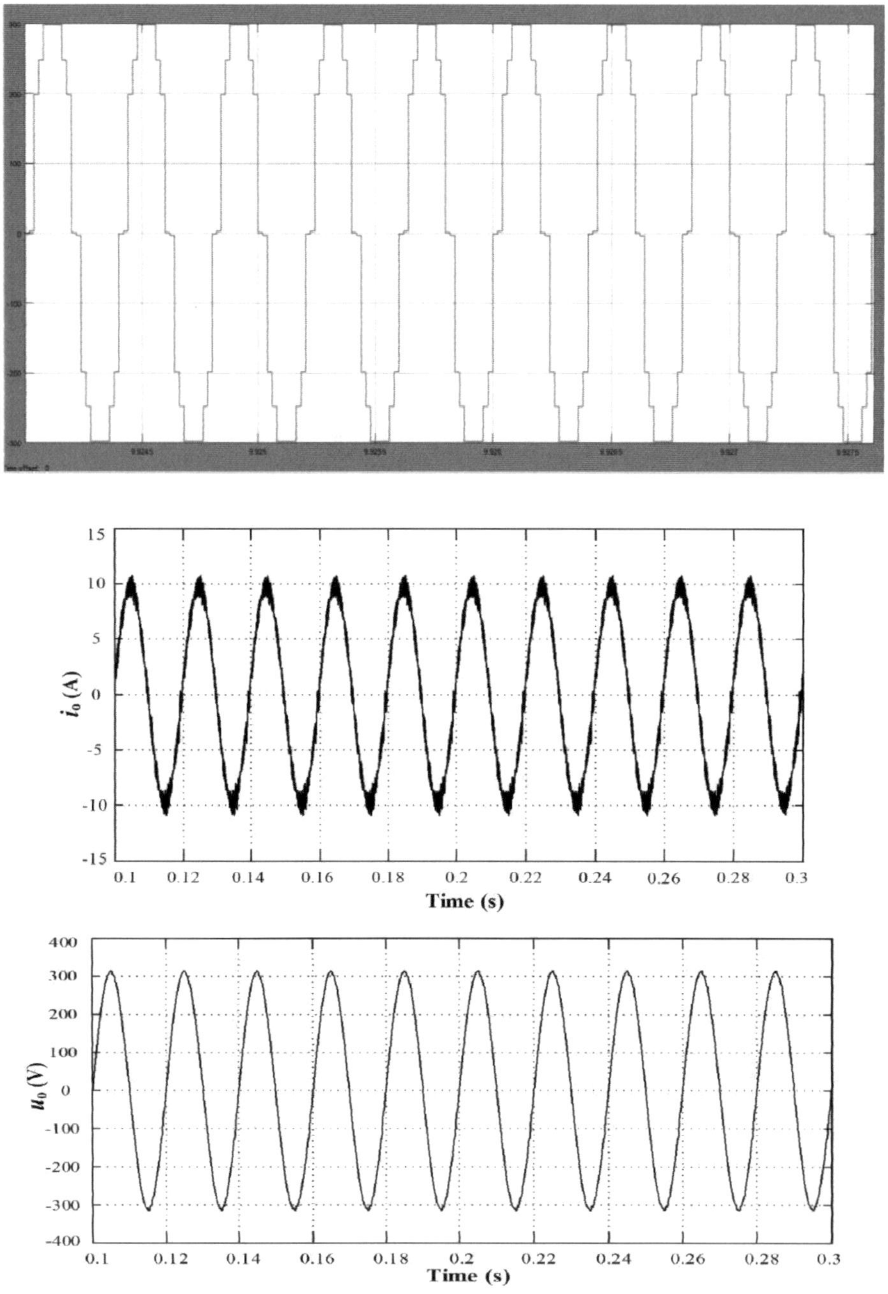

Figure 5: Simulation results of output voltage (a) 7L output voltage, (b) Inductor current, (c) Filtered output voltage

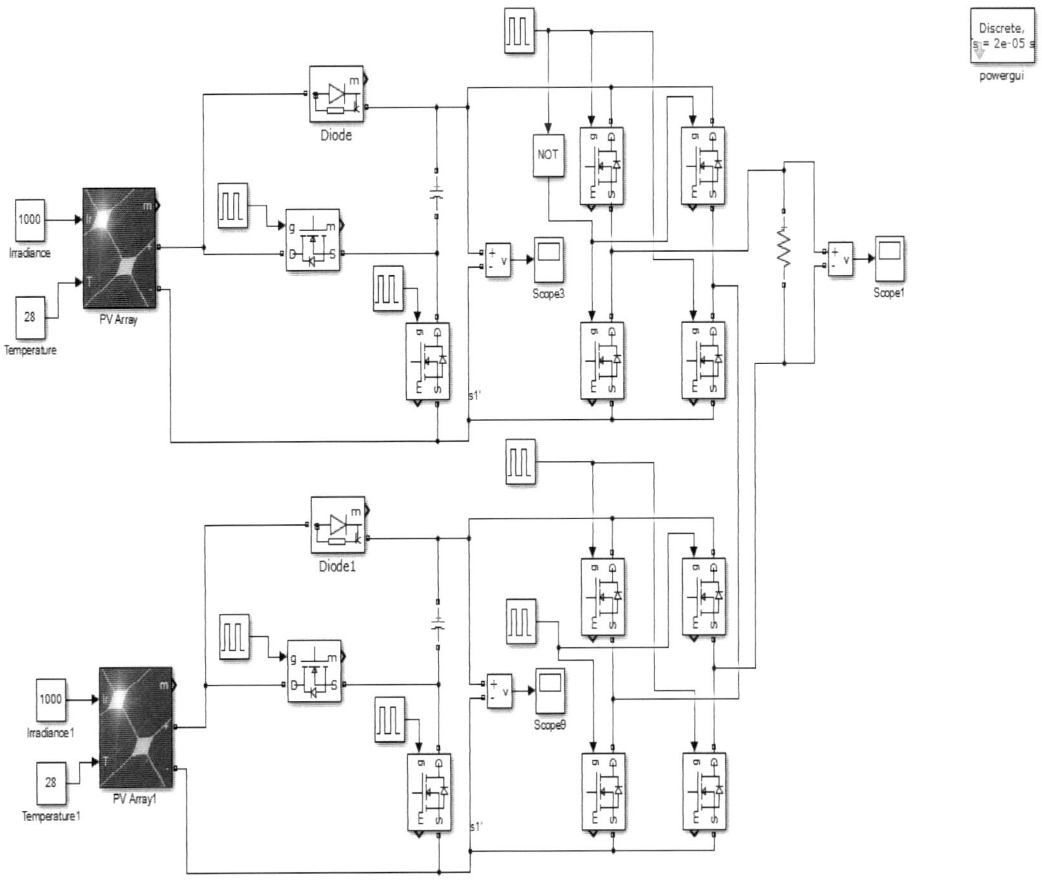

Figure 6: Simulation model of solar powered seven level inverter

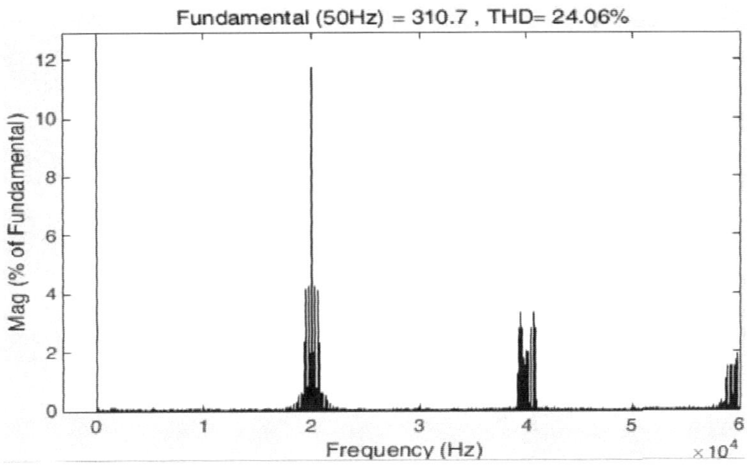

(a)

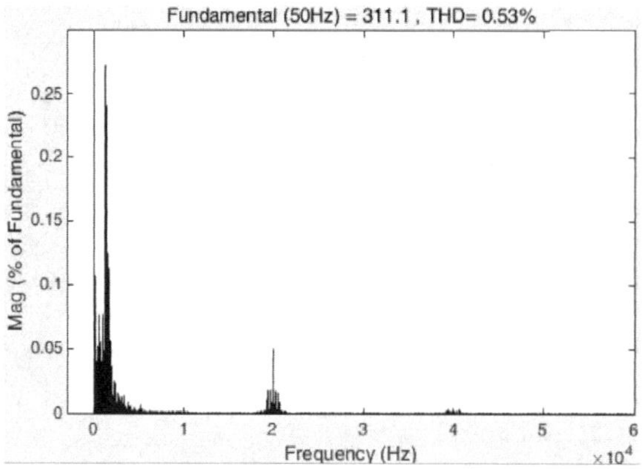

(b)

Figure 7: Harmonic spectra of output voltage (a) 7L output voltage, (b) Filtered output voltage

4. Conclusion

This paper proposed a novel operated seven-level inverter dependent on solar powered with exchanged capacitors. The arrangement uses the shared condenser component to create two degrees of voltage E and 2E and extends the excess states. The topology can also recognize seven output voltage ranges with reduced number of switch transistors and gate drivers resulting in ease. The seven-level inverter waveform solar driven THD and filtered output voltage are 24.06 percent and 0.53 percent respectively. Results of replication demonstrated the practicality of the planned seven stage inverter powered by the solar.

References

[1] A Nazar Ali, D Sivamani, R Jaiganesh M Pradeep, Solar powered air conditioner using BLDC motor, IOP Conference Series: Materials Science and Engineering, vol. 23,2019
[2] V Venkatesh, A Nazar Ali,, R Jaiganesh. V Indiragandhi , Extraction and conversion of exhaust heat from automobile engine in to electrical energy Energy‖, IOP Conference Series: Materials Science and Engineering, vol. 23,2019 .
[3] A Ali Nazar, R Jayabharath, MD , Udayakumar, An ANFIS Based Advanced MPPT Control of a Wind-Solar Hybrid Power Generation System,international review of modelling and simulations. vol..7, no. 4, pp. 638– 643,2014.
[4] Shyam D, Premkumar K, Thamizhselvan T, Nazar Ali A, Vishnu Priya M(2019),Symmetrically Modified Laddered H-Bridge Multilevel Inverter with Reduced Configurationally Parameters ,International journal of engineering and advanced technology ,Vol 9, issue 1,2019
[5] A.Nazar Ali and R. Jayabharath, "Performance Enhancement of Hybrid Wind/Photo Voltaic System Using Z Source Inverter with Cuk-sepic Fused Converter." Research Journal of Applied Sciences, Engineering and Technology7.19 pp. 3964-3970,2014
[6] AT SankaraSubramanian,PSabarish, A Nazar Ali , A power factor correction based canonical switching cell converter for VSI fed BLDC motor by using voltage follower technique‖IEEE International Conference on Electrical, Instrumentation and Communication Engineering (ICEICE), pp.1-8.2017
[7] C.Kalavalli, S.R.Paveethra, S.Murugesan, Dr.A.Nazar Ali, Design And Implementation Of High Efficiency H6 PV Inverter With Dual Axis Tracking, International Journal of Scientific & Technology Research, Vol 9, issue 02, pp. 4728-31, 2020
[8] .R.Paveethra, C.Kalavalli, S.Vijayalakshmi, Dr.A.Nazar Ali, D.Shyam, Evaluation Of Voltage Stability of Transmission Line With Contingency Analysis, International Journal of Scientific & Technology Research, Vol 9,issue 02, pp. 4018-22,2020
[9] Premkumar, Kamaraj et al. (2018), Antlion Algorithm Optimized Fuzzy PID Supervised On-line Recurrent Fuzzy Neural Network Based Controller for Brushless DC Motor, Electric Power Components and Systems, 45, 20, pages.2304-2317.
[10] Rajkumar, B. Karthikeyan and S. Senthil Kumar , Fuzzy Logic Control of Fast Transient Response DC-DC Converter. Middle-East J. Sci. Res., 23 (9): 2256-2263, 2015.

[11]K Premkumar et al. (2018), Novel bacterial foraging-based ANFIS for speed control of matrix converter-fed industrial BLDC motors operated under low speed and high torque, Neural Computing and Applications, 29, 12, pages.1411–1434.
[12]Premkumar et al. (2015), GA-PSO optimized online ANFIS based speed controller for Brushless DC motor, Journal of Intelligent & Fuzzy Systems, 28, 6, pages.2839-2850.
[13]A.Nazar Ali and R. Jayabharath. "Performance Enhancement of Hybrid Wind/Photo Voltaic System Using Z Source Inverter with Cuk-sepic Fused Converter." Research Journal of Applied Sciences, Engineering and Technology7.19 (2014): 3964-3970.
[14]S. S. Dobakhshari, J. Milimonfared, M. Taheri, and H. Moradisizkoohi, ''A quasi-resonant current-fed converter with minimum switching losses,'' IEEE Trans. Power. Electron., vol. 32, no. 7, pp. 353–362, Jan. 2017.
[15]A.Nazar ali, Dr.R.Jayabharath, R.Shanthi Priyadharshini. —A Single phase high efficient transformer less inverter for PV Grid connected power system using ISPWM technique." International Journalof Applied Engineering Research (IJAER) Volume 10, No.9 (2015), pp.7489-7496
[16]A Nazar Ali, R Jeyabharath, MD Udayakumar‖ Cascaded Multilevel Inverters for Reduce Harmonic Distortions in Solar PV Applications‖ Asian Journal of Research in Social Sciences and Humanities, Vol 6,pp.703-715.
[17]A.Nazar Ali and R. Jayabharath. "Performance Enhancement of Hybrid Wind/Photo Voltaic System Using Z Source Inverter with Cuk-sepic Fused Converter." Research Journal of Applied Sciences, Engineering and Technology7.19 (2014): 3964-3970.
[18]M D Udayakumar et al 2019 IOP Conf. Ser.: Mater. Sci. Eng. 623 012018

IOP Conf. Series: Materials Science and Engineering **906** (2020) 012015 doi:10.1088/1757-899X/906/1/012015

Applications of Internet of things on Smart Grid

P Paul Divakar[1*], G Vijaya Lakshmi[2] and L Devi[3]

[1] Department of Physics, Sir C R Reddy College, Eluru, India
[2] Department of Electronics, Sir C R Reddy College, Eluru, India
[3] Department of Computer Science, Sir C R Reddy College, Eluru, India

*Email: *putlapaul1965@gmail.com, gudurivijayalakshmi.crr@gmail.com devi.labba@gmail.com*

Abstract. Internet of Things devices in public and private spaces has changed our lifestyle. For example, home automation applications that support smart devices, locks, keys, surveillance systems, and Internet-connected devices has change the way we interact with our living spaces. These systems are widely adopted; the Internet of Things digitally expresses concern on safety and security. This technology makes life easier, safety as its priority. The firewall is also very important for the Internet of objects. The algorithms designed for users should be unique and safety. The basic principle of the Internet of Things is based on the integration of information networks to make people's lives easier. In this way, the objects can connect to each other and share information. The development of the Internet of Things has triggered by technology. A smart grid is an electricity network based on digital technology that is used to supply electricity to consumers via two-way digital communication. This system allows for monitoring, analysis, control and communication within the supply chain to help improve efficiency, reduce energy consumption and cost, and maximize the transparency and reliability of the energy supply chain. The smart grid was introduced with the aim of overcoming the weaknesses of conventional electrical grids by using smart net meters. Smart Grid technologies contribute to manage IoT energy efficiently, which the present ones lack in the existing framework. The two-way communication between connected devices and hardware which can sense and respond to user demands makes the IOT smart grid better. A smart Grid with the help of these technologies is more resilient and less costly than the current power infrastructure

1. Introduction

Internet of Things (IoT) is a network of devices which can sense, accumulate and transfer data over the internet without any human intervention. IoT is the idea of the fusion of the "real world" with the "digital world", making the individual be in constant communication and interaction, either with other people or objects. With the Internet of things, objects can be activated and controlled remotely through an existing network infrastructure creating opportunities for integration between the physical world and computer systems. IoT is not science fiction anymore, the power of internet connectivity has gone beyond Smartphone's and computers. Now, the internet is been used in all smart devices which are aimed to resolve actual real-world problems, from smart watches to driverless.

2. Internet of Things for Human Life

Smart electrical grids are the most important building block in the development of technology. Networks provide access to quality energy with low budgets and costs. Smart grids for use ensure efficiency. Smart grids, economic and social applications provide development Fig [1]. One of the most important goals of smart networks is establishment of micro-networks that allow more distributed generation and energy storage at the end-user side. The expansion of power generation from micro grid applications and renewable energy sources, the systems and energy storage technologies that will serve as required energy buffers have become very important in order to meet the increasing power demand. Better and faster communication devices and technologies for the electricity grid, preventing most of the electrical circuits and voltage drops. Smart grids are very important for the development of technology and economic innovations. In order to make the existing electricity grid a smart grid, the design and implementation of a new communication infrastructure for the grid are two important research areas. Smart grids provide quality and appropriate electrical energy requirements. Internet of Things technology is the user. This technology receives the quality and efficient energy it needs from smart grids. It enables technology security in energy controllers and communication systems. It is the new age for users who can able to find and want to use the IoT applications. These applications need safe energy with the Internet of Things. And smart grids infrastructure technology and solution can provide energy with efficiency.

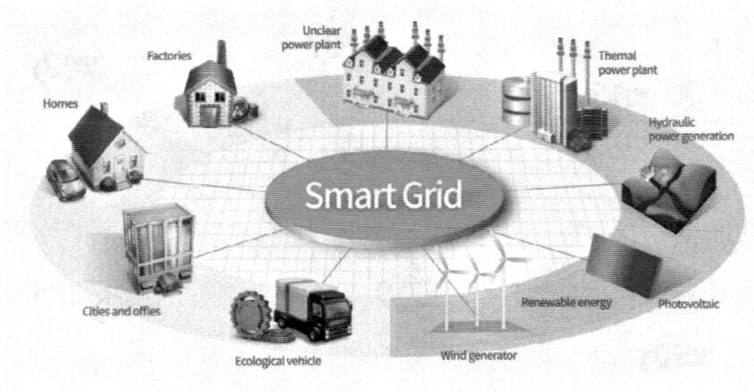

Figure 1. Smart Grid IOT Applications.

2.1. A Smart grid IoT application allows:

Smart Grid technologies all contribute to efficient IoT energy management solutions that are currently lacking in the existing framework [4]. What makes the IoT Smart Grid better is two-way communication between connected devices and hardware that can sense and respond to user demands. These technologies mean that a Smart Grid is more resilient and less costly than the current power infrastructure.

2.1.1. *Mini Smart Grids*

The basic part of electronic devices is microprocessors. From the most frequently used to the most rarely used device, the brain of most of them is the microprocessor. And these microprocessors are very sensitive. So the microprocessor really works as the brain of the device [3]. This information is not just a similar for the brain to the microprocessor. The microprocessor working principle is almost like a human's brain. Smart grids, save energy, reduce costs and increases reliability. It provides energy independence by connecting everyone to an abundant, affordable, clean, efficient and reliable electrical energy anywhere, anytime. The most preferred type of these smart grids in the Internet of Things technology is **mini smart grids**. Smart grids have an important place for the Internet of Things that developed to easier for human life.

2.1.2. *Communication and Networking in Smart Grids*

Provided communication infrastructure is important for the use of smart grids. Communication is important for the development and operation of smart grids [2]. Consumers can minimize energy expenditure by adjusting their smart home appliance operations to avoid peak home hours and instead use renewable energy Fig [2]. The basis of these applications is the intelligent network developed. The complexly developed network keeps the benefits of smart networking in the background [1]. Therefore, there is a way to provide the most efficient energy and infrastructure to provide high efficiency to the user. The most common way is to choose the right communication and network in smart grids.

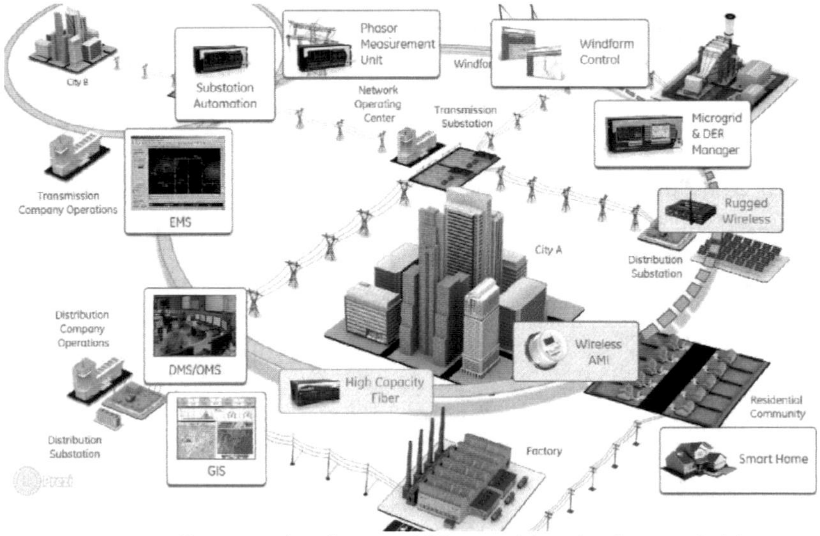

Figure 2. Communication and Networking in Smart Grids.

3. Conclusion

It is the new age for users who can able to find and want to use the Internet of Things applications. These applications need safe energy with the Internet of Things. And smart grids infrastructure technology and solution can provide energy with efficiency.

References

[1] Guo Dengfeng, Xu Shan and Kun, "The Internet of Things hold up Smart Grid networking technology", North China Electric, vol. 2, pp. 59-63, 2010.
[2] Songshu Fang, Li Lanxin and Shen Jie, "Smart Grid technology review", Power System Technology, pp. 23-29, Aug. 2009.
[3] Wang Fang Fang, "Smart Grid technology content and comparison", High Technology and Industrialization, pp. 9-13, May 2009.
[4] Qing Song, Jing Tang and Xiao Feng, "Domestic and international situation and the development of Smart Grid analysis", Electrical electrical, pp. 1-4, Mar. 2010.
[5] HY Li and Gui chao, "Application of the Internet of Things and trends", Fujian PC, pp. 1-2, Sep. 2010.

IOP Conf. Series: Materials Science and Engineering **906** (2020) 012016 doi:10.1088/1757-899X/906/1/012016

Analysis of Back To Back (BTB) Converter Control Strategies in Different Power System Applications

Bandla Pavan Babu[1], Indragandhi V[2]
[1]ResearchScholar, [2]Associate Professor
[1,2] School of Electrical Engineering, VIT Vellore, Tamilnadu, India.

Abstract Power Quality (PQ) is an important term in many industries and other electrical power consumers. Maintaining of power quality during the operation of industries is an important task as the PQ issues may arise and cause the deviation in normal mode operation. In order to maintain the stability and mitigate the PQ issues, Back to Back (BTB) are mostly used in different power systems majorly in industries. This paper provides a major review on the BTB converters role in different power system configurations which are playing important role in facing the power quality issues such as voltage sags occurred in the system due to any faulty condition or short circuits, harmonic distortions, and some other abnormal conditions. Also in this paper, the various control strategies implemented in the BTB converter at their respective power system configurations are presented which will be helpful in developing the new methods or modifications in using BTB converters for power system applications.

Keywords- Power Quality, Power Quality issues, Back to Back (BTB) converter and Control Strategies.

1. Introduction

The Utilization of power converters has developed largely especially in industries such as generation systems, Uninterrupted Power Supplies (UPS), effective transmission of power between energy sources and also capability of fault tolerance. Generally, low quality of power may result in additional losses in power, change in behaviour of the machines in industries, abnormal operations (especially effect on voltage and frequency parameters), harmonic distortion, and possibility of interference in communication systems at the transmission lines etc. This results in the need of power converters in different levels of power systems for maintaining power quality with proper control strategies [1] .

Three phase AC/DC/AC which also comes under the back-to-back (BTB) converters are operated with a proper control strategies which are extensively used in Power and Industrial applications, especially for the control of Voltage in case of High Voltage DC Transmission (HVDCT), Electrical machines, Flexible AC Transmission systems (FACTS) and Unified Power Quality Conditioners (UPQC), and some other devices [2-3]. These BTB converters with the proper control strategies protect the system by avoiding the disturbances which effect on power quality. Also this BTB converter has a high demand in controlling the power quality issues like Voltage sags under faulty or short circuited conditions, Grid operating in weak conditions etc. This BTB converter topology utilizes DC link at its intermediate station for the process of power conversion i.e., from AC/DC/AC in addition to the other shunt connected filters at AC and DC side [4-6].

BTB converters of high power ratings are employed in industries, distribute the electrical power between the numerous power semiconductor switches. In order to improve the power quality, advancements in topologies of the BTB converters such as multi-level of voltage in the converter stations are made to enhance the redundancy, reliability, and for better voltage control [7]. With this multi-level, voltage stress, power losses, harmonic content at voltage and current waveforms can be reduced.

In BTB converters, another important property is the most often power flow manipulation and balance among the two stations (AC sides) by using a capacitor across the DC link. Dynamic Control strategies for processing the power provides stability more at the DC link which results in choosing a less sized capacitor across the DC link without any affect at the converter stations [8-11]. In BTB applications, mostly electrolytic capacitors are used across the DC link so that the life time of converters can be increased with its natural characteristics.
Finally in this paper, we are presenting BTB converters at different power system configurations and the major control strategies implemented at respective power system configurations such as the Pulse Width Modulation (PWM), Space Vector-Pulse Width Modulation (SV-PWM), Proportional Integral (PI) controllers are analyzed.

2. Basic Back to Back (BTB) Converter Configuration

In industries, 3-ϕ AC/DC/AC converters with adjustable Voltage and adjustable frequency operations have gained a broad application. Diode based Voltage source Inverter (VSI) is the most commonly preferred configuration for Uninterruptible power supplies (UPS), adjustable/variable speed drives and some other industrial utilizations. As shown in figure 1, it is one of the less cost and performs more reliable operation with the usage of diode rectifier. But the drawback is that in input line currents, this diode rectifier generates extreme distortions.

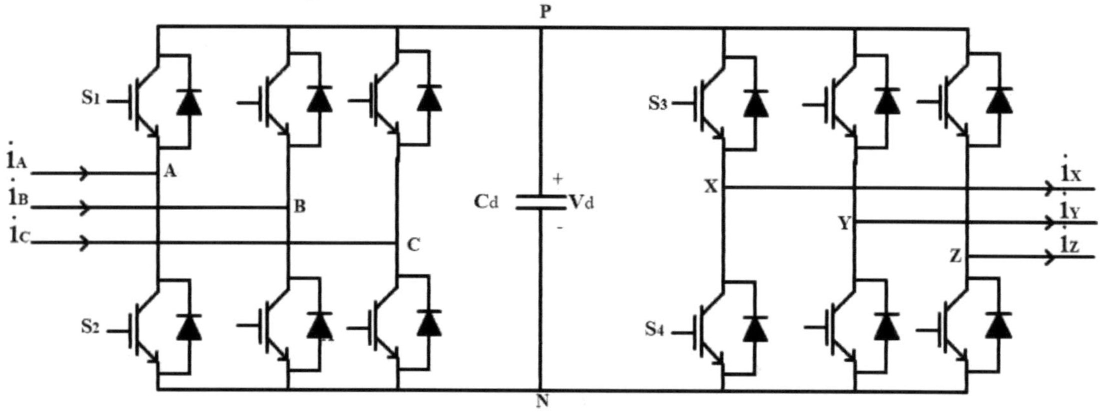

Figure 1. Three leg – Two level BTB converter

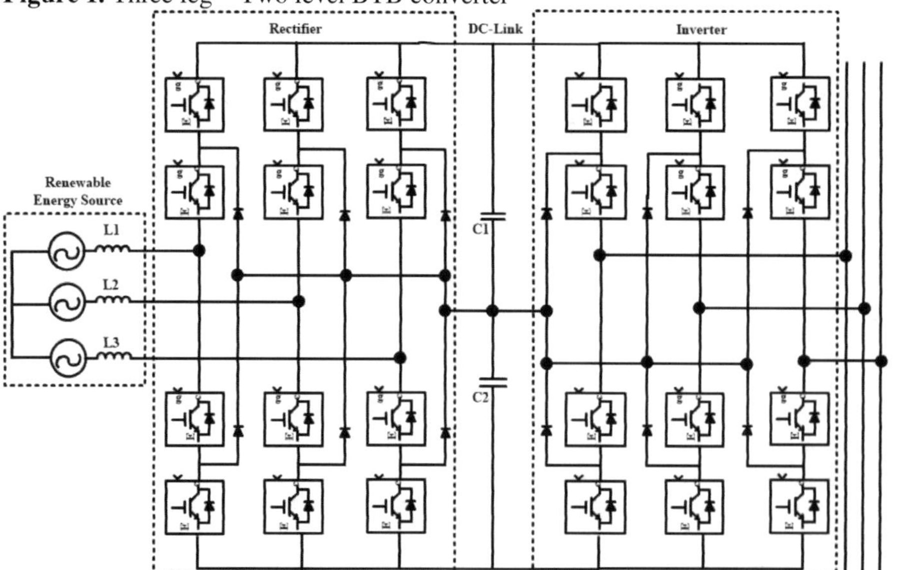

Figure 2. Three level BTB converter

In order to overcome this problem a Pulse width Modulation (PWM) based voltage source converter as shown in Fig.1, is used by replacing the diode rectifier. This BTB three leg-two level VSI lacks a twelve number of controllable switches i.e., Insulated Gate Bipolar Transistors (IGBTs) are used for controlling the voltage of higher ratings. Along with this, capacitor connected across the DC link responsible for increase in the effective operation of the converter in addition to increase in cost. Similarly figure 2, represents the three leg-three level BTB converter consisting of 24 switching devices (each converter station has 12 switches) for controlling the voltage with capacitors C_1 and C_2 connected across the DC link [1-2]. Various BTB configurations connected in power systems and their control strategies proposed are presented in this paper.

3. Back to Back Converter configuration in different power systems

3.1 BTB in HVDCT System Configuration

For many of the commercial applications, there are continuous growth in technologies especially in transmission techniques like HVDCT and FACTS. These transmission techniques has undergone many research and developments in the past few years and depended on the thyristors, controllable power semiconductor switches and other Voltage source converters (rectifier and inverter). Due to the increase in growth of the power semiconductor switches in the area of the power electronics, the full control over the voltage can be achieved even for higher operating levels as shown in figure 3. The major contribution of control over the voltage is due to the thyristors in the converters .i.e., either at the rectifier and inverter stations in case of HVDCT as shown in figure 4. There exist two type of converters which are mostly preferred i.e., Voltage Source Converter (VSC) and Current Sources Converters (CSC) in which the PWM is implemented at the Inverter end. VSC and CSC have the ability to withstand the sudden voltage and current respectively due to its inherent properties of the elements used in it [3].

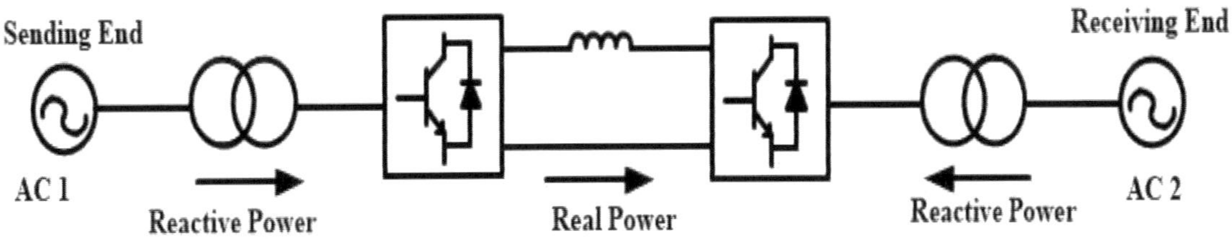

Figure 3. HVDC system based on VSC technology built with IGBTs.

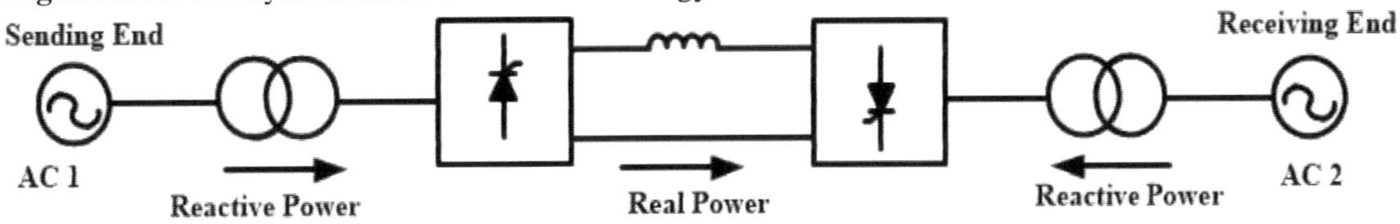

Figure 4. HVDC system based on CSC technology with thyristors.

With the use of Micro Grids (MGs), the overall effect on the environment can be reduced by exempting the use of fossil fuels for the meeting the power demand from the customers. These MGs consist of the distributed generations (DGs) which have a rapid growth in their developments will have distinct characteristics with respect to the changes in environment which are unpredictable. This leads to the cause of the power quality (PQ) issues which greatly affect the MG. An active power conditioner (APC) which consists of a three phase BTB converter as shown in figure 5, connected between the two MGs which enhancing by the PQ by limiting the disturbances/PQ issues at both ends. This APC can compensate the active and reactive power as per the dynamic behaviour of the MG and due to this APC has become a major part in the MGs [4-6].

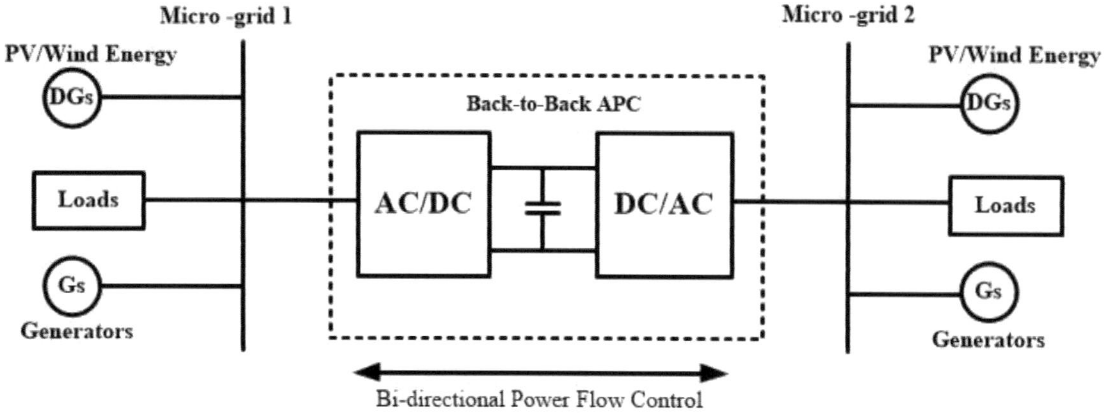

Figure 5. MG with BTB Active Power Conditioner.

3.2 *BTB in flywheel applications*

Nowadays on Electric Vehicles (EVs), a considerable research done recently. The evolution of robust and efficient propulsion is a fundamental task for the EVs facility. But, even though sophisticated engines and advanced electrical power trains exist, the main issue, the long-term energy storage, has not been resolved. In order to reduce the power density if the batteries, flywheels, super capacitors and their combinations are used. As shown in figure 6, the flywheel is split in to two levels of voltage over stator winding. Where Low Voltage (LV) side are associated to the battery and High Voltage (HV) side of flywheel allied to motor wheel. A bidirectional system exists which consists of two modes of operation i.e., acceleration mode (Battery-Wheel) and regenerative braking mode (Wheel-Battery).

The function of the flywheel in the mode of acceleration is basically to afford the power variant needed by motor wheel; hence the smooth operation can be performed by the battery. Whereas in the mode of regenerative, motor wheel functions as a generator and the flywheel will store the energy during this regenerative mode. In order to optimize and regulate the flux between these modes of operations, BTB converters which consist of the power electronic components mostly the fully controlled power semiconductor switches are used [7-9].

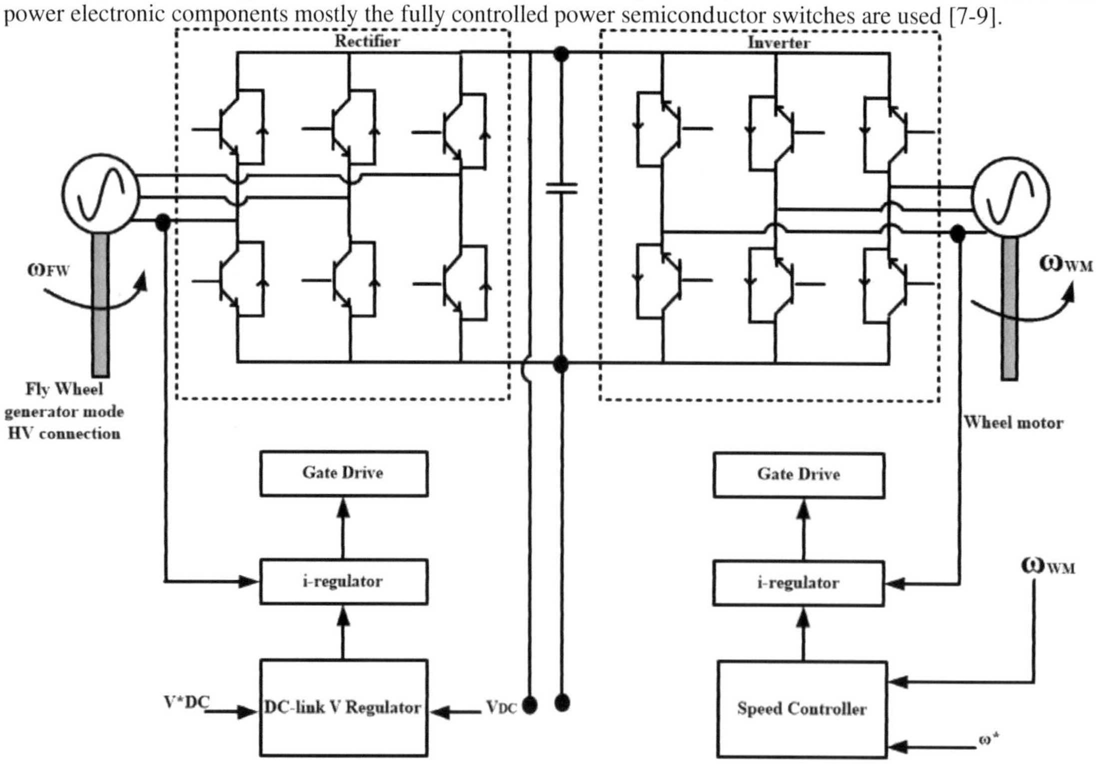

Figure 6. Acceleration mode of BTB Converter

3.3 BTB and Three Phase Induction Motor connected to a grid

In High power applications, multilevel based BTB converters are being mostly used nowadays. Numerous topologies have been used in multilevel converters (inverters) such as a flying capacitor (FC), Neutral point clamped (NPC) and cascaded H Bridge (CHB) types. With major combinations and modifications, a t-type BTB converter as shown in figure 8, is more effective than the other types of converters discussed. This T-type converter is used in industries where the most often used machine is the three phase induction motor so that the full control can be achieved with a medium switching range of frequencies [18-20].

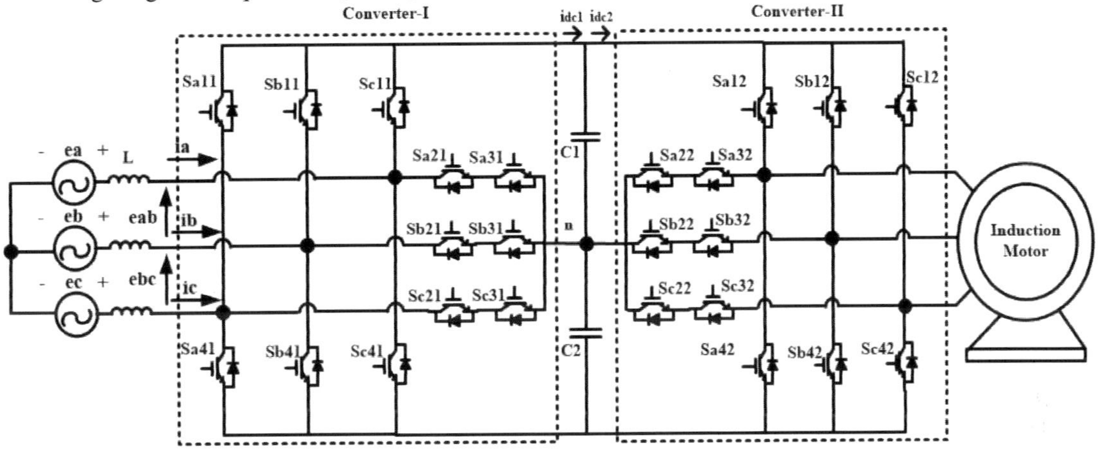

Figure 7. T-type three levels BTB Converter

3.4 BTB in Wind Power system Configuration

With the use of renewable energy sources such as wind energy systems for electrical power generation, the concept of stability is one of the important tasks. Due to the environmental conditions, the PQ issues like Voltage unbalance and dips may rise in the system which can be controlled by using the Power Electronic systems consisting of the BTB converters as shown in figure 8. This BTB Converter compensates the reactive power and active power which ultimately control the Voltage dips and unbalance caused in the systems due to the climatic changes in wind Power generation. A Local control level (LCL) filter as shown in Fig.6, connected at the VSC 1, used to limit the harmonics which are caused during the abnormal conditions [28-29].

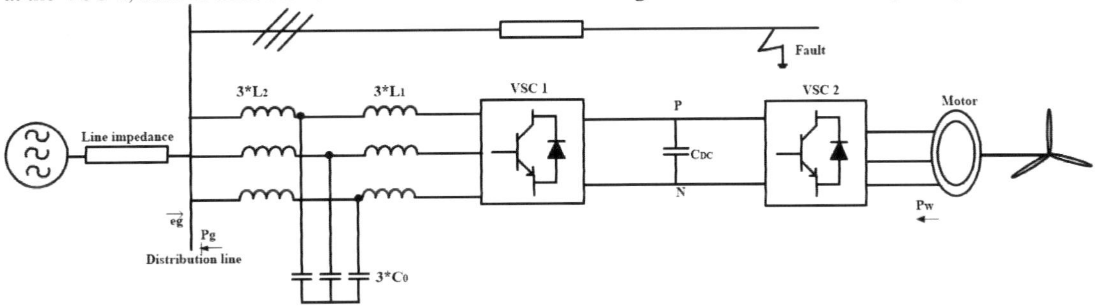

Figure 8. Back-to-back VSC connected to the utility distribution line

1.4 BTB and Double Fed Asynchronous Machines (DFAM) connected to a grid

In view of the sustainable energy sources development, the balanced usage of electrical energy signifies having the capability of network transportation for a long distance as well as to the storage plants which are adequate. Approximately most of the pump turbine groups functioning recently consist of Synchronous Motor-Generators (SMGs), working at the fundamental frequency and at constant speed. With the use of the variable speed MGs and especially by using the DFAM generators with a slip range of 0.1 to 0.15 around synchronous speed, the overall efficiency and performance of such groups can be effectively improved [30].

Some of the groups are equipped with the cyclo-converter cascade which are now replaced with BTB-3-level VSC (inverter) neutral point clamped (NPC). Regulating reactive power can be achieved with this type of

IOP Conf. Series: Materials Science and Engineering **906** (2020) 012016 doi:10.1088/1757-899X/906/1/012016

converter in the rotor cascade. Where as in case of the cyclo converter cascade there exists disturbances such as sub harmonics which are can be overcome by using the BTB converter as shown in figure 9.

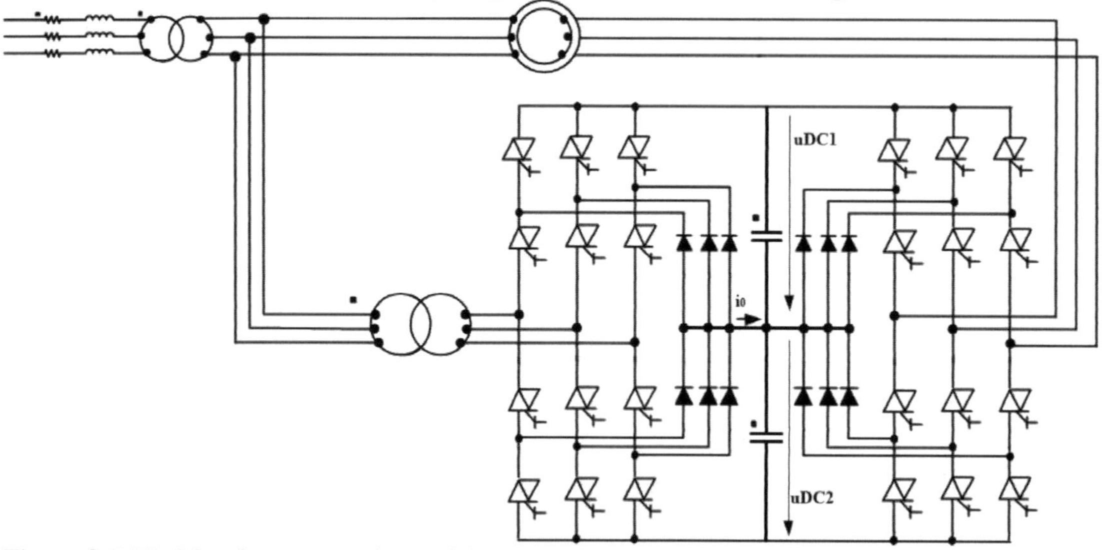

Figure 9. BTB-3 level rotor cascade supplying a DFAM

Table 1. Classification of BTB converter based on different level of control

Function	Converter Description	Area of Function	Control method(s)/ Control strategies	Reference
Voltage Control	3-phase 3-level VSC	Power Rectifier	space vector pulse width modulation (SVPWM)	[1]
Voltage balancing	3-level Neutral point clamped Converter	Inverter	Nearest-Three-Vector (NTV) Modulation	[2]
Voltage Control	Direct and Indirect Matrix Converter	DC-link	outer speed control loop and two inner current control loops	[5]
voltage control	3-Phase, 3-leg VSC	DC-link	optimal ac-line current regulation and dynamic dc-link voltage regulation (DDVR)	[6]
switching harmonic currents	3-phase 2-level VSC	DC-Link	optimal PWM shift angle	[7]
Harmonic Distortion	Three single Phase, 9-leg multilevel VSC	Inverter	SV-PWM	[8]
Voltage and Current	3-Phase, 3-leg VSC	Dc-Link and Inverter	linear multivariable	[10]
voltage control	Two 3-leg VSC	DC-Link	SV-PWM	[11]
voltage control and Harmonic Distortion	Grid Connected Converter	Dc-Link and Inverter	adaptive proportional integral (PI) controller	[12]
Voltage and Current	1-phase 3-leg CSC	Rectifier and Inverter	Cartesian space	[13]
voltage control	two single phase, 3-leg modules	DC-link	SV-PWM	[15]
Harmonic Distortion	two parallel, 1-phase 3-leg converters	Inverter	four-carrier PWM	[16]
Reactive power compensation	3-Phase, 3-leg,9-Switch VSC	PCC	closed loop proportional integral control	[17]
Oscillation	1-phase, X-type CSC	Dc Link	Pulse Width Modulation	[18]

voltage control and Harmonic Distortion	3-phase 4-wire, 3-leg VSC	Inverter	Pulse Width Modulation	[21]
voltage control	1-phase,3-leg VSC	DC Link	differential- and common current (power)-based state-feedback control	[22]
voltage control	single-phase four-leg VSC	DC Link	SV-PWM	[23]
voltage control	unidirectional neutral-point-clamped (NPC) leg and two bidirectional NPC legs VSC	DC Link	Pulse Width Modulation	[24]
voltage control	1-phase, 2-3leg VSC	DC Link	SV-PWM	[25]
voltage control and Harmonic Distortion	1-phase, 8-leg VSC	DC Link	SV-PWM	[26]
voltage control	3-phase 3-level NPC VSC	DC Link	DSP and FPGA	[28]
Reduces power losses and improves the efficiency of power converter.	5- channel, 3 L-NPC VSC	AC ends	variable speed pumped storage unit (VSPSU)	[29]

4. Classification of Back to back converter control strategies

4.1 3-level BTB Converter using Space Vector Pulse Width Modulation (SVPWM)

From [1], a new model of power rectifier circuit for BTB-3-level converter was proposed. The control is achieved at the DC link voltage by developing a control algorithm using MATLAB/Simulink. Along with this using dSPACE DS1104 board a prototype is developed. In this model, voltage and current distortions can be reduced by the converter. In addition to that DC link voltage can also be adjusted by this type of converter. From the power rectifier circuit results obtained from simulation by using SVPWM, the DC link voltage is controlled suitably and the distortions in current waveform can be reduced and hence overall performance in conversion of energy is done.

4.2 3-level BTB Neutral point Clamped (NPC) Converter

From [2], the limitations of NPC particularly, the control of current in order to balance the voltage was explained. This mathematical model applicable to converters operating with higher number of levels which is also very useful for the finding out the limitations of modulation techniques. Nearest three Vector (NTV) modulation techniques presented in this paper provides a better performance output voltage, Electro Magnetic Interference (EMI) and reducing the switching losses. All these issues are important especially while operating at higher ratings. Also this mathematical model helps in analysing the conditions such as balance during a stand-alone operation of 3-level Converter.

4.3 Micro Grids with BTB-Active power conditioner(APC)

For [6], MGs applications, a three phase BTB-APC with voltage control strategy across DC link were presented. Due to dynamic behaviour of the MGs, the demand of active and reactive power compensation has been increased, which can be done by the BTB-APC. In this APC, two voltage regulation methods across the DC links are presented which improves the stability and power quality of the MGs. In this paper, regulation of ac line current method is capable of achieving the voltage regulation across the DC link at 60Hz frequency supply. Along with this it can reduce the variations of current in ac line.

4.4 9-leg BTB Converter using Level Shifted (LS) PWM

In [8], three phase BTB; 9-leg configuration was presented which consists of three 1-ϕ, 3-leg converter with three banks of DC link and shared legs without a circulating current and isolation transformer. SVPWM and LS-PWM methods are implemented to control the voltage across the DC link with proper developments made in PWM techniques. Along with this, a comparison of 9- level, 5-level and NPC configurations are also presented in this paper.

4.5 BTB LCL filter based Converter using state space model

In [10], a modern unified model and a procedure for controlling the BTB converter was presented in which all the interactions in dynamic nature between two ends of AC and at intermediate DC station particularly at capacitor are also presented in the form of state space. In a single linear model the BTB converter is presented in which the DC capacitor and LCL filters are together linked. In order to reduce the changes in the DC link voltage at the time of transients in active power flow, a linear Multi-variable control strategy was proposed in this paper. This strategy is also helpful to accurately control the real power flow. Even for the large variations, the size of the capacitor across the DC link has to be chosen large enough to overcome the variations and maintain the stability.

4.6 6-leg BTB Converter using SVPWM Technique
From [11], BTB topologies for the applications of non-regenerative which consist of two 3-leg converters with an equal voltage across the DC link which are in series connection are presented in this paper. In this proposed system, controllable switches, drivers and overall cost can be reduced. SVPWM at the inverter end and voltage vector redundancies across the DC links are presented in this paper.

4.7 Grid connected BTB converter using adaptive proportional integral (API) controller
In [12], an anti-windup API controller based voltage controller across the DC link is proposed. With the proposed API controller, a fast dynamic response after step rise of input current and dc link voltage can be obtained. Along with this after step rise there will be no overshoot across DC link voltage and less THD under low grid current while performing the stable operation.

4.8 3-leg BTB Converter series connected H-Bridge using SVPWM
From [15], two 3-leg converters which are connected in series with H bridge converter are investigated. This is formed by the single phase BTB multilevel converter of high power ratings. SVPWM strategy is used in this proposed model which is capable of reducing the switching losses, stress in switches and harmonic distortions. Along with this, it can also reduce the lower voltages across DC link and voltages are generated with more steps.

4.9 4-leg BTB Converter using universal active power filters (UAPFs)
In [23], based on single phase, 4-leg BTB converters, two UPAFs topologies are proposed for performing the compensation as well as mitigating the disturbances in the current which are caused by the loads (nonlinear) and power supply voltage disturbances. With the proposed topologies, Total Harmonic Distortion (THD) can be reduced wherever required and also can lower the average frequency of switching.

4.10 Transformer based 3-leg BTB converter using SVPWM
In [25], single phase BTB topology for UPQC and UPS applications are proposed which comprises of transformer and a 6 level converter (termed as 6LT). This 6LT allows power supply with a constant voltage and frequency to the load connected. It also improves the power factor by lowering the Harmonic Distortion and the losses in semiconductor switches.

4.11 3-leg,3-level BTB Neutral Point Clamped (NPC) Converter
In [24], unidirectional single phase 3 level NPC converter is proposed. This proposed converter has less in number of power semiconductor switches than the traditional converters. By using Current control loop and Phase locked loop (PLL), the utilization and its synchronization method was simulated in this paper. Between the voltage and current for synchronization at the grid, the PWM strategy depends on the direction of the current flow on the voltage synthesis. With the use of diodes which not only reduces the numbers of controllable switches but also results in less in drive circuits. This leads to reduction in complexity of the configuration.

4.12 BTB Electric Railway Application
From [27], 3-ϕ to 1-ϕ BTB topology based on multi converter is proposed, which can enhance the quality of power in electrified railways. Flexibility in control at the inverter helps in removing the neutral sections which also helps in removal of negative sequence components (currents) during the abnormal operations. With the use of H-bridge based BTB converter, control actions can be easily performed and neutral drift problem can be overcome. At the rectifier station, PWM and DC voltage control results in control of unbalanced voltage. By comparing the traditional power system at the tractions, the solution of mitigating the power quality problems can be identified.

4.13 BTB Neutral Point Clamped (NPC) Converter in Wind Turbines
In [28], BTB converter is divided in to two subsystems which are interred operating. Those are power electronic and control systems. In the first subsystem, thermal components study, filter components in grid, passive

components design, capacitors at DC bus and assembly of converter are discussed. Whereas in second sub-system, coordination between the processors are discussed.

5. Conclusion
This paper presents the overall review and discussions of the BTB converters at different power system configurations. Also numerous converter control strategies which are implemented at various stages in order to control the system from any disturbances such as voltage unbalance, voltage dips, ripples, DC link voltage variations and harmonics etc. Along with the discussion of implemented various control strategies, Classification of BTB converter based on different level of control were listed in this paper which will be helpful in in developing the new methods or modifications in using BTB converters for power system applications.

6. References
[1] Phankong, Nathabhat, et al. "Design of Power Rectifier Circuit for Three-Level Back-to-Back Converter." *Energy Procedia*, vol. 56, Jan. 2014, pp. 574–583.

[2] J. Pou, R. Pindado, D. Boroyevich and P. Rodriguez, "Limits of the neutral-point balance in back-to-back-connected three-level converters," in *IEEE Transactions on Power Electronics*, vol. 19, no. 3, pp. 722-731, May 2004, doi: 10.1109/TPEL.2004.826528.

[3] N. Flourentzou, V. G. Agelidis and G. D. Demetriades, "VSC-Based HVDC Power Transmission Systems: An Overview," in *IEEE Transactions on Power Electronics*, vol. 24, no. 3, pp. 592-602, March 2009, doi: 10.1109/TPEL.2008.2008441.

[4] V. Khadkikar, "Enhancing Electric Power Quality Using UPQC: A Comprehensive Overview," in *IEEE Transactions on Power Electronics*, vol. 27, no. 5, pp. 2284-2297, May 2012, doi: 10.1109/TPEL.2011.2172001.

[5] T. Friedli, J. W. Kolar, J. Rodriguez and P. W. Wheeler, "Comparative Evaluation of Three-Phase AC–AC Matrix Converter and Voltage DC-Link Back-to-Back Converter Systems," in *IEEE Transactions on Industrial Electronics*, vol. 59, no. 12, pp. 4487-4510, Dec. 2012, doi: 10.1109/TIE.2011.2179278.

[6] C. Tang, Y. Chen, Y. Chen and Y. Chang, "DC-Link Voltage Control Strategy for Three-Phase Back-to-Back Active Power Conditioners," in *IEEE Transactions on Industrial Electronics*, vol. 62, no. 10, pp. 6306-6316, Oct. 2015, doi: 10.1109/TIE.2015.2420671.

[7] L. Shen, S. Bozhko, G. Asher, C. Patel and P. Wheeler, "Active DC-Link Capacitor Harmonic Current Reduction in Two-Level Back-to-Back Converter," in *IEEE Transactions on Power Electronics*, vol. 31, no. 10, pp. 6947-6954, Oct. 2016, doi: 10.1109/TPEL.2015.2511304.

[8] A. C. N. Maia, C. B. Jacobina, N. B. de Freitas and M. A. Vitorino, "Investigation of Three-Phase AC–DC–AC Multilevel Nine-Leg Converter," in *IEEE Transactions on Industry Applications*, vol. 52, no. 5, pp. 4156-4169, Sept.-Oct. 2016, doi: 10.1109/TIA.2016.2574771.

[9] A. C. N. Maia, C. B. Jacobina and G. A. de Almeida Carlos, "A New Three-Phase AC–DC–AC Multilevel Converter Based on Cascaded Three-Leg Converters," in *IEEE Transactions on Industry Applications*, vol. 53, no. 3, pp. 2210-2221, May-June 2017, doi: 10.1109/TIA.2016.2646343.

[10] A. Rodríguez-Cabero, F. H. Sánchez and M. Prodanovic, "A unified control of back-to-back converter," *2016 IEEE Energy Conversion Congress and Exposition (ECCE)*, Milwaukee, WI, 2016, pp. 1-8, doi: 10.1109/ECCE.2016.7854799.

[11] N. B. de Freitas, C. B. Jacobina, N. S. de Moraes Lima Marinus and N. Rocha, "AC–DC–AC Single-Phase Multilevel Six-Leg Converter With a Reduced Number of Controlled Switches," in *IEEE Transactions on Power Electronics*, vol. 33, no. 4, pp. 3023-3033, April 2018, doi: 10.1109/TPEL.2017.2707064.

[12] M. Merai, M. W. Naouar and I. Slama-Belkhodja, "An Improved DC-Link Voltage Control Strategy for Grid Connected Converters," in *IEEE Transactions on Power Electronics*, vol. 33, no. 4, pp. 3575-3582, April 2018, doi: 10.1109/TPEL.2017.2707398.

[13] Y. Liu, Y. Sun, M. Su, X. Li and S. Ning, "A Single Phase AC/DC/AC Converter With Unified Ripple Power Decoupling," in *IEEE Transactions on Power Electronics*, vol. 33, no. 4, pp. 3204-3217, April 2018, doi: 10.1109/TPEL.2017.2707443.

[14] Devassy, S. (.1,2)., and B. (. 1.). Singh. "Control of Solar Photovoltaic Integrated UPQC Operating in Polluted Utility Conditions." *IET Power Electronics*, vol. 10, no. 12, pp. 1413–1421. *EBSCOhost*, doi:10.1049/iet-pel.2016.0999. Accessed 29 May 2020.

[15] A. De P. D. Queiroz, C. B. Jacobina, N. B. de Freitas, A. C. N. Maia and V. F. M. B. Melo, "Single-phase AC-DC-AC multilevel converter based on H-bridges and three-leg converters connected in series," *2017 IEEE Energy Conversion Congress and Exposition (ECCE)*, Cincinnati, OH, 2017, pp. 2129-2136, doi: 10.1109/ECCE.2017.8096421.

[16] N. Rocha, A. E. L. da Costa and C. B. Jacobina, "Parallel of Two Unidirectional AC–DC–AC Three-Leg Converters to Improve Power Quality," in *IEEE Transactions on Power Electronics*, vol. 33, no. 9, pp. 7782-7794, Sept. 2018, doi: 10.1109/TPEL.2017.2771464.

[17] C. N. Jibhakate, M. A. Chaudhari and M. M. Renge, "Reactive Power Compensation Using Induction Motor Driven by Nine Switch AC-DC-AC Converter," in *IEEE Access*, vol. 6, pp. 1312-1320, 2018, doi: 10.1109/ACCESS.2017.2778291.

[18] L. A. L. de Azevedo Cavalcanti Costa, M. A. Vitorino and M. B. de Rossiter Corrêa, "Improved Single-Phase AC–DC–AC Current Source Converter With Reduced DC-Link Oscillation," in *IEEE Transactions on Industry Applications*, vol. 54, no. 3, pp. 2506-2516, May-June 2018, doi: 10.1109/TIA.2018.2799186.

[19] A. Alizade and J. B. Noshahr, "Evaluating noise and DC offset due to inter-harmonics and supra-harmonics caused by back-to-back converter of (DFIG) in AC distribution network," in *CIRED - Open Access Proceedings Journal*, vol. 2017, no. 1, pp. 629-632, 10 2017, doi: 10.1049/oap-cired.2017.0045.

[20] J. King, J. Berry and N. Murdoch, "Steady-state modelling for the integration of a bi-directional AC–DC–AC flexible power link," in *CIRED - Open Access Proceedings Journal*, vol. 2017, no. 1, pp. 445-449, 10 2017, doi: 10.1049/oap-cired.2017.0941.

[21] A. C. N. Maia, C. B. Jacobina, N. B. de Freitas, A. de Paula Dias Queiroz and E. R. C. da Silva, "Three-Phase Four-Wire AC–DC–AC Multilevel Topologies Obtained From an Interconnection of Three-Leg Converters," in *IEEE Transactions on Industry Applications*, vol. 54, no. 5, pp. 4728-4738, Sept.-Oct. 2018, doi: 10.1109/TIA.2018.2843352.

[22] A. Rodríguez-Cabero, M. Prodanovic and J. Roldán-Pérez, "Full-State Feedback Control of Back-to-Back Converters Based on Differential and Common Power Concepts," in *IEEE Transactions on Industrial Electronics*, vol. 66, no. 11, pp. 9045-9055, Nov. 2019, doi: 10.1109/TIE.2018.2873518.

[23] P. L. S. Rodrigues, C. B. Jacobina, M. B. de Rossiter Corrêa and I. R. F. M. P. da Silva, "Single-Phase Universal Active Power Filter Based on Four-Leg AC–DC–AC Converters," in *IEEE Transactions on Industry Applications*, vol. 55, no. 2, pp. 1639-1648, March-April 2019, doi: 10.1109/TIA.2018.2876027.

[24] R. P. R. de Sousa, N. S. de Moraes Lima Marinus, C. B. Jacobina and N. Rocha, "A Unidirectional Single-Phase AC–DC–AC Three-Level Three-Leg Converter," in *IEEE Transactions on Industry Applications*, vol. 55, no. 2, pp. 1708-1716, March-April 2019, doi: 10.1109/TIA.2018.2879794.

[25] N. B. de Freitas, C. B. Jacobina, B. S. Gehrke and M. F. Cunha, "Transformer-Based Single-Phase AC–DC–AC Topology for Grid Issues Mitigation," in *IEEE Transactions on Industry Applications*, vol. 55, no. 4, pp. 4001-4011, July-Aug. 2019, doi: 10.1109/TIA.2019.2912547.

[26] P. Leal Serafim Rodrigues, C. Brandão Jacobina and N. B. de Freitas, "Single-phase universal active power filter based on ac–dc–ac converter with eight controlled switches," in *IET Power Electronics*, vol. 12, no. 5, pp. 1131-1140, 1 5 2019, doi: 10.1049/iet-pel.2018.5391.

[27] L. Li, M. Wu, S. Wu, J. Li and K. Song, "A Three-Phase to Single-Phase AC-DC-AC Topology Based on Multi-Converter in AC Electric Railway Application," in *IEEE Access*, vol. 7, pp. 111539-111558, 2019, doi: 10.1109/ACCESS.2019.2933949.

[28] E. J. Bueno, S. CÓbreces, F. J. RodrÍguez, Á. HernÁndez and F. Espinosa, "Design of a Back-to-Back NPC Converter Interface for Wind Turbines With Squirrel-Cage Induction Generator," in *IEEE Transactions on Energy Conversion*, vol. 23, no. 3, pp. 932-945, Sept. 2008, doi: 10.1109/TEC.2008.918651.

[29] K. Desingu, R. Selvaraj, T. R. Chelliah and D. Khare, "Effective Utilization of Parallel-Connected Megawatt Three-Level Back-to-Back Power Converters in Variable Speed Pumped

Storage Units," in *IEEE Transactions on Industry Applications*, vol. 55, no. 6, pp. 6414-6426, Nov.-Dec. 2019, doi: 10.1109/TIA.2019.2936365.

[30] A. Hodder, J. -. Simond and A. Schwery, "Unbalanced DC-Link voltage regulation in a back-to-back 3-level PWM converter for a double-fed induction motor-generator," in *IEE Proceedings - Electric Power Applications*, vol. 152, no. 6, pp. 1477-1481, 4 Nov. 2005, doi: 10.1049/ip-epa:20045170.

[31] Kumar Neelamsetti Kirn, and Indragandhi Vairavasundaram. "Analysis on Various Optimization Techniques Used for Load Frequency Control in Power System." *Serbian Journal of Electrical Engineering*, vol. 15, no. 3, Jan. 2018, pp. 249–273. *EBSCOhost*, doi:10.2298/SJEE1803249K.

[32] Kumar, N. Kirn and Gandhi, V. Indra. 'Implementation of Fuzzy Logic Controller in Power System Applications'. 1 Jan. 2019 : 4115 – 4126.

[33] Pavan Babu Bandla, Indragandhi V., RajaSingh R. and Kannan Ramani 2018 Design Approach of Grid Coupled Solar Inverter *MATEC Web Conf* **225** 06001.

[34] Pavan Babu, B., and V. Indragandhi. "A Review on Dynamic Voltage Restorer in Power Systems Concerned to the Issues of Power Quality." *IOP Conference Series: Materials Science and Engineering*, vol. 623, no. 1. *EBSCOhost*, doi:10.1088/1757-899X/623/1/012015. Accessed 31 May 2020.

[35] Indragandhi, V., et al. "Implementation of a DC Micro-Grid for House Hold Applications." *2018 8th International Conference on Power and Energy Systems (ICPES), Power and Energy Systems (ICPES), 2018 8th International Conference On*, Dec. 2018, pp. 141–145. *EBSCOhost*, doi:10.1109/ICPESYS.2018.8626889.

Energy Optimal Intelligent Switching Mechanism for Induction Motors with Time Varying Load

R. Raja Singh[1], C. Thanga Raj[2], Ryszard Palka[3], V. Indragandhi[1], Marcin Wardach[3], and Piotr Paplicki[3]

[1] Advanced Drives Laboratory, Department of Energy and Power Electronics, Vellore Institute of Technology, India.
[2] Hydropower Simulation Laboratory, Department of Water Resources Development & Management, Indian Institute of Technology Roorkee, India.
[3] Department of Power Systems and Electrical Drives West Pomeranian University of Technology in Szczecin, Poland.
Correspondence: rrajasingh@gmail.com

Abstract. In medium and small process industries Induction Motor (IMs) enabled star-delta energy conservation scheme is a common practice. In such industrial applications, the Power Quality Disturbances (PQD) must be considered for hassle free operation. This paper is an attempt in the above direction on star-delta energy conservation applied to IMS (2.2 kW and 37.5 kW) operated with time varying loads. In apart to machine's performance evaluation, variation of switchover point for uninterrupted operation and optimized energy savings under various electrical perturbations is realized. Moreover, safe restarting feature is equipped in controller for resuming the machine during unplanned shutdown or any interruption.

Keywords: energy conservation; power quality disturbances; star-delta scheme; switchover point; squirrel cage induction motor

1. Introduction

Energy conservation (EC) is a key requirement for any industry, and most industries are following it economically. Processing industries, textile mills, tiles mills, paper mills, leather mills, etc., generally prefer IMs because of their reliability and robustness, but these engines run out inefficiently during part load operation. Electrical machines are designed and around the full load to achieve maximum output and high power factor. Therefore, the part-load system results in inefficiency and inadequate power factor [1]. The electrical grid with many partially charged motors, causes inadequate power factor and causes instability of voltage. Optimally, the efficiency and the power factor of the motor will be improved. Therefore, an energy-optimal strategy is mandatory for all industries without affecting machine performance. Although there are plenty of energy conservation techniques available, due to investment constraints, most industrialists show less interest in adopting energy-efficient technologies In this field, several researchers are developing various economy energy conservation scheme that receive considerable attention [3]. Kothals Altes et al. initially introduced the induction motor star delta scheme for decreasing magnetizing current [4]. The energy-saving aspect of the scheme was later attempted by Alger et al. with a harmonic distortion analysis [5]. Taylor, et al. achieves optimum load-based motor torque demand by automatic star-delta configuration [6]. Gjota et al. modified engine windings to improve IM performance [7]. Ferreira, and so on. Miloje, et al. predicted a power factor enhancement star delta energy conversion scheme using PIC microcontroller [8–9]. In another path, through the adoption of drive technology, different energy conservation schemes are implemented. Advancing control techniques and electronic power topologies has energy-efficient operation on industrial drives, but these electronic power controls are very expensive. In addition, electronic power switches could be the origin of harmonics, leading to device instability and girding [10]. In addition, the overall power electronic drive system efficiency is reduced due to losses in switching and conduction [11]. Through controlling the flux using various control techniques the efficiency of the IMs is

accomplished with energy conservation [12–16]. The issues regarding energy saving (e.g. through increased efficiency) have also been discussed in papers related to other types of machines, e.g. linear induction machines [17] or hybrid excited machines [18–19].
On the other hand, these electrical drives in any industry are influenced by various electrical perturbations, and causes process interruptions and equipment failure [20]. Severity of PQD in star-delta scheme based EC is more as motor windings receive less voltage at star mode. This paper, therefore, analyses the star-delta energy conservation system in IM under different PQDs including under-voltage, over-voltage, voltage sag, voltage unbalance, under-frequency, over-frequency, and single phasing. Considering these PQDs, a strategic approach for continuous operation is provided through automatically readjusting the switch over point. Moreover, safe restarting feature is embedded in the proposed controller for resuming the machine during unplanned shutdown or any interruption. (i.e.) Conventional star-delta energy saving scheme resume only in star connection, however, star connected motor fails to develop sufficient torque at heavy loads. Focusing on this direction, the last operating mode is recorded for safe restarting in the proposed strategy. A 37.5 kW induction motor spinning machine of a medium scale textile industry is considered as a case study. The motivation of this work is to enforce the proposed controller for providing reliable energy conservation in small and medium scale process industries where conventional star-delta starter is employed.
In overall, the paper contains the description of experimental arrangement in section 2, besides, a case study with star delta based energy conservation has been carried out using the load curve of textile spinning mill in section 3. Effect of power quality disturbances on Y-Δ scheme based IM in section 4, summary of PQD with an intelligent switching mechanism in section 5 and concluded in section 6.

2. Experimentation

Laboratory tests are carried out on the basis of theory and empirical evidence to discover the effects of the proposed energy saving strategy and to analyze the test machine's behavior. Systematic operation is performed to obtain the results for analysis.

2.1. Experimental Arrangement

Dynamic performance of a 2.2 kW IM is analyzed experimentally and a 37.5 kW IM is simulated using Matlab/Simulink software. Experimental arrangement (Fig. 1) consists of a 2.2 kW squirrel cage induction motor (SCIM) coupled with dynamometer loading arrangement (DC generator, excitation system, and electrical load). Machine parameters are given in Table 1. An AC programmable power supply (AMETEK 3000 Lx) is used to inject various electrical disturbances on test machine and the electrical quantities are measured recorded using three-phase power quality analyzer (PQA – Fluke 435).

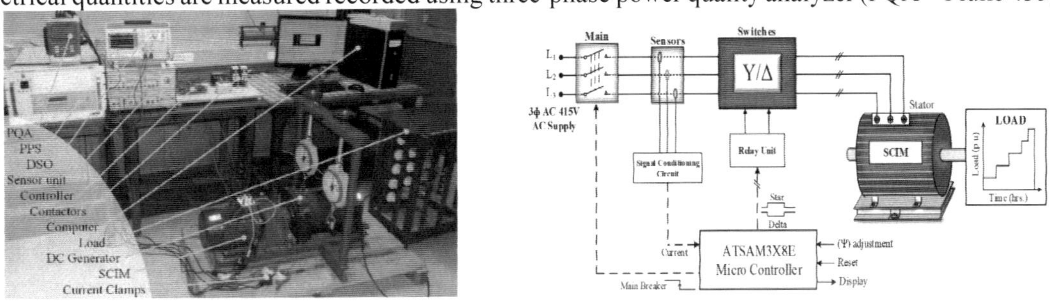

Figure 1. Experimental arrangement for Y-Δ Scheme (**a**) Experimental Arrangement (**b**) Schematic diagram of Y-Δ Scheme.

Similarly, the output quantities (speed, torque, and temperature) are measured using tachogenerator, electro dynamo meter arrangement and thermal imager (FLUKE Ti-32) respectively. The control strategy consists of sensor unit (Hall Effect current and voltage sensors) with signal conditioning circuit to provide the digital signals to controller. Atmel micro-controller (ATSAM3X8E) with opto-coupler isolated driver and relay unit is used to drive the contactors with respect to load variation. Two contactors are used for switching the winding connections alternatively (star or delta), additionally a main contractor is used to energize or isolate the system from supply voltages.

Table 1. Machine particulars.

Particulars	Induction Machine	DC Machine
Rated Power	2.2 kW	2.2 kW
Rated Voltage	415 V	220 V
Rated Current	4.7 A	12 A
Rated Frequency	50 Hz	–
Rated Speed	1410 rpm	1500 rpm
No. of poles	4 poles	4 poles
Type	Squirrel Cage	Shunt
Cooling Type	Fan cooling	Fan cooling

2.2. Experimental Arrangement
As per IEEE 112TM-2004 benchmark [21], an experiment is performed with the 2.2 kW SCIM (test machine). A load-test is conducted on the test machine with star and delta contact to obtain preliminary results. Using the typical equations (1) to (5) in the collected experimental data, the energy and efficiency load curves are plotted, shown in Fig. 2. Efficiency, energy saving and power factor of the star connected motor is found to be higher than the delta connection during the light load operation, but otherwise during the high load operation. The machine operating with star at light load region is considered as Mode-I and running with the delta at high load region is considered as Mode-II, which enhances the energy-saving and strengthen the machine's performance. Matlab/Simulink software is used to validate the proposed scheme and also for generalizing to the higher machine ratings. Efficiency-load and energy-load profile intersecting point of the star and delta curve is considered to be a switch-over point denoted by Ψ. concerning the machine rating and its parameters, the switchover point varies.

$$Efficiency\ [\%] = \frac{P_o}{P_i} \cdot 100 \tag{1}$$

$$Mechanical\ output\ power\ (P_o) = \omega \cdot T \tag{2}$$

$$Electrical\ input\ power\ (P_i) = \sqrt{3} U_L I_L \cos\varphi \tag{3}$$

$$Energy = P_i \cdot time\ period \tag{4}$$

$$Power\ Factor\ (\cos\varphi) = \frac{P_i}{S_i} \cdot 100 \tag{5}$$

where, ω is speed of the machine in rad/s, V_L is the input line voltage in volts, T is the output load torque in newton meter, I_L is the input line current in amps and S_i is the apparent power in volt ampere.
This paper's primary objective is to testify the tangible benefits of the Y-Δ energy conservation technique in test machine with minimal capital expenditure, less downtime and resource requirement. Implementation of the Y-Δ energy conservation technique under part-load operations indicates that the power factor, energy-saving, and efficiency are improved significantly. The test machine is operated with a linear step load (Fig. 5b) using conventional and proposed energy saving strategy. All the input electrical quantities including the machine's speed are shown in Fig. 3 and Fig. 4.

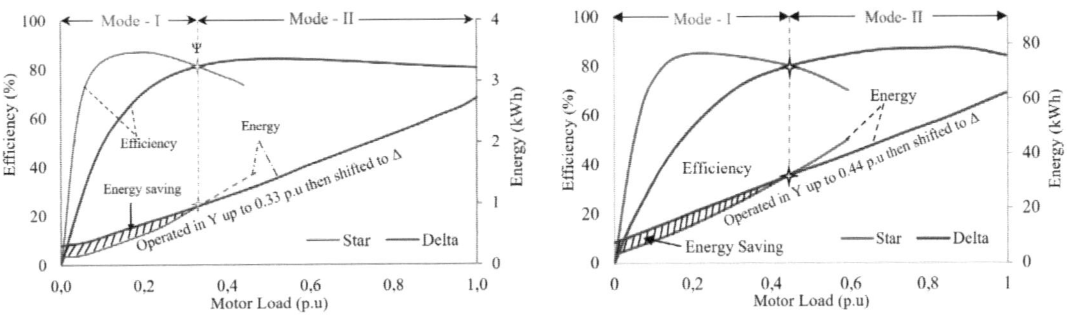

(a) (b)

Figure 2. Efficiency and energy profile of star and delta connected SCIM; (**a**) 2.2 kW IM, (**b**) 37.5 kW IM.

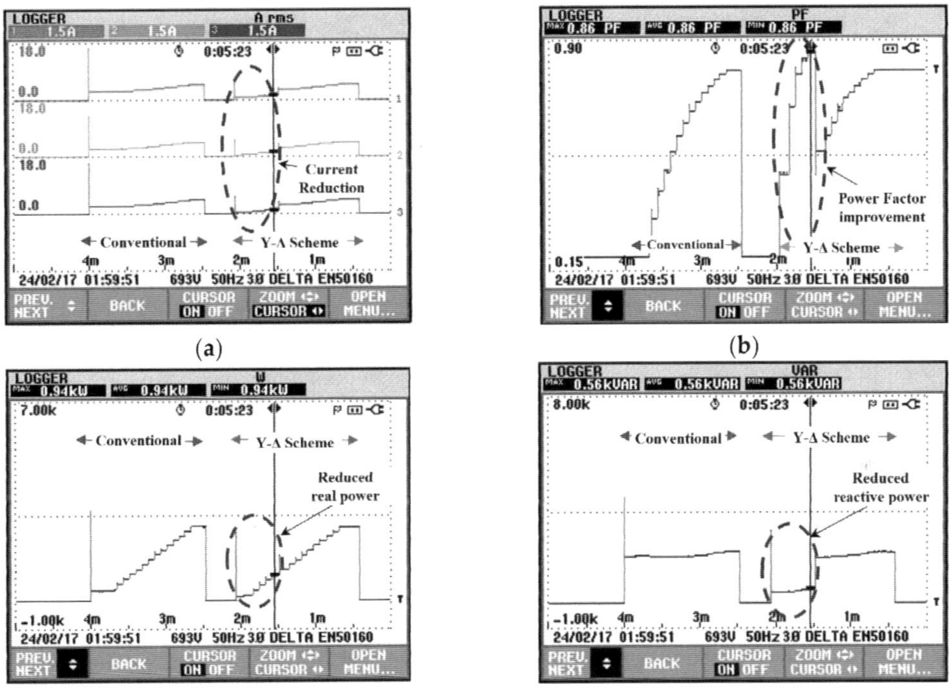

(c) (d)

Figure 3. Electrical parameters of SCIM with conventional Δ connected and Y-Δ EC strategy: (**a**) Input Current; (**b**) Input Power factor; (**c**) Input Real power; (**d**) Input Reactive Power.

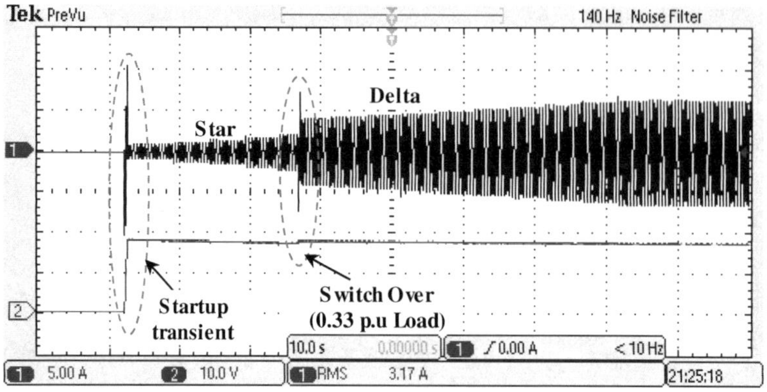

Figure 4. Current and Speed of SCIM with Y-Δ EC strategy process.

Analysis findings are as follows:

2.2.1. *Current*
Starting the machine with the star connection, provides lesser current transient 33.47 percent lower than direct on-line starting. In addition, the switch-over transient is lower than the starting transient (1.32 p.u.) shown in Fig 3a and Fig 4.

2.2.2. *Power factor*
At light loads, the machine's power factor with the proposed strategy is greater than the conventional method. The no-load power factor is 43.37 percent higher in the proposed scheme shown in Fig. 3b

2.2.3. *Real power*
Compared to the conventional method shown in Fig. 3c, the proposed technique conserves 3.68 % more energy, ultimately leading to improved efficiency. Also, it is noted that the amount of energy-saving varies in accordance with the rating of the machine.

2.2.4. *Reactive power*
Adopting the proposed strategy, the consumption of reactive power is significantly minimized. Operating the machine for a complete cycle (i.e.) no load to full load, the consumption of reactive power in the proposed star-delta scheme is 21.82 percent lesser than the conventional direct scheme is depicted in Fig. 3d. Excess reactive power demand, which is accountable for iron losses, winding losses, and inappropriate loading of the machine.

2.2.5. *Speed*
At switchover point (0.33 p.u load), the speed of the star-connected system is 4.67 percent lower than the delta-connected system. Nevertheless, it is 0.52 percent higher than the rated speed of the test machine (146.53 rad/sec).

3. Case study on textile spinning mill with star-delta energy conservation strategy
A spinning drive motor in the textile mill is accounted for energy conservation studies. The textile industry has different processes such as spinning, carding, combing and so on, amongst these process, the spinning drive motor consumes more than 50% of total energy [1]. Mechanically acquiring thread from cotton fibers is called the process of spinning, shown in Fig. 5a. The thread quantity on the spindle increases gradually from zero to the total permissible limit during this spinning process (from hollow bobbin to full bobbin), and therefore the load on the spinning drive motor increase to rated value from no-load value, depicted in Fig. 5b. Acquiring the Y-Δ energy conservation strategy in the system of

spinning, the drive motor operates at light load region with star connection and at high load region with delta connection.

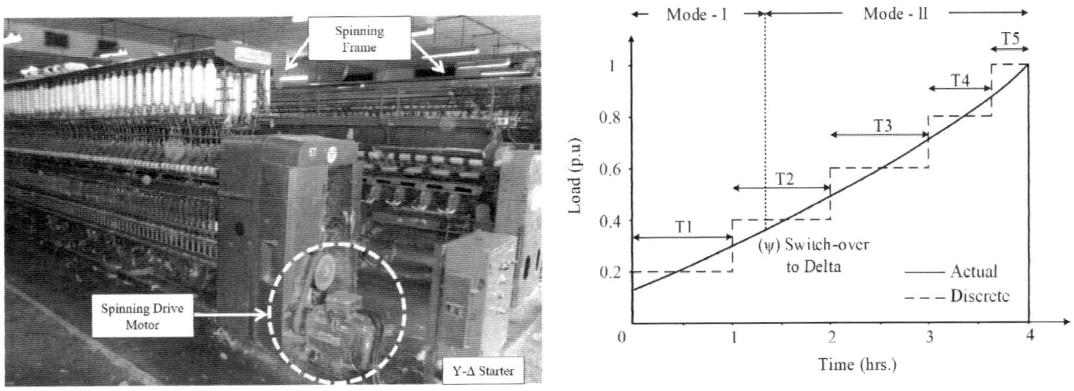

Figure 5. Textile spinning frame with the load diagram: (**a**) Spinning frame with drive motor; (**b**) Load diagram of spinning drive motor.

Considering the spinning frame process cycle in the textile mill (360 workdays per year with 5 processes cycle per day and 4 hrs per cycle process), annual energy saving is calculated using equation 6 and shown in Table 2. From the results, it is observed that Y-Δ scheme offers 3.68 % and 5.97 % of energy conservation in 2.2 kW and 37.5 kW IM, respectively. In addition, under the star-delta scheme, IM ensures better power factor improvement and thermal effect reduction.

$$E_s = (E_{c\Delta} - E_{cY}) * H_T \tag{6}$$

$$E_{c\Delta} = \sum_{n=1}^{\propto} \left\{ \frac{(x_{n+1} - x_n)(y_{\Delta\, n+1} + y_{\Delta\, n})}{2} \right\} \tag{7}$$

$$E_{cY} = \sum_{n=1}^{\propto} \left\{ \frac{(x_{n+1} - x_n)(y_{Y\, n+1} + y_{Y\, n})}{2} \right\} \tag{8}$$

Where, E_s is the total energy saved in kWh, $E_{c\Delta}$ is energy consumed per process with conventional delta (Δ) connection in kWh, E_{cY} is total energy consumed per process using star-delta (Y-Δ) scheme in kWh, H_T is total working hours per annum, n is number of samples, x is process time in hours, y_Δ is power consumed with conventional delta connection in kW, y_Y is power consumed using star-delta scheme in kW.

Moreover, the temperature of test machine is measured for analyzing the thermal effect at light load with star and delta winding connections. Since the proposed system is mainly meant for small and medium scale industries, the rating of the machine is lower where fan type cooling is mainly used [IEC 60034-1] and the cooling of motor directly proportional to shaft revolution. Corresponding to load diagram (Fig. 5b), test machine is operated from no load to 0.33 p.u. load in 1.2 hours with star and delta connection individually and snapshot of test results is shown in Fig. 6a and 6b. It is observed that, machine with star connected produces 13.6 °F less temperature than the machine with delta connection. As the temperature produced in machine is proportionate to the voltage and current in windings, SCIM with star connection produces less thermal effect than delta connection. These variances of thermal characteristics are common for other higher capacity machines (37.5 kW)

Table 2. Energy conserved in IM using Y-Δ scheme with machine constants.

Machine rating	Switchover point (ψ)	Energy Consumed kWh / year		Energy Saving kWh / year	Total cost saving / year ($)[#]
	p.u load	Y-Δ Scheme	Normal (Δ)		
2.2 kW*	0.33	9721	10091	371	39
37.5 kW**	0.44	231990	246739	14749	1520
Machine constants					
2.2 kW*	$R_1 = 3.678$, $R^{|}_2=1.3$, $X_1 = X^{|}_2 = 7.813$, $X_m = 88.57$				
37.5 kW**	$R_1 = 0.0823$, $R^{|}_2=0.0503$, $X_1 = X_2^{|} = 0.2274$, $X_m = 8.51$				

* Experimentation & Simulation; ** Simulation; # Considered 0.103 $ per kWh [22].
Considered as class A for empirical distribution of leakage reactance [21]
R_1 and X_1 are the resistance and the reactance of the stator in ohms,
$R^{|}_2$ and $X_2^{|}$ are the resistance and the reactance of rotor referred to the stator in ohms,
X_m is the magnetizing reactance in ohms.

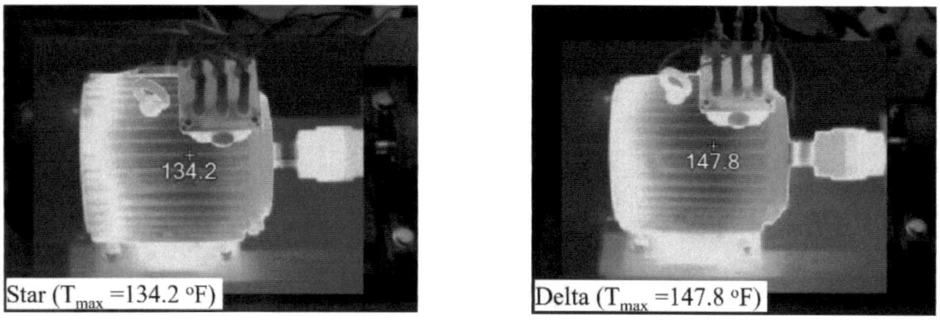

Figure 6. Thermal Effect of 2.2 kW IM in light load region (Mode-I): (**a**) Y-Δ scheme, (**b**) normal operation (Δ).

4. Electrical perturbations on Y-Δ scheme enabled IM

Most aged industries are concerned about issues of power quality including voltage sag, under-voltage, over-voltage, voltage unbalance, under-frequency, over-frequency, single phasing and harmonics. Such fluctuations in power performance are usually caused by non-linear loads, system failures, direct line starting of motor, imbalance loads, power conversion switching, uneven distribution of single phase loads etc. [20–23]. Power quality disturbances (PQD) have a detrimental effect on the industrial drive motor and the degree of impact on the motor is dependent on its connections and span. Fortunately, the use of power quality analyzers to measure PQD easily as they are longer and slower lasting events [1]. This section analyses the effects of various PQD in Y-Δ schemed IM at light loads and compared the results with conventional delta connected IM. Experimentally, power quality disturbances are created and analyzed using AC programmable power supply (AMETEK 3000Lx) and three phase power quality analyzer (Fluke - 435) respectively. To validate the obtained experimental results and to generalize with higher machine rating, Matlab/Simulink software is used.

4.1. Voltage sag

Voltage sag is defined as the momentary variation in voltage magnitude between 10 and 90 percent of nominal value from half cycle to one minute [24]. Voltage sag is created due to various factors such as short circuit faults, starting of motors in same bus, single line to ground faults, etc., Voltage sag is the most common problem in industries, that reduces the efficiency and life span of machine. Also, sag affects the quality of end product in process industries [25]. During voltage sag, due to small time constant, the electromagnetic torque reduces in proportional to square of sag depth (rms value) without changing mechanical torque. This reduction of electrical torque force to operate the motor at a new operating point with speed reduction, thereby exerted torque on motor may fail to drive the load [26]. The test machine is experimentally analyzed by allowing it to experience 10 to 60 percent of three phase single event symmetrical voltage sag for 1 sec, shown in Fig. 7. Similarly, 2.2 kW and 37.5 kW SCIM are analyzed using Matlab/Simulink software with the same percentage of voltage sag for generalization.

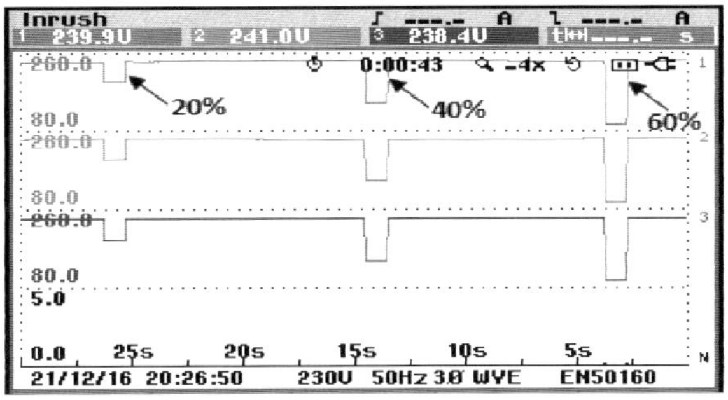

Figure 7. Symmetrical voltage sag (20%, 40% and 60%).

It is observed that, voltage sag induces current transients, but comparatively lesser than the starting and switchover transients. In case of star connected IM, sag creates heavy flow of current (transient) due to magnetic saturation, whereas, the transient is lower in delta connected IM due to absence of magnetic saturation. Similarly, torque of star connected IM gets affected three times more than the delta connected induction motor. Also, the star connected IM experiences more drop in speed than the delta connected motor shown in Fig. 8, which could reduce the quality of the end product and cause heavy loss to industries.

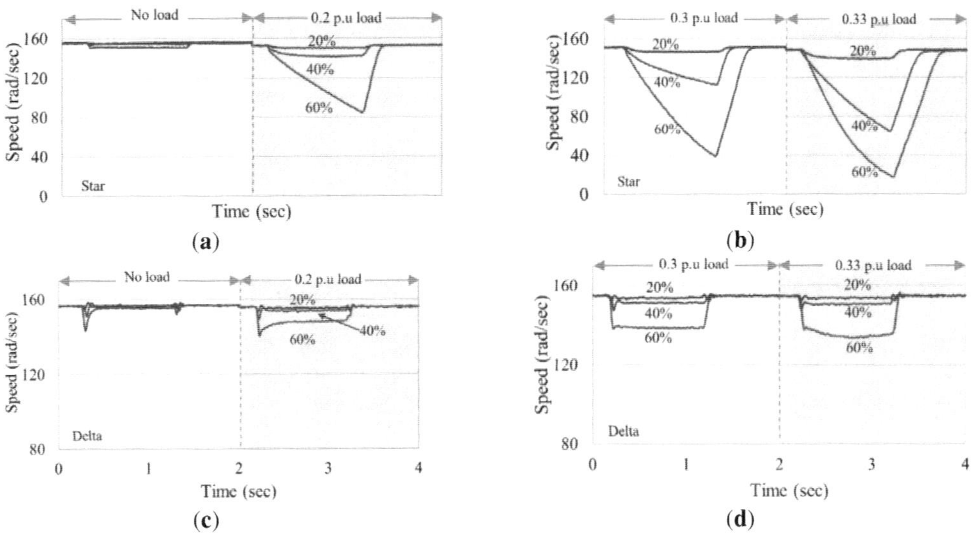

Figure 8. Speed variation of 2.2 kW IM during voltage sag (20%, 40% and 60%): (**a**) Star at No load and 0.2 p.u. load; (**b**) Star at 0.3 p.u. and 0.33 p.u. load; (**c**) Delta at No load and 0.2 p.u. load; (**d**) Delta at 0.3 p.u. and 0.33 p.u. load.

Assuming if 20 percent of speed variation is allowed in a process industry, 2.2 kW SCIM with Y-Δ scheme gets affected at 0.27 p.u. and 0.8 p.u. load with 40 % and 60 % of voltage sag respectively.

Similarly, the 37.5 kW IM operating with Y-Δ scheme gets affected at 0.22 p.u. load on 60 percent voltage sag. Contrastingly, the delta connected IM does not get affected by same percentage of voltage sag, shown in Fig. 9. Overall, IM with Y-Δ scheme is more affected than the delta connected IM. Hence, the switchover point needs to be adjusted for a smooth operation. From the obtained test results, the new switchover point is determined and enforced in the controller.

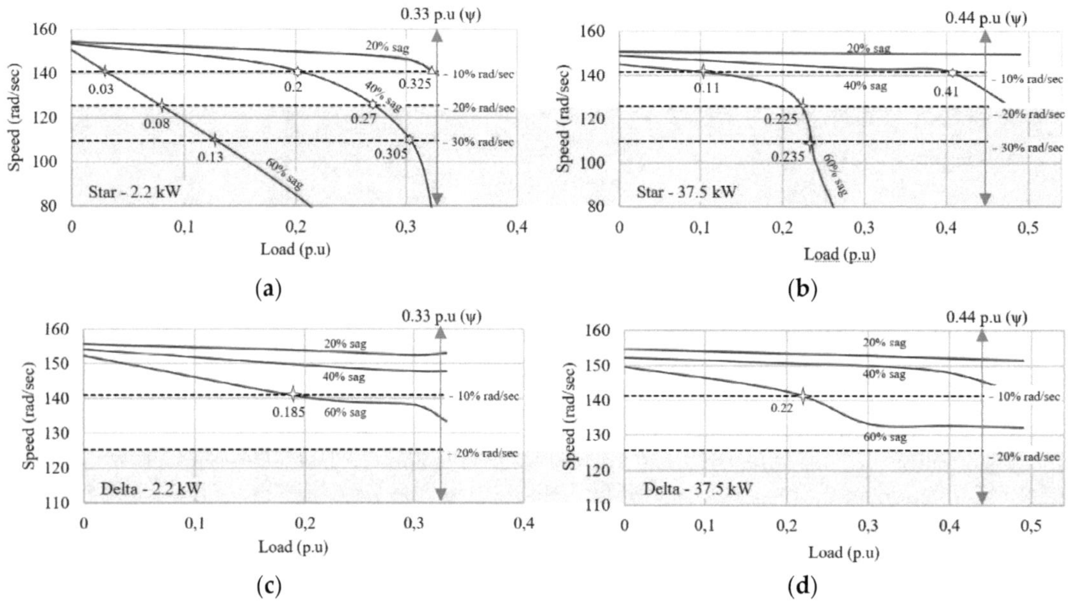

Figure 9. Speed variation of induction motors during voltage sag (20%, 40% and 60%): (**a**) Star connected 2.2 kW IM; (**b**) Delta connected 2.2 kW IM; (**c**) Star connected 37.5 kW IM; (**d**) Delta connected 37.5 kW IM.

4.2. Voltage Variation

An under voltage and over voltage is defined as the variation in rms voltage greater than 10 percent of its nominal value for longer than 1 min [24]. Under voltage fault is the main cause of overloaded circuits and poor voltage regulation that impacts on motor efficiency and lifespan. This under voltage in IM reduces the speed and overheats the windings which extend to insulation deterioration and winding failure. Also, the under-voltage fault lengthens the starting time of machine and raise the current flow in windings [27]. Similarly, overvoltage is the consequence of switching off large loads, tap changing of transformers and poor voltage regulation. Even though the occurrence of overvoltage is not frequent as under voltage, impacts of overvoltage on industrial motors are higher than the under-voltage fault [27]. Performance characteristics of lightly loaded 2.2 kW SCIM have been experimentally analyzed with 20 percent voltage variation on both star and delta connection individually, shown in Fig. 8 and using Matlab/Simulink software the 2.2 kW and 37.5 kW SCIM are analyzed.
The analyses have been performed at light loads (closer to the switch over point), because the aim of this paper is to analyze test machine under various electrical disturbances and modify the switch over point accordingly. The change in current and speed is the main consequence of voltage variation, hence, throughout the course of voltage variation, speed and current are monitored precisely.

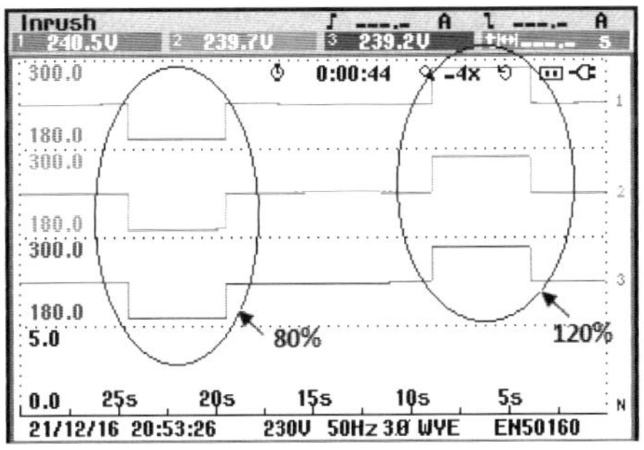

Figure 10. Voltage variation (80% and 120%).

During under voltage (20 %), the speed of star connected IM gets dropped to zero (halted) while trying to run beyond 0.32 p.u. load. Wherein delta connected IM is not affected with the same percentage of under voltage fault. Similarly, during over voltage, IM with Y-Δ scheme provides more beneficial effects such as, improvement in power factor and increase in energy conservation. Delta connected IM with over voltage create more thermal gradient that could even damage motor's stator windings. During voltage variation (under voltage or over voltage) on Y-Δ scheme based IM and normal (Δ) connected IM in the light load region, electromagnetic torque (*T*) and starting torque (T_{st}) of IM are obtained using the steady state torque equations (9) to (12) [28] and result are summarized in Table 3.

$$T = \frac{3}{\omega_s} \frac{{U_{L-L}}^2 \left(\frac{R_2^{|}}{s}\right)}{\left(\frac{R_2^{|}}{s}\right)^2 + \left(X_2^{|}\right)^2} \tag{9}$$

$$s = \frac{\omega_s - \omega_m}{\omega_s} \tag{10}$$

$$T \propto s\,U^2 \tag{11}$$

$$T_{st} = \frac{3}{\omega_s} \frac{{U_{L-L}}^2 R_2^{|}}{\left(R_1 + R_2^{|}\right)^2 + \left(X_1 + X_2^{|}\right)^2} \tag{12}$$

Where, U_{L-L} is the line to line voltage; s is the slip; ω_m is the motor speed and ω_s is the synchronous speed.

4.3. Voltage Unbalance

Voltage unbalance is one of the principal causes of power quality disturbances and it is evaluated in terms of voltage unbalance factor (VUF). The VUF is defined as the ratio of the negative sequence voltage (U_n) to the positive sequence voltage (U_P), usually expressed in percentage as in eqn. (13),(14) and (15) [24].

$$VUF\ (\%) = \frac{U_n}{U_p} \times 100 \tag{13}$$

$$U_p = \frac{1}{3}(U_a + \alpha U_b + \alpha^2 U_c) \tag{14}$$

$$U_n = \frac{1}{3}(U_a + \alpha^2 U_b + \alpha U_c) \tag{15}$$

where, $\alpha = e^{j120^\circ} = \cos 120^\circ + j\sin 120^\circ$, U_a, U_b, U_c are the phase voltage.

Effects of unbalanced voltages are equivalent to the superposition of negative-sequence voltage on the positive-sequence voltage. In the supply voltage, the positive sequence component is the direct function of the motor's efficiency and power factor. Similarly, negative sequence component influences negative sequence current flowing through motor windings that affects the motor torque and tends motor to draw higher current for equivalent mechanical load. Overall, voltage unbalance causes inefficiency and derating on induction motors and higher level of voltage unbalance can result in shaft vibration, failure of winding, failure of bearings, etc., [29]. To analyze machine's performance, induction motor in star and delta connection are operated with unbalance supply. In the laboratory, different value of voltage unbalance is generated using the AC programmable power supply, shown in Fig. 11. Negative sequence and positive sequence component are maintained same for both star and delta connected IM, mentioned in Table 3. The magnitude of voltage unbalance is computed using equation (13), (14) & (15), as well, measured using three phase power quality analyzer. Electrical and mechanical quantities are measured using power quality analyzer for estimating the efficiency and losses and thermal effect on IM is measured using thermal analyzer. Similarly, positive sequence torque (T_p), negative sequence torque (T_n) and net torque (T_{net}) are calculated using the steady state torque equation (16), (17) & (18) [28]. The results are tabularized in Table 3.

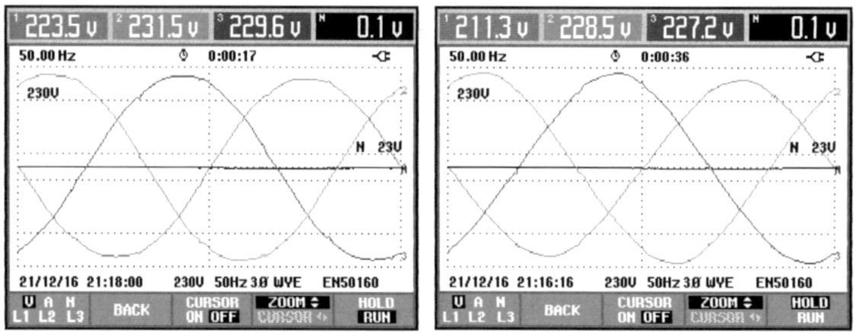

Figure 11. Voltage Unbalance (2.5% and 5%).

Investigating the effect of voltage unbalance results that, speed reduction is more in Y-Δ schemed IM than normal Δ connected IM. Similarly, Y-Δ schemed IM has experienced more shaft vibration and thermal gradient that indirectly entails the life span of machine. Overall, Y-Δ schemed IM are affected more than the delta connected IM during voltage unbalance.

$$T_{net} = T_p + T_n \tag{16}$$

$$T_p = \frac{3}{\omega_s} \frac{U^2_{p(L-L)}\left(\frac{R_2^{|}}{s}\right)}{\left(\frac{R_2^{|}}{s}\right)^2 + \left(X_2^{|}\right)^2} \tag{17}$$

$$T_n = \frac{3}{\omega_s} \frac{U^2_{n(L-L)}\left(\frac{R_2^{|}}{2-s}\right)}{\left(\frac{R_2^{|}}{2-s}\right)^2 + \left(X_2^{|}\right)^2} \tag{18}$$

4.4. Frequency Variation

Frequency variation is the deviance of fundamental frequency from its specified nominal value [32]. At any instant, the frequency of system depends on balance of load and generation. Frequency changes occur due to load switching, faults in transmission system, large block of deload or change in load, large generation going offline, control system tuning issues, etc., [31]. As the supply frequency is directly proportional to motor speed, frequency variation directly influences the speed of motor. For experimental analysis, a lightly loaded 2.2 kW SCIM with individual star and delta connection is energized using AC programmable power supply and for generalizing, two machines (2.2 kW and 37.5 kW) are analyzed using Matlab/Simulink software.

Similar to voltage analysis, frequency variation analysis on test motor is executed closer to switchover point. Speed and current fluctuation are the main consequences of frequency variation. During frequency variation, speed of Y-Δ schemed IM gets affected more than delta connected IM. Since the cooling effect of fan cooled induction motor (usually preferred in small and medium scale industries) is the direct function of rotational speed, the temperature gradient is more in star connected IM than the delta connected IM. Besides, Y-Δ scheme based IM draws less current with under frequency fault which results in loss reduction. On the whole, Y-Δ scheme based IM has less loss, but in energy conservation point of very less energy conservation is attained during under frequency. Likewise, during over frequency, the energy conservation and performance of star and delta connected IM has improved. From overall analysis, it is observed that, under frequency affects system more than the over frequency especially, the star connected IM. The percentage variation from the nominal values are summarized in Table 3.

4.5. Frequency Variation

The loss of one phase in three phase supply is termed as single phasing. An induction motor can suffer a single phasing failure due to any defects such as safety system breakdown, fuse broken, driver breaking, jumpers burning, substation failures, transmission transformer failures, etc., [30]. During single phasing fault, current flow increases on other two phases which creates voltage unbalance in larger magnitude that extends to winding damage [32]. Similarly, during single phasing, stationary machines fails to develop starting torque due to higher influence of negative sequence component in net torque development [33]. Considering this power quality problem, the 2.2 kW Y-Δ schemed IM is analyzed experimentally. Additionally, 2.2 kW and 37.5 kW Y-Δ scheme based IM are analyzed through Matlab/Simulink software for generalizing purpose. Single phasing fault is created experimentally using AC programmable power supply, during single phasing fault, magnitude of voltage unbalance is calculated using equation (13) and measured using three phase power quality analyzer.

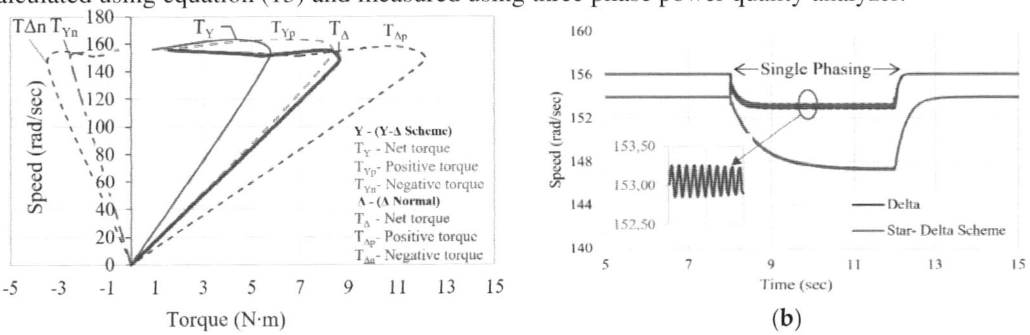

Figure 12. Speed and Torque variation during single phasing on 2.2 kW IM: (**a**) Speed-torque profile with single phasing; (**b**) Speed variation during single phasing at 0.2 p.u.

In additional to this, during single phasing, the positive sequence torque (T_p), negative sequence torque (T_n) and net torque (T_{net}) are computed using the steady state torque equations (16), (17) & (18) and plotted in Fig. 12a. From obtained result, it is realized that unbalance is created in large magnitude in Y-Δ schemed induction motor than the conventional delta connected induction motor. This consequence to high current flow in the stator windings resulting high thermal gradient. Also, it produces torque fluctuation and vibration in large magnitude. Long run of induction motor under single phasing fault could damage the bearing and other parts of the machine. During single phasing fault, speed droop and recovery time are higher in Y-Δ schemed induction motor as shown in Fig. 12b. Moreover, the test machine with Y-Δ scheme fails to drive the load above 0.3 p.u. load, but in delta connected induction motor stalling has not occurred. Also, the variation in the input and output parameters including temperature and vibration is higher than the nominal value in Y-Δ schemed test machine, whereas variation of quantities is in an acceptable limit up to 0.33 p.u. load in delta connected test machine. The percentage variation of input and output quantities from the nominal value during single phasing fault at light loads with star and delta connection are summarized in Table 3.

5. Summary of electrical perturbations

The influence of various electrical perturbations on induction motor under Y-Δ energy saving strategy is thoroughly investigated and compared to conventional delta connected IM. Thc results of investigations are as follows,

(i) Voltage Sag: Induction motor under Y-Δ energy saving strategy experiences speed drop in large magnitude, approximately three times larger than the conventional delta connected system, (Fig. 8). This creates an interruption in process, equipment damage and production losses. Also, sag malfunctions the protection system and even collapse the control system. Hence, this effect should be accounted and the switching point must be varied for continuous service.

(ii) Voltage Variation: Induction motor under Y-Δ energy saving strategy gets affected more than motor with delta connection. Drop in speed and rise in current are higher in case of Y-Δ scheme. The prolonged operation of the machine with under-voltage could interrupt the process and causes enormous loss of output.

Wherein, In the case of an over-voltage malfunction, the device has the advantages of energy conservation. Therefore, to attain uninterrupted operation during voltage variation the switchover point should be adjusted.

(iii) Voltage Unbalance: Y-Δ energy saving strategy experiences speed drop and more vibration affecting the process and causing enormous loss of production. As well as, fault increases the magnitude of line current and de-rates the machine. Therefore, adjusting the switching point for trouble-free service during voltage unbalance is mandatory.

(iv) Frequency Variation: As the motor speed is directly proportional to the input supply frequency, the motor speed is directly influenced by the frequency variation. During under-frequency fault, the line current of IM with Y-Δ scheme increased and there is no energy conservation during under-frequency fault. However, during over-frequency fault, the Y-Δ schemed IM conserves significant amount of energy. Therefore, the fitting of switching point in accordance to the frequency variation is essential.

(v) Single Phasing: The repercussion of single phasing fault in Y-Δ schemed IM is higher than other electrical perturbations. Motor fails to drive the load during single phasing failure and generates more losses for industries. Thus, the machine must be halted during single phasing failure instead of modifying the switchover point.

Table 3. Summary of various PQD on 2.2 kW and 37.5 kW induction motor in percentage variation.

	Operating Characteristics	Results for 2.2 kW IM at 0.32 p.u. load				Results for 37.5 kW IM at 0.4 p.u. load			
		80% (Under voltage)		120% (Over voltage)		80% (Under voltage)		120% (Over voltage)	
		Y-Δ scheme	Normal (Δ)	Y-Δ scheme	Normal (Δ)	Y-Δ scheme	Normal (Δ)	Y-Δ scheme	Normal (Δ)
Voltage Variation	Speed	↓ 7.9	↓ 0.13	↑ 4.16	↑ 0.81	↓ 0.74	↓ 0.33	↑ 0.40	↑ 0.133
	Torque	↑ 17.14	↓ 32.09	↓ 32.89	↓ 7.78	↑ 0.02	↓ 0.04	↓ 0.12	↓ 0.012
	Current	↑ 59.87	↓ 20.14	↓ 5.19	↑ 63.80	↑ 26.54	↓ 16.26	↓ 6.36	↑ 36.53
	Power factor	↓ 5.68	↑ 32.07	↓ 6.81	↓ 22.64	↓ 7.86	↑ 30.03	↓ 4.49	↓ 1.01
	Starting torque	↓ 39	↓ 37.33	↑ 44.08	↑ 47.22	↓ 34.85	↓ 35.96	↑ 44.13	↑ 39.22
	Magnetic noise	Increases	No change	No change	Increased	--			

	Operating Characteristics	Results for 2.2 kW SCIM							
		0.2 p.u. load				0.32 p.u. load			
	VUF	2.5 %		5 %		2.5 %		5 %	
	Un (volts) / Up (volts)	5.86 / 234.04		11.73 / 232.85		5.86 / 234.04		11.73 / 232.85	
		Y- Δ Scheme	Normal (Δ)	Y- Δ Scheme	Normal (Δ)	Y- Δ Scheme	Normal (Δ)	Y- Δ Scheme	Normal (Δ)
Voltage Unbalance	Speed	↓ 0.55	↓ 0.34	↓ 0.96	↓ 0.61	↓ 0.86	↓ 0.34	↓ 1.44	↓ 0.82
	Torque	↑ 33.03	↑ 20.32	↑ 38.72	↑ 35.24	↑ 15.36	↑ 27.4	↑ 26.94	↑ 38.33
	Power factor	↓ 0.41	↓ 3.59	↓ 2.41	↓ 2.12	↑ 1.04	↓ 0.62	↑ 0.23	↓ 2.68
	Temperature	↑ 0.16	↑ 0.69	↑ 10.42	↑ 1.00	↑ 3.10	↑ 0.69	↑ 7.84	↑ 3.56

	Operating Characteristics	Results for 37.5 kW SCIM							
		0.2 p.u. load				0.4 p.u. load			
	VUF	2.5 %		5 %		2.5 %		5 %	
	Un (volts) / Up (volts)	5.597 / 224.4		11.62 / 219.8		5.597 / 224.4		11.62 / 219.8	
		Y- Δ Scheme	Normal (Δ)	Y- Δ Scheme	Normal (Δ)	Y- Δ Scheme	Normal (Δ)	Y- Δ Scheme	Normal (Δ)
Voltage Unbalance	Speed	↓ 0.13	↓ 0.11	↓ 0.25	↓ 0.20	↓ 0.22	↓ 0.18	↓ 0.54	↓ 0.40
	Torque	↑ 2.58	↑ 25.8	↑ 11.48	↑ 84.28	↑ 2	↑ 7.37	↑ 3.04	↑ 28.37
	Power factor	↓ 4.25	↓ 18.36	↓ 14.89	↓ 16.97	↑ 2.32	↓ 8.16	↑ 4.65	↓ 38.77
	Temperature	--							

	Operating Characteristics	Results for 2.2 kW IM at 0.32 p.u. load				Results for 37.5 kW IM at 0.4 p.u. load			
		95% (Under Frequency)		105% (Over Frequency)		95% (Under Frequency)		105% (Over Frequency)	
		Y- Δ Scheme	Normal (Δ)	Y- Δ Scheme	Normal (Δ)	Y- Δ Scheme	Normal (Δ)	Y- Δ Scheme	Normal (Δ)
Frequency Variation	Speed	↓ 5.44	↓ 5.31	↑ 7.61	↑ 6.40	↓ 4.95	↓ 4.92	↑ 5.02	↑ 4.99
	Torque	↓ 2.71	↓ 18.74	↓ 23.34	↓ 18.95	↓ 0.257	↓ 0.281	↓ 0.171	↓ 0.244
	Current	↑ 10.35	↑ 20.22	↑ 10.34	↓ 4.58	↑ 5.52	↑ 10.81	↑ 5.55	↓ 12.01
	Power factor	↓ 0.61	↓ 9.29	↓ 0.44	↑ 12.84	↓ 2.24	↓ 1.01	↓ 1.12	↑ 1.01
	Starting torque	↑ 7.26	↑ 11.07	↓ 4.86	↓ 1.82	↑ 10.80	↑ 12.81	↓ 9.17	↓ 10.67
	Slip	↓ 6.52	↓ 26.86	↓ 19.63	↓ 17.45	↓ 6.43	↓ 12.28	↓ 0.52	↓ 4.76

	Operating Characteristics	Results for 2.2 kW IM				Results for 37.5 kW IM			
		0.2 p.u. load		0.32 p.u. load		0.3 p.u. load		0.4 p.u. load	
		Y-Δ Scheme	Normal (Δ)	Y-Δ Scheme	Normal (Δ)	Y-Δ Scheme	Normal (Δ)	Y-Δ Scheme	Normal (Δ)
Single phasing	Speed	↓ 2.45	↓ 0.95	Stall	↓ 0.95	↓1.20	↓ 0.40	Stall	↓ 0.73
	Torque	↑ 187.57	↑ 26.41	Stall	↑ 13.39	↑ 133.1	↑ 89.0	Stall	↑ 57.4
	Avg. line current	↑ 135.96	↑ 68.67	↑ 293.9	↑ 75.37	↑ 159.2	↑ 144.4	↑ 316.19	↑ 162.92
	Temperature °F	↑ 17.68	↑ 17.68	Stall	Stall	--			

All the above results are in percentage

↑ represents increase from its nominal value

↓ represents decrease from its nominal value

Overall performance of induction motor (2.2 kW and 37.5 kW) under Y-Δ energy conservation scheme and induction motor with normal Δ connection is analyzed under the influence of various power quality disturbances. Consequence of the analysis is tabularized with percentage variation from nominal value in Table 3.
Hence, by considering all these power quality effects in the existing Y-Δ energy conservation scheme an option for readjusting switchover point (ψ) is provided in the proposed control strategy. In view of Fig. 2a and [8], the 2.2 kW IM in star mode can be operated up to 0.33 p.u., but on consideration of various power quality disturbances, IM cannot be operated up to 0.33 p.u. load. Similarly, 37.5 kW IM cannot be operated up to 0.44 p.u. load during PQD. Therefore, a new strategy is developed in controller to provide uninterrupted operation by readjusting the switchover point (ψ), discussed in subsequent subsection. Percentage variation in energy conservation under various PQD from nominal value and change in switchover point (ψ) corresponding to the PQD are presented in Table 4.

Table 4. Percentage variation in energy conservation under various PQD.

PQD	Results for 2.2 kW IM		Results for 37.5 kW IM	
	Switch over point (Ψ)	% Variation	Switch over point (Ψ)	% Variation
Nominal	0.33	--	0.44	--
Under voltage (20%)	0.25	↓ 74.89	0.24	↓ 64.74
Over voltage (20%)	0.61	↑ 140.3	0.64	↑ 108.5
Under frequency (5%)	0.12	↓ 10.6	0.08	No EC
Over frequency (5%)	0.58	↑ 3.475	0.85	↑ 40.38
Unbalance (5%)	0.32	↓ 46.93	0.39	↓ 77.4
Single phasing	0.16	↓ 14.33	0.19	No EC

All the above results are in percentage
↑ represents increase from its nominal value
↓ represents decrease from its nominal value

6. Intelligent Switching Mechanism Intelligent Switching Mechanism

In addition to existing energy conservation scheme discussed in [5], control strategy is embedded with safe restarting and switchover point readjustment feature. This feature facilitates to continue the process without any interruption and to increase the lifespan of machine.

6.1. Safe restarting

From the investigation of the EC system based on a Y-Δ scheme, if the Y-Δ scheme enabled machine is deliberately disturbed by any grid disruptions or by an operator accidentally, it starts again in star contact only. Nevertheless, the star-connected induction motor does not resume at high loads due to low starting torque (about one-third of the based delta motor). Starting the IM with a star contact often increases the start time, leads to thermal stress and lifespan reduction. Therefore, during the process, the IM must resume in the star contact or delta contact by considering the recess load point from past interruption. Hence, the previous operating mode (Y and Δ) storage facility is installed in the microcontroller to resume the motor in star or delta connection. This function is mandatory for the Y-Δ scheme enabled machine to establish the required starting torque. For example, induction motor is interrupted while running at 60 percent of its full load, star-connected machine fails to resume due to insufficient starting torque shown in Fig. 13 with stripped lines, but the machine with delta connection resumes in 23 millisecond shown in a continuous line.

IOP Conf. Series: Materials Science and Engineering **906** (2020) 012017 doi:10.1088/1757-899X/906/1/012017

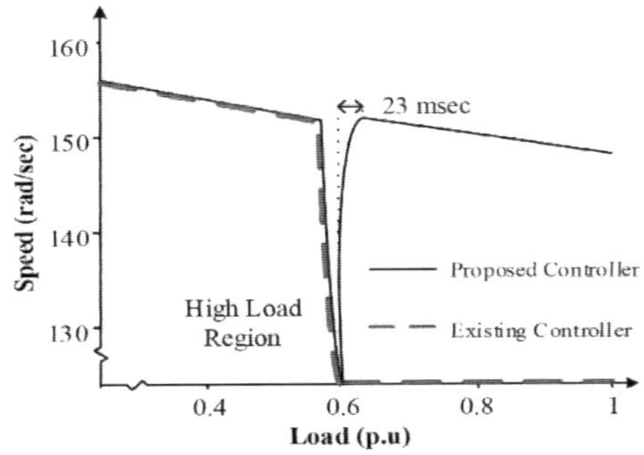

Figure 13. Speed profile of test machine at 0.6 p.u. load during restart.

6.2. Switchover point readjustment

The motor loses its stability during various electrical disturbances and causes loss of production. From the analysis of various power quality issues in Y-Δ based energy conservation scheme, readjustment of switchover point (ψ) is mandatory for obtaining more energy conservation as well as to provide uninterrupted service. Therefore, the preconfigured value of the switchover point will be changed automatically based on recurring electrical disruptions. Automatic readjustment of switchover point is depending upon the quality of power received in industry and it can be readjusted externally by considering the power quality impact on process and end product. Therefore, it is possible to conserve more energy without interrupting the process using the proposed control strategy shown in Fig.14.

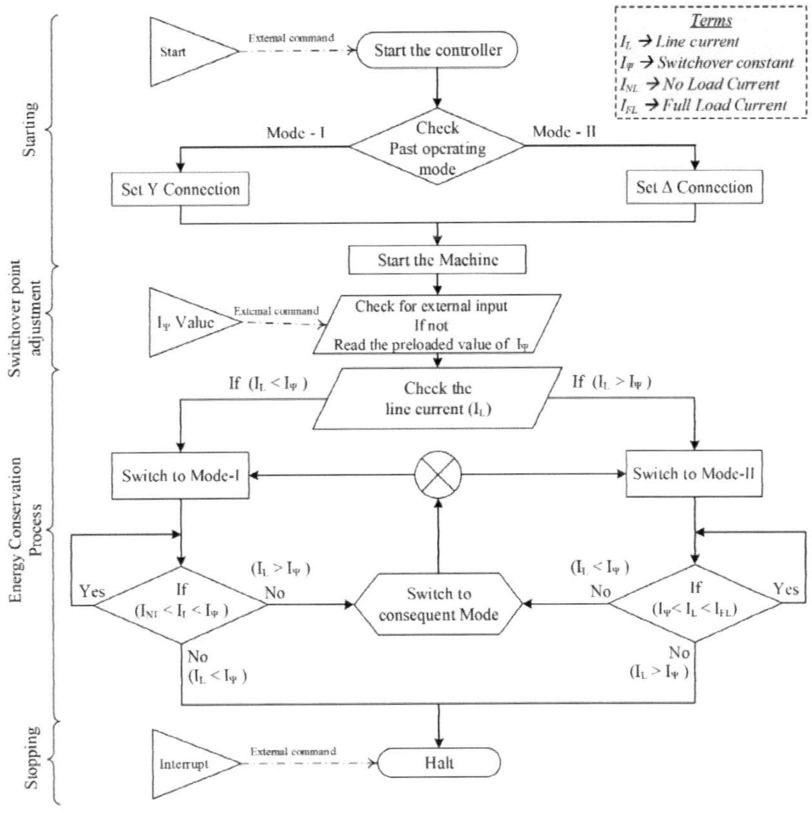

Figure 14. Proposed Intelligent Switching Mechanism for induction motor.

7. Conclusion

Economic energy conservation is an important scenario for any industry, but reliability of preferred energy conservation method is also very important. Hence, the performance of an IM under Y-Δ EC scheme is evaluated and outlined in this paper with the effects of different energy value disruptions. The analyzed results convey that the effects of electrical perturbation cause serious problems in production with an increase in operating costs, motor losses and motor failure. Considering these effects on induction motor, an optimal energy saving strategy is developed with memory storage and switchover point readjustment features, in which, the switchover adjustment may depend upon the quality of power received in the industry and the sensitivity of the end product. Therefore, no matter how much resources are used to conserve energy, proactivity analyzing and responding to anticipation risk is extremely important.

References

1. N. Kumar, T. R. Chelliah and S. P. Srivastava, "Energy conservation study on induction motors using MATLAB/Simulink for enhancing electric machinery courses," Proceedings of IEEE International Conference on Teaching, Assessment, and Learning for Engineering (TALE) 2012, Hong Kong, 2012, pp. H4B-10-H4B-16, doi: 10.1109/TALE.2012.6360358.
2. Gnacinski, P. Energy saving work of frequency controlled induction cage machine. *Energy Conversion and Management*, **2007**; 48, 919–26. DOI:10.1016/j.enconman.2006.08.002.
3. Kothals-Altes, W. Motor winding; U.S. Patent 1 267 232, May 21, 1918.
4. Jannati, M.; et. al, A review on Variable Speed Control techniques for efficient control of Single-Phase Induction Motors: Evolution, classification, comparison. *Renewable and Sustainable Energy Reviews*, Available online 16 November 2016, ISSN 1364-0321.
5. Alger, P.L.; Hughes, A. New 3-phase winding of low m.m.f. harmonic content. *Proceedings of the Institution of Electrical Engineers*, **1971**, 118, 607-608. DOI: 10.1049/piee.1970.0293
6. Taylor, N.R.; Taylor, P.A. Automatic load seeking control for a motor; U.S. Patent 4 413 218, Nov. 1, 1983.
7. Gjota, R. Winding arrangement of a stator and/or rotor of a three-phase generator or electromotor with improved performances. U.S. Patent 4 710 661, Dec. 1, **1987**.
8. Ferreira, F.J.T.E.; De Almeida, T. Method for in-field evaluation of the stator winding connection of three-phase induction motors to maximize efficiency and power factor. *IEEE Transactions on Energy Conversion* **2006**, 21, 370–379. DOI: 10.1109/TEC.2006.874248.
9. Kostic, M.M.; Radakovic, J. Induction Motors with YY / Δ Connection Change for Efficiency and Power Factor Increasing at Partial Loads. *Journal of Electrical Engineering*, **2006**, 19, 85–98.
10. Blinov, A.; Vinnikov, D.; Jalakas, T. Loss calculation methods of half-bridge square-wave inverters. *Electronics and Electrical Engineering*, **2011**; 7, 9–14. doi:10.5755/j01.eee.113.7.604.
11. Marouani, K.; Khaldi, M.; Khoucha, F.; Kheloui, A. Switching losses and harmonic currents evaluation of PWM techniques for VSI-fed dual stator induction motor drive. *2009 17th Mediterranean Conference on Control and Automation*, Thessaloniki, **2009**, pp. 1492-1497.
12. Sakthivel, V.P.; Subramanian, S. On-site efficiency evaluation of three-phase induction motor based on particle swarm optimization. *Energy*, **2011**, 36 (3), 1713-1720, ISSN 0360-5442.
13. Lu, B.; Habetler, T.G.; Harley, R.G. A survey of efficiency-estimation methods for in-service induction motors. *IEEE Trans Ind Appl*, **2006**, 42 (4), pp. 924–933.
14. Sivarani, T.S. Jawhar J.S., Cumar, A.K. Intensive random carrier pulse width modulation for induction motor drives based on hopping between discrete carrier frequencies. *IET Power Electron.*, **2016**, 9, pp. 1–10.

15. Singh, R.R.; Thanga Raj, C. Enforcement of cost-effective energy conservation on single-fed asynchronous machine using a novel switching strategy, *Energy* **2017**, 126, pp. 179-191, https://doi.org/10.1016/j.energy.2017.03.003.
16. Mohan, H.; Singh, R.R.; Agrawal, K.; Kumar, B.A.; Thanga Raj, C. Energy conservation on induction motors drives with immune control strategy using full spectrum simulator, *2017 International Conference on Computer Communication and Informatics (ICCCI)*, Coimbatore, **2017**, pp. 1-6. doi: 10.1109/ICCCI.2017.8117800
17. Woronowicz, K.; Palka, R. Optimized thrust control of linear induction motors by a compensation approach. *International Journal of Applied Electromagnetics and Mechanics* **2004**, 19(1-4), pp. 533-536.
18. Wardach, M.; Bonislawski, M; Palka, R.; Paplicki, P.; Prajzendanc, P. Hybrid Excited Synchronous Machine with Wireless Supply Control System. *Energies* **2019**, 12, 3153, doi:10.3390/en12163153.
19. Paplicki, P.; Wardach, M.; Bonisławski, M.; Palka, R. Simulation and experimental results of hybrid electric machine with a novel flux control strategy. *Arch. Electr. Eng.* **2015**, 64, 37–51.
20. Christopoulos, G.A.; Zafiris, A.; Safacas, A.N. Energy savings and operation improvement of rotating cement kiln by the implementation of a unique new drive system. *IET Electr. Power Appl.*, **2016**, 10 (2), pp. 101–109.
21. IEEE Standard Test Procedure for Polyphase Induction Motors and Generators. IEEE Standard 112TM-2004, 2004; 1-109. DOI: 10.1109/IEEESTD.1991.114383.
22. EIA report. Average Price of Electricity to Ultimate Customers by End-Use Sector. October 2015 and 2014 (Cents per Kilowatt-hour) electric power monthly data for October 2015. Release date: December 24, 2015.
23. Manish Kumar Saini, Rajiv Kapoor. Classification of power quality events – A review. *International Journal of Electrical Power & Energy Systems* **2012**; 43: 11-19. DOI:10.1016/j.ijepes.2012.04.045
24. IEEE Recommended Practice for Monitoring Electric Power Quality. IEEE Standard 1159TM-2009, 2009; 1-76. DOI: 10.1109/IEEESTD.1995.79050
25. Rolán, A.; Córcoles, F.; Pedra, J.; Monjo, L.; Bogarra, S. Testing of three-phase equipment under voltage sags. *IET Electr. Power Appl.*, **2015**, 9 (4), pp. 287–296.
26. Milanovic, J.V.; Aung, M.T.; Vegunta S.C. The Influence of Induction Motors on Voltage Sag Propagation — Part I: Accounting for the Change in Sag Characteristics. *IEEE Transactions on Power Delivery*, **2008**, 23, 1063–1071. DOI: 10.1109/TPWRD.2007.915846
27. Wallace, A.; Von Jouanne, A.; Wiedenbrug, E.; Ünsal, A.; Andrews, P. The Effects of Nonideal Voltage and Winding Conditions on the Performance of Induction Motors. *Electr. Power Components Syst.*, 2001, 29, 543–554. DOI:10.1080/153250001300338772.
28. Dubey, G.K., Fundamentals of electrical drives. **2002**, CRC press.
29. Reineri, C.A.; Gómez, J.C.; Balaguer, E.B.; Morcos, M.M. Experimental Study of Induction Motor Performance with Unbalanced Supply. *Electr. Power Components Syst.*, **2006**, 34, 817–829. DOI:10.1080/15325000500488636.
30. NEMA Standards Information Guide for General Purpose Industrial AC Small and Medium Squirrel-Cage Induction Motor Standards. NEMA Standards Publication Condensed MG 1-2007.
31. Bonnett, A.H. The impact that voltage and frequency variations have on AC induction motor performance and life in accordance with NEMA MG-1 standards. *Pulp and Paper Industry Conference* (Cat. No.99CH36338), **1999**. DOI: 10.1109/PAPCON.1999.779341.
32. Agamloh, E.B.; Peele, S.; Grappe, J. Induction Motor Single-Phasing Performance Under Distribution Feeder Recloser Operations. *IEEE Transactions on Industry Applications*, **2014**, 50 (2), pp. 1568-1576. DOI:10.1109/TIA.2013.2279188

33. Kersting, W.H. Causes and Effects of Single-Phasing Induction Motors. *IEEE Transactions on Industry Applications*, **2005**, 41, 1499–1505. DOI: 10.1109/TIA.2005.857467.

Study of RCD on Industrial Commercial and Residential Electrical Safety – A Hazard Awareness

Preethi Devadas[1], Rathi Abeth[2], Shruthi Singh[3], Abeth Siluvairajan[4], Raja Singh[5]

[1]JIS Electrical Trading LLC, Dubai.
[2]Sahara Educations, Dubai.
[3]Department of Electronics and Communication, PSNCET, Tirunelveli, India.
[4]Department of Artificial Intelligence, Dubai Electricity and Water Authority, Dubai.
[5]Department of Energy and Power Electronics, VIT Vellore, India.
Correspondence: rathiabeth@gmail.com

Abstract. Every human relies on electricity to power the operation and comfort of their businesses and homes. And they expect their electrical installations to be secure, to protect people from the risk of shock or electrocution, and to protect buildings against the risk of electrical fire. Although the municipal laws on electrical health and defensive equipment continue to improve, many risks still remain today. Hence the paper reflects on the real development in the electrical safety industry through innovative engineering technologies covering all sorts of industrial, business, residential safety. Thus, during the electrical faults, the paper studies the function and importance of highly sensitive residual current device (RCD) and its defensive techniques. RCD is a protective device that detect the electrical faults and switch off the electricity automatically.

Keywords: electrical shock; fire hazards; electrical safety; RCD

1. Introduction

Electromechanical devices could create an electrical shock and fire hazards by their own existence. During the construction of these devices one of the most critical aspects is to add effective protection against electric shocks and fire hazards to them. Residual Current Device (RCD) that can be referred to as a life-saving system intended to protect from lethal electric shock if anyone hit something live, like a bare cable. In addition, which can provide electronic fire safety. RCDs provide a specific degree of security that regular circuit breakers and fuses can't do. For example, when mowing the grass, you cut through the cable and inadvertently hit the uncovered active wires or a defective device overheats allowing electrical current to spill to earth. The electrical current flowing through circuits is tracked by the residual device for securing it. During the tracking process if electricity is detected flowing in unexpected direction, automatically RCD can turn off the particular circuit reducing the risk of serious injury and death.

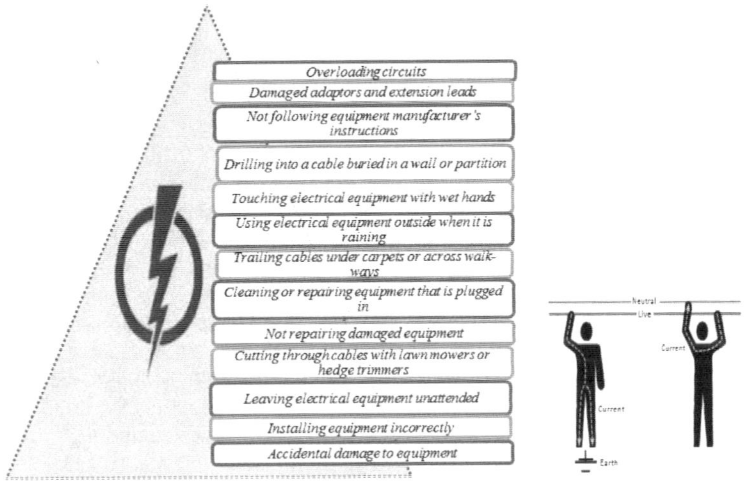

Figure 1. Electrical hazards and current flow in direct contact

IOP Conf. Series: Materials Science and Engineering **906** (2020) 012018 doi:10.1088/1757-899X/906/1/012018

Electrical devices like television, battery chargers, toasters, computers, electrical cookers, washing machines used to draw electrical power. In a house electrical installation comprised of three components (i) fuse or circuit breakers in main switch, (ii) lighting point and the permanent wiring to the power, and (iii) related power outlets, lights, switches on load side. However, there are certain possible dangers mentioned in Fig 1, even the fuse and circuit breakers are installed to the residential load.

1.1 Significance of electric shock protection for human life safety

The influence of electric shock on human body is complex, as they depend on both the time it flows and the current value. Ventricular fibrillation is the key reason people die from electric shock when the heart fails its usual pulse of contraction. The electric could happen directly, when direct contact with bare conductor or when contacting the equipment with leakage current. For instance, a current may be felt in the body of as small as 0.5 mA (0.0005 A), but usually the affected person will not be shocked. Currents between 0.5 mA and 10 mA are likely to be uncomfortable, triggering involuntary muscle contractions, but typically do not have any harmful effect. A 10 mA current (0.01 A) is considered the let-go threshold. When current is greater than this, it is not being possible to loosen the grip on an electrically active device. Currents that last for a long time between about 10 mA and 50 mA may cause severe breathing trouble, disrupt heart function, and muscle contractions. Considerable currents of more than 50 mA (0.05 A) can results burns, breathing problems, cardiac arrest and damages to tissue. With the size and length of the current the probability of this occurring increases. Alternating current (a.c.) is high risky than direct current (d.c.) because it affects the heart's natural rhythm more often.

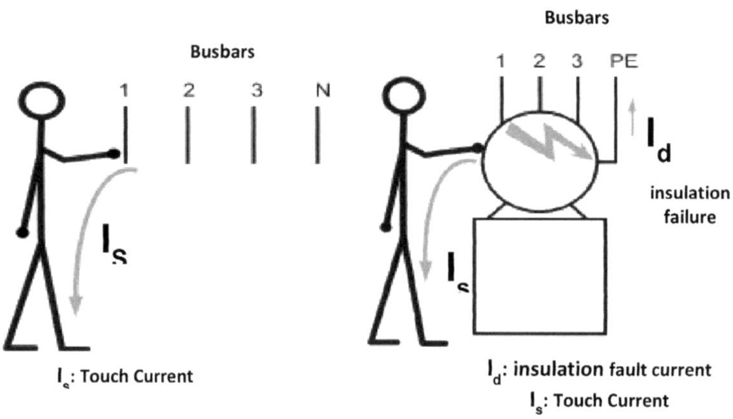

Figure 2. Effect of electric shock during direct and indirect contact

Standard EN 61140[1] discuss the fundamental safety rules for high voltage system and low voltage systems. Especially, hazardous live parts which is permitted and not permitted to accessible has been explained. Similarly, in IEC / HD 60364 standard defined specific rules on protection against electrical shock for low-voltage systems and in HD 60364-4-41[2] low voltage electrical installation is described. Protection against basic contact and fault protection must be employed for safety purpose. Indirect contact protection is mostly rectified through auto cut-off supply, which is the most common form of defence against indirect contact. In the case of a line-to-earth failure, hazardous voltage can appear at the current-using equipment's metal enclosure, and due to indirect contact there is a risk of electrocution (Figure 2). An RCD can be employed as uncoupling device. These systems technically provide safety precautions in case of direct and indirect contact with the live conductor. A person can be rescued from fatal electric shock through only the reliable high sensitivity RCD by providing residual current rated does not exceeding 30 mA, that is assumed as ventricular fibrillation threshold [3]. In many applications contemporary current-use device uses power converters for controlling the power absorbed and regulate the speed of the motor. Power converters are used at industries as well as household appliances. Nowadays, residual current devices (RCDs) are widely adopted to protect from electric shock in low-voltage power systems.

1.2 Necessity of preventing electrical faults and fire hazards

Electric fires in cables and connections are caused by electric arcs, earth leakage current, overloads and short circuits. Two types of faults are possible to occur within the electrical circuit. First if a live component comes in contact with the earthed metal works causes earth fault. It usually happens if a washing machine motors fails with insulation failure. Second, when a live conductor comes in contact with the neutral conductor may leads to short circuit current. This faults should be quickly detected and terminated because it may cause drastic heating and damages the electrical installation will result in set fire. The country wise registered accident data with their current are shown in Figure 3.

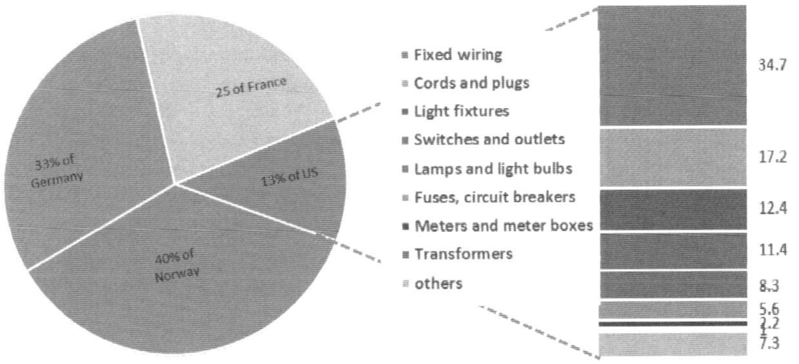

Figure 3. Proportions of electrical fires **[4]** and causes of electrical distribution **[7], [8]**

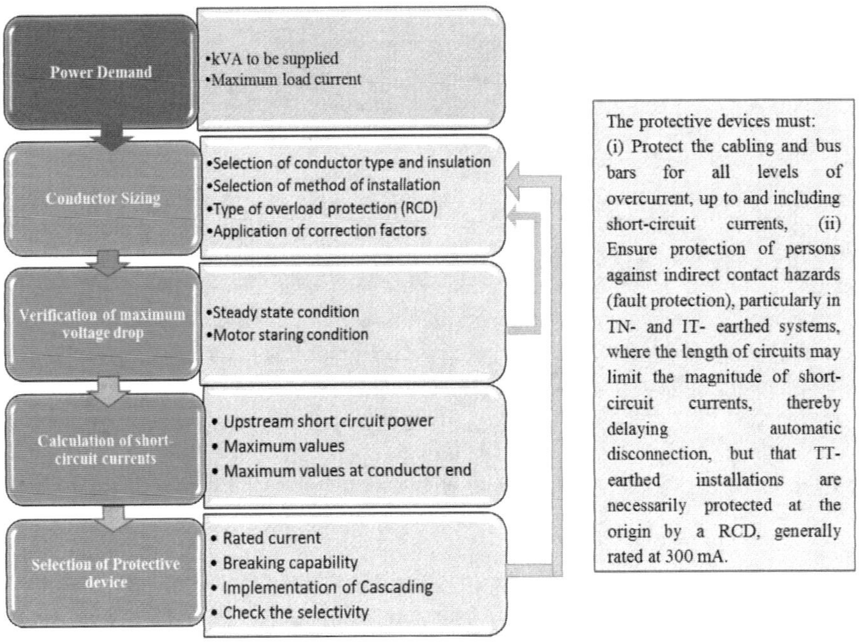

Figure 4: Flow chart of selection of overcurrent protective device

Every year in Europe 2,000,000 fires are registered, causing more than 4,000 deaths, and 70,000 people are hospitalized because of serious injuries. On average, buildings account for 90 percent of fires in the EU. Electricity is a source of domestic fires which is reported quite frequently in [3]. Every year there are 280000 electrical fires in homes [4]. An electrical installation organization [4] reported the electrical fires proportions of different countries based up on the region and the investigative methods; 13 percent in the United States, 25 percent in France, 33 percent in Germany and 40 percent in Norway. In addition, according to the National Fire Protection Association's new statistics [5]; 41200 home structure fires are only due to 'electrical delivery' each year. Earlier figures compiled by FEMA [6] revealed very similar results: the fifth-ranking cause of fires was electrical distribution, the fourth-ranking cause of fire deaths and the second-ranking cause of property damage. The proportions of electrical fires and the causes of electrical distribution are depicted in Figure 3 and the preventive methods are imparted in Figure 4.

2. Fault protection devices for Electrical Hazards

In general, there are various types of the safety devices used in electrical and electronic circuits. Among these, safety devices used to avoid electrical hazards are fuses and circuit breakers, residual current circuit breaker (RCCB), residual current circuit breaker with overload protection (RCCBO), earth leakage circuit breaker (ELCB), and residual current device (RCD) as shown in Figure 5.

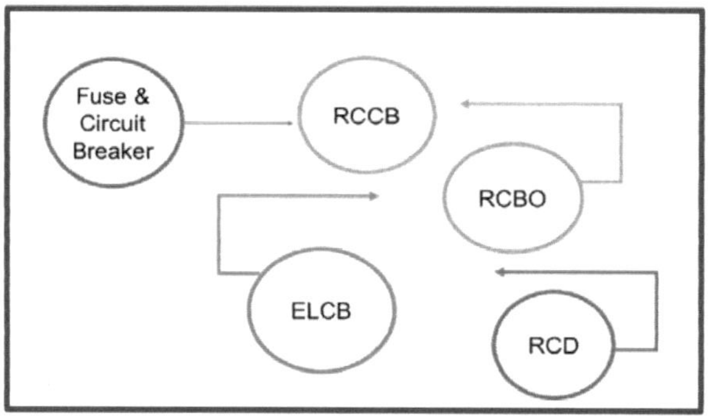

Figure 5: Different types of protection devices

2.1 Fuses and circuit-breakers:

Contrary to common opinion, the fuse in an appliance socket is not intended to protect the device from overload but to protect the flexible wire that provides it from overheating in the event of failure (i.e. short circuit or earth failure). It should be consistent about what and not what a wire fuse does. If a device develops a fault the following problems may happens: A current that is too big causes the fuse to melt, (ii) prevents current flow, (iii) prevents flex overheating and causes fire, (iv) prevents further damage to the equipment. An integrated electromagnetic trip switch could secure the device itself. Consequently, as a result of direct contact, fuses and circuit breakers cannot provide protection against the very tiny electrical currents that pass through the body to earth. Residual current devices can afford this safety as defined in this technical article provided they have been selected correctly.

2.2 Residual current circuit breaker (RCCB):

It is a mechanized switch together with the tripping feature of residual current connected in it. Essentially, it can only sever the circuit when leakage current to the earth. But, short circuit fault or over current cannot be detected by the RCCBs they need to be attached to a fuse or MCB (Miniature Circuit Breaker) in line. RCCBs usually have a break up power and fault-making of 1 kA. This means that if it is a fault to Earth they can accommodate a 1 kA fault on their own. The Wiring Laws allow other tools to provide protection for a line to neutral short circuits and overloads. The system used for protection against short circuits could boost the RCCB's short circuit rating when working together. RCCBs provide protection against earth leakage but a big point to bear in mind when installing them it should always be mounted in combination with a properly designed SCPD (Short Circuit Protective System).

2.3 Residual current circuit breaker with overload protection (RCBO):
It is also a residual current system with a built-in MCB. RCBO is equivalent to residual current circuit breaker and MCB. Residual current circuit breaker together with overload protection have a significant act in (i) short circuit current and overload protection, (ii) earth fault current protection. The RCBO can be used in each one of the circuit, so that if fault occur in one circuit it does not affect the subsequent circuits.

2.4 Earth leakage circuit breaker (ELCB):
This circuit breaker is mainly designed mainly for industrial requirement. It allows the selectivity between different circuit breakers through adjusting the threshold of tripping delay and residual current. Its main function is to protect from ground breaker fault and works with combination with circuit breaker. ELCB follows the IEC 62020, standard which focus on monitoring and isolation of the network. It also shields human lives from electric shocks. The primary usage is for the manufactures who need the tripping time in the production chain.

2.5 Residual current device (RCD):
An additional direct contact protection is provided by RCD. It measures the current size of the neutral and live conductors. Also the circuit of the residual current device is design to easily detachable following the size of the threshold size. The RCDs reacts within 0.04 seconds and it is rated 30mA. If anyone touches the RCD circuit directly one do experience the nonlethal electric shock.

3. Critical review on high sensitivity residual current device (RCD)

High sensitive residual current device uses residual current for its operation at rated 30 mA or less, if offers many protections against the electrical hazards from interaction with live parts. Different reasons to fall in fault is considered which includes (i) improper maintenance, (ii) abrasion and bending of connecting leads (iii) insulation fault (natural tear and wear) (iv) carelessness (v) water immersion and (vi) unintentional touching. This rapid tripping device identify the residual currents to the earth and detach the power supply automatically in order to protect human or animal lives.

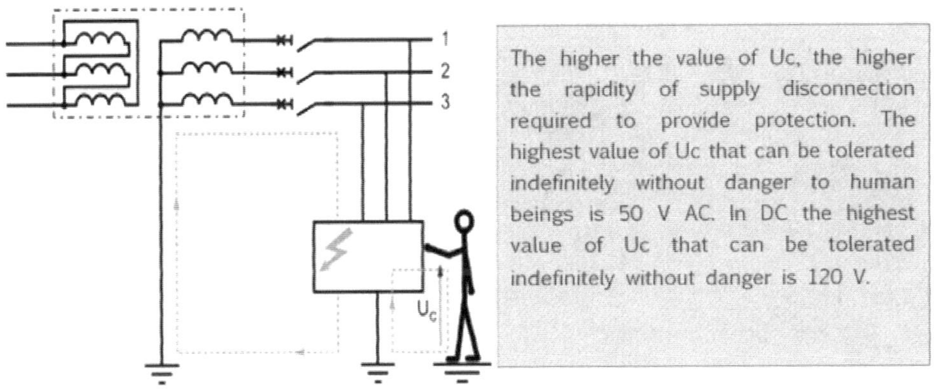

Figure 6: Illustration of dangerous touch voltage Uc

RCD devices calculates the residual current and identify the difference in the input current and output current leaving the system with earthed source. To provide the rapid security, it follows the IEC 61008 and IEC 61009 standard. Therefore, it offers extra protection where standard protection entails failures like human error, damaged or old insulation etc. It can interrupt the current even in the device failure and terms as ultimate protection device. For instance, in the case of greenhouse

IOP Conf. Series: Materials Science and Engineering **906** (2020) 012018 doi:10.1088/1757-899X/906/1/012018

possibility of proportion of electrocutions is high, wherein the people are in contact with earth. A low track of resistance can be deployed by damp ground for the passage of electric shock through the body. In other way, RCD can be attached with the fuse box to have in the existing system. Moreover, there are plug-in RCDs available that are cost-effective, and can literally save lives. Residual current instruments (RCDs) are usually used to protect against over direct and indirect interactions.

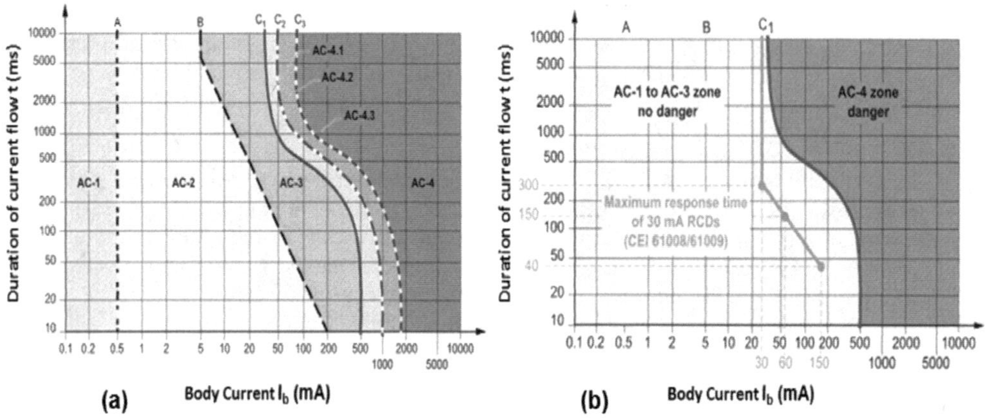

AC-1 zone: Imperceptible, AC-2 zone: Perceptible, AC-3 zone: Reversible effects: muscular contraction, AC-4 zone: Possibility of irreversible effects, AC-4-1 zone: Up to 5% probability of heart fibrillation, AC-4-2 zone: Up to 50% probability of heart fibrillation, AC-4-3 zone: More than 50% probability of heart fibrillation, A curve: Threshold of perception of current, B curve: Threshold of muscular reactions, Ci curve: Ventricular fibrillation unlikely to happen, C2 curve: Threshold of 5% probability of ventricular fibrillation, C3 curve: Threshold of 50% probability of ventricular fibrillation

Figure 7: (a) Body sensitivity curve and (b) High sensitivity RCD tripping curve

3.1 Principle of Residual Current Device

The flux produced in the toroidal magnetic core at any moment depends on the sum of the vector currents, if one path of current (I_1) moment is considered positive, then the other direction of current (I_2) will be negative. The detailed explanation is shown in Figure 8.

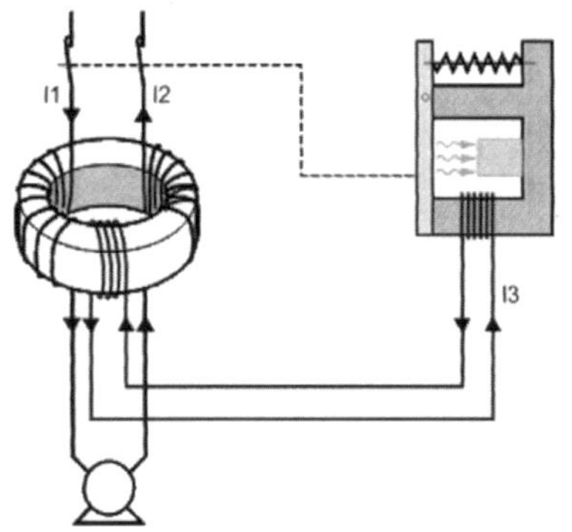

In a normally healthy circuit, there will be no flux in the magnetic core, and zero e.m.f. in its coil. An earth-fault current Id will pass through the core to the fault, but will return to the source via the earth, or via protective conductors in a TN-earthed system. The current balance in the conductors passing through the magnetic core therefore no longer exists, and the difference gives rise to a magnetic flux in the core. The difference current is known as the "residual" current and the principle is referred to as the "residual current" principle.

Figure 8: Working principle of RCD

3.2 Technologies of Residual Current Device

Two technologies such as voltage Independent (VI) technology and voltage dependent (VD) technology are implemented by the RCDs. In which technology, whatever the line voltage may, but the voltage drop is < 50 V, the RCD will migrate in the event that the residual current exceeds the operating value. Using these technologies, When the voltage of the line is too low, the RCD would be able to sense but not to move, because the electrical circuit and the tripping devices must be driven. The supply line's minimum voltage is 50 V, to allow for tripping as shown in Figure 9.

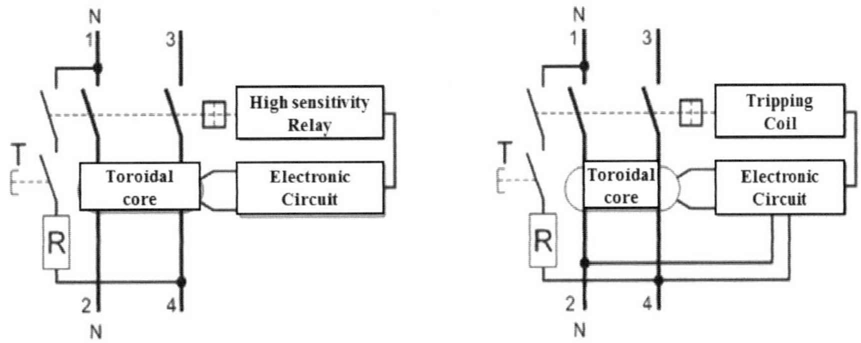

In the Voltage Independent (VI) technology, the resultant alternating flux in the core induces an e.m.f. in its coil, so that a current I3 flows in the tripping device operating coil. If the residual current exceeds the value required to operate the tripping device either directly or via an electronic relay, then the associated circuit breaker will trip. The energy necessary to trip mechanism comes directly from residual current, independent of line voltage.

With the Voltage Dependent (VD) technology, the summation current transformer measures residual current, an electronic circuit detects the tripping level and then sends an order to tripping unit to open the protected circuit. In this case, the power supply of electronic circuit and energy for tripping unit come from the line voltage.

Figure 9: Voltage Independent and Voltage Dependant Technology

3.3 Types of RCDs

An appropriate safety device with a suitable level of sensitivity should be introduced and applied to identify a line to earth fault and insulation fault which may leads to fire hazard or electrical shock. The operating algorithm of the protective system depends on the power network, special signals, detection of nonlinear sinusoidal alternating earth faulty current, DC current, distorted voltages, current comprising component of low frequency system, detection of an arc failure, higher order harmonics, power electronic converter circuits [7] – [23]. Different types and its description is depicted in Figure 10.

The international standards [24] – [26] consider the following types of RCDs in accordance of their sensitivity and the shape of waveform:

- AC-type: only intended for residual sinusoidal currents of 50/60 Hz,
- A-type: intended for waveforms similar to AC-type RCDs and residual DC-pulsing currents with a smooth DC portion up to 6 mA;
- F-type: designed for waveforms, the same as for A-type RCDs (but the smooth DC portion can be increased up to 10 mA) and mixed-frequency residual currents; the RCD must be provided from one single phase.
- B-type: designed for residual currents: designed for sinusoidal waveforms up to 1000 Hz, pulsating DC, smooth DC, and also mixed frequency. It can be concluded that nowadays A-type RCDs are the most common RCDs used in low voltage systems [27, 28].

Figure 10: Different types of RCDs

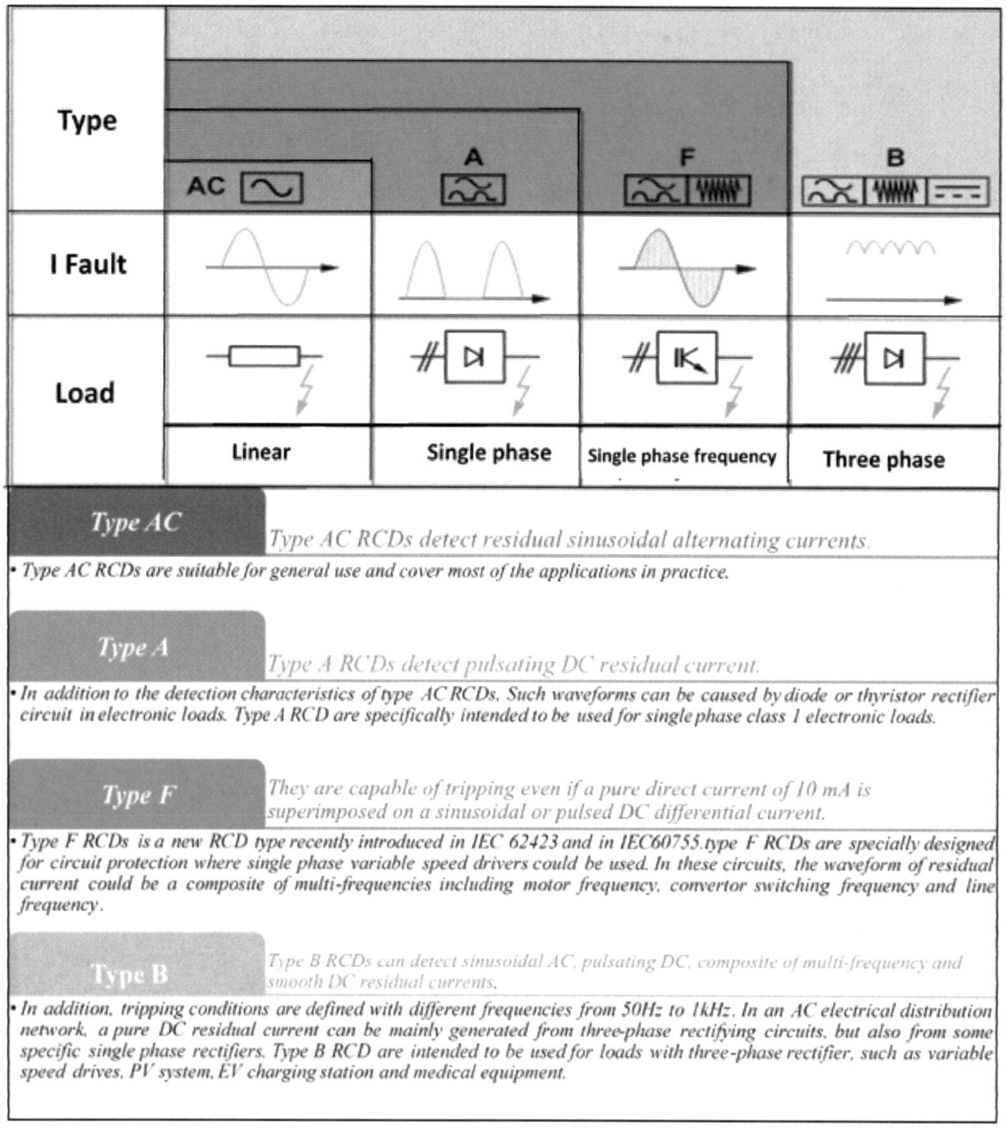

3.4 Selectivity of residual current device

Selectivity of residual current device is accomplished either by delaying time or by splitting circuits which are then separately or jointly protected, or by combining the two methods together. Such selectivity protects any RCD tripping, rather than the one instantly upstream of a fault detection. With the device presently offered, selectivity at different distribution levels is achievable, systems for automated disconnection in case of indirect contact danger are placed on distribution panel boards and on security of electrical appliances, along with added safety against direct interaction hazards.

3.4.1 Selectivity among RCDs

RCD selection is accomplished by leveraging the different levels of uniform sensitivity: 30 mA, 100 mA, 300 mA and 1 A as shown in Figure 11.

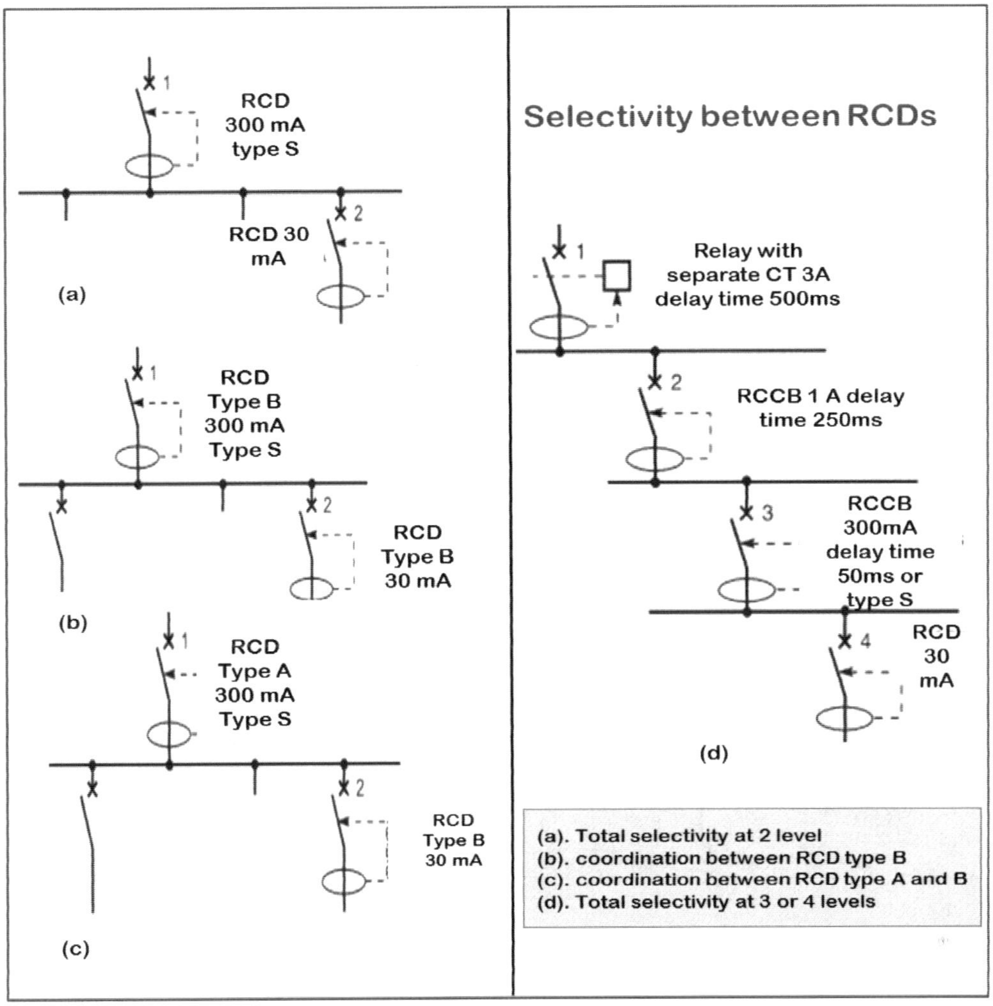

Figure 11: Different levels of selectivity with RCDs

a) *Selectivity at 2 levels (Figure 10a.)*: Level-1: RCD time delayed configuration of either I (industrial devices) or S (domestic devices) for defect safety. Level-2: High sensitivity RCD on at-risk outlet socket products (washing machines, heater, etc.).

b) *Selectivity at 2 levels (Coordination with RCDs type B Figure 10b)*: If earth-leakage DC fault current is present, an RCD type B must be used to protect against electric shock. The upstream RCD should not be blinded by the possible DC residual current in this case and should have its usual protection when any residual fault current exists in some other part of the circuits. For instance, in Figure 10(b), the 30mA RCD type B at level-2 might have had a maximum DC tripping threshold of (2*I), according RCD brand standard IEC 62423. This ensures that type B 30mA RCD should require residual current to pass through nearly 60mA DC before tripping and the upstream RCD will not lack any of those output with both the existence of this significantly high DC level. Therefore, it is always suggested using a level-1 RCD type B to prevent any blinding effect from DC current, as shown in Figure 10(b).

c) *Selectivity at 2 levels (another possibility Figure 10c.)*: Many class A RCDs are eligible in the industry to never be susceptible to residual DC current of up to 60mA. A 30 mA RCD type B could be used upstream, but with no risk of ignorance, as seen in Figure 10c. Its type

AC and type A prevention traits are guaranteed though in the existence of 60 mA smooth DC residual current.

d) *Selectivity at 3 or 4 levels (Figure 10d.)*: Level 1: RCD time-delayed (setting III), level-2: RCD time-delayed (setting II), level-3: RCD time-delayed (setting I), or type S, level-4: RCD instantaneous. The upstream RCCB architecture must abide by the selectivity laws and take into consideration all downstream ground leakage currents.

4. Advanced Trends in RCD Type B for Frequency Converter

Safety is the important aspects in VSD, the design and development of USD has various factor including protection, site, rating, cooling, emc filter, etc., among them the protection of VSD is the important as it could same the VSD from power interruption and the same for the other systems connected to it intentionally or un intentionally from electrical hazard. Also consideration should be taken to reducing hazards arising from fairly predictable mishandling of the variable velocity drives that may occur during their lifetime. And the other is defect safety that safeguards the user from an electrical shock in coarse of simple fault. Usually, in the variable speed drives the protections are provided by protective connection in earth or by the use of plastic enclosure.

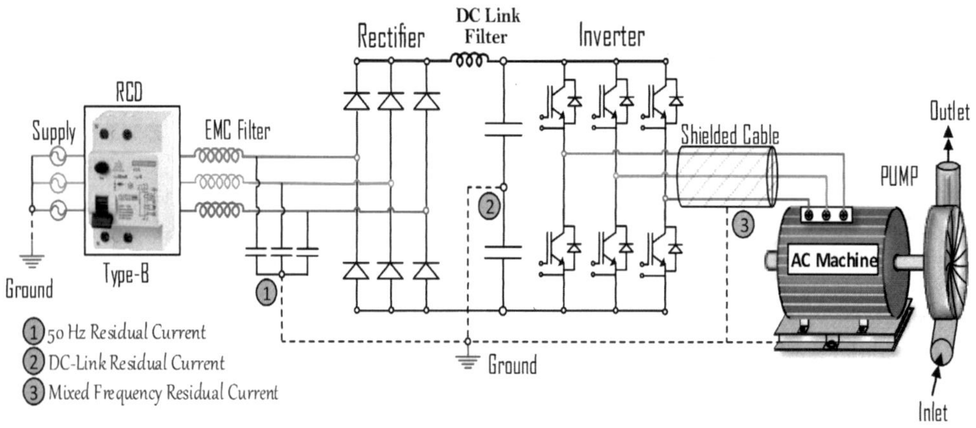

Figure 12: RCD for Frequency Converter

Additionally, protective galvanic insulation is provided for variable sped drives of the power circuit components and attainable control components. It is to ensure that no harmful voltage will appear on the control lines. This would potentially cause lethal contact with the control lines, and also destroy the equipment. The IEC / EN61800-5-1 international / European standard describes the electric shock protection to variable speed drives. Enclosure rating of FC offers protection from damage or loss to the contact. A better cage rating than IP-21 avoids personal injury associated with the contact. Compliance FC's enclosure rating provides protection from contact damage or injury. A better enclosure ranking than IP-21 avoids contact-related personal injury. In order to protect against contact hazards, compliance with national accident prevention regulations (such as BGV-A3, which is mandatory for electrical equipment in Germany) [29] is also essential.

The noise source in inverter voltage that produces a pulse-shaped output voltage with very quick acceleration and deceleration periods. This voltage is fed to motor, and the motor results in a common-mode current through parasite ground capacity. The current of common mode will return to its source after closing, the DC-link, shown in figure 12. One main element in keeping electromagnetic interference under control is through regulating the return path of the common-mode current. There are common mode condensers inside the VSD This includes VSD condensers and ground / earth loops. Common-mode capacitors can be used as decoupling condensers in the RFI circuit, or in the DC-link. By using a shielded motor wire and attaching the motor end of the power cord to the motor frame and attaching the VSD end to the VSD frame, the standard mode current is preferably returned to the DC link by the common mode capacitor. The common-mode current that passes from the supply of the mains is unwanted because it can create interference on other devices attached to the mains. Therefore, this current needs to be limited e.g. by using RFI filters. When utilizing unshielded motor cables, just a part of the common-mode current passes via the variable speed drives' frames and common-mode condensers, creating pressure on main grid.

The variable-speed drive systems are the most controversial from the perspective of RCD spectrum and operation. Such circuits produce complex nonlinear earth fault currents [30], the range of which depends on the pulse width modulation (PWM) frequency and individual motor velocities. For practical applications, the PWM frequency may be within the range of a few to more than 20 kHz. The larger the PWM frequency, the weaker the RCD-operation situation. Such currents comprising the frequency domain differentiate tripping properties of the most typical RCDs from those for the nominal 50/60 Hz [31]. The outcome of the easiest RCD style test described in [32] indicates that their threshold for tripping depends entirely on the magnetic properties of the tripping circuit configuration and current sensor. The elevated tripping frequency of RCDs, which can be harmful for the efficiency of electrical shock protection [33], and the authors recommend a redesign of the RCD structure. Upon correction the correct function of the evaluated RCD is achieved but only for a frequency not much higher than the central wavelength. The problem of RCD operation in the existence of harmonic currents is demonstrated in the guide [34]; it illustrates strongly that the selection of RCDs for circuits with skewed earth fault current needs a detailed review of the existing earth fault shapes. The selected RCD should be capable of responding to them, while at the same time being resistant to natural earth-leakage currents. For these circuits the manufacturer [34] also recommends different RCD types. An advanced RCD sort (B-type) is studied in [35], but is a fairly costly RCD and is used mainly for limited applications. The standard related to RCD is shown in Figure 13.

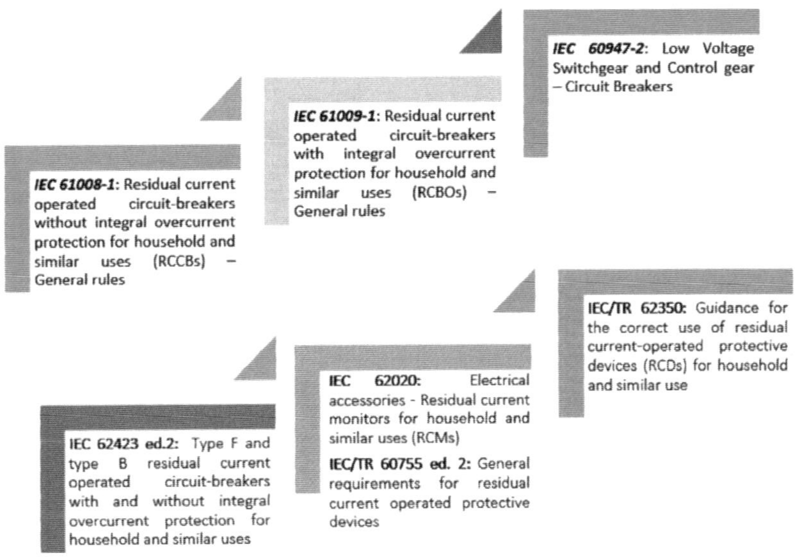

Figure 13: Standards related to residual current device.

An RCD is the only solution on a TT system to protect against indirect contacts because the dangerous fault current is too low for overcurrent protection devices to detect. The TN-S and IT systems are also an easy solution. RCD applications are briefly shown in Figure 14.

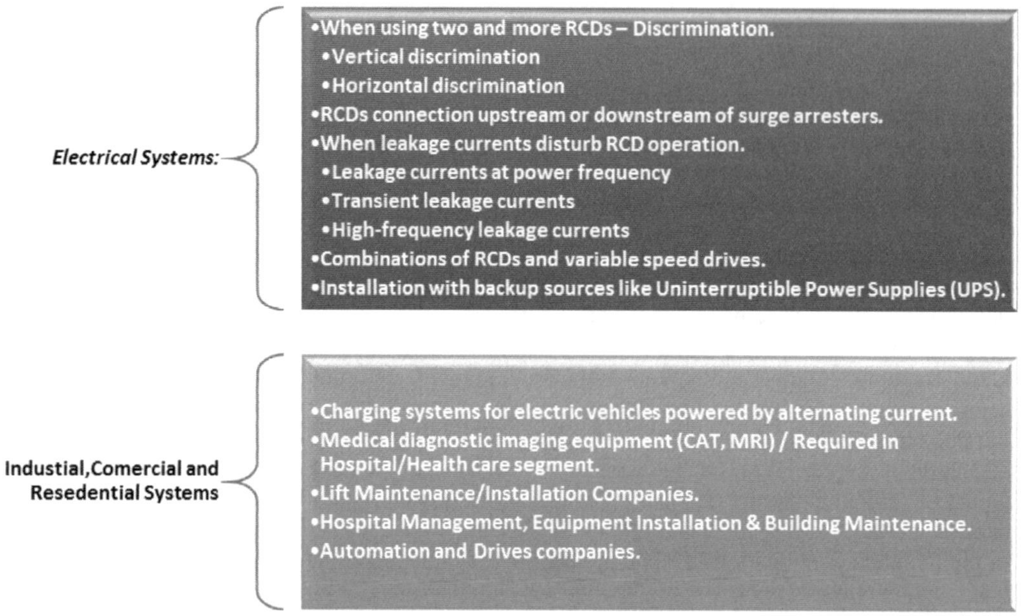

Figure 14: Special application of RCD

5. Conclusions

Electrical devices are frequently handled with currents and voltages which is inherently harmful to animals, humans and structures. Those dangers can be caused by physical interaction, overloading, short circuiting, and loss of component, or impact of heat or moisture. An RCD can sense low currents of leakage that can pass through a person's body. The paper has studied on RCD for electrical safety as well electrical hazards and its caused that are also described in detail. Moreover, during the electrical faults, the paper studies the function and importance of highly sensitive residual current device (RCD) and its defensive techniques. These residual current device is a protective device, which detect the electrical faults and switch off the electricity automatically. The ensuing possible dangers have to be eliminated by precautionary planning and architecture combined to fault analysis and detection system. Therefore, the study helps the readers to understand the importance of electrical safety and give awareness about the electrical hazards.

6. References:

[1] IEC. IEC 61140:2016. Protection Against Electric Shock–Common Aspects for Installation and Equipment; International Electrotechnical Commission: Geneva, Switzerland, 2016.
[2] CENELEC. HD 60364-4-41:2017. Low-Voltage Electrical Installations–Part 4-41: Protection for Safety–Protection Against Electric Shock; European Committee for Electrotechnical Standardization: Brussels, Belgium, 2017.
[3] IEC. IEC TS 60479-1:2016 Effects of Current On Human Beings and Livestock. Part 1: General Aspects; International Electrotechnical Commission: Geneva, Switzerland, 2016.
[4] Protection Against Electric Shock and Electric Fault "https://www.electrical-installation.org/enwiki/Fires_of_electrical_origin".
[5] Marty Ahrens "Home structure fired" from Nation home protection association issued October 2019
[6] V. Babrauskas at the 7th international Fire & Materials conference, 2001, San Francisco, USA, pp39-50. Proceedings are available from the publisher, Interscience Communications, UK
[7] Liu, Y.; Xie, X.; Hu, Y.; Qian, Y.; Sheng, G.; Jiang, X. A novel transient fault current sensor based on the PCB Rogowski coil for overhead transmission lines. Sensors 2016, 16, 742.
[8] Granizo, R.; Blánquez, F.R.; Rebollo, E.; Platero, C.A. A novel ground fault non-directional selective protection method for ungrounded distribution networks. Energies 2015, 8, 1291–1316.
[9] Bao, G.; Gao, X.; Jiang, R.; Huang, K. A novel differential high-frequency current transformer sensor for series arc fault detection. Sensors 2019, 19, 3649.
[10] Yang, K.; Zhang, R.; Chen, S.; Zhang, F.; Yang, J.; Zhang, X. Series arc fault detection algorithm based on autoregressive bispectrum analysis. Algorithms 2015, 8, 929–950.
[11] Sułowicz, M.; Ludwinek, K.; Tulicki, J.; Depczyński, W.; Nowakowski, Ł. Practical adaptation of a low-cost voltage transducer with an open feedback loop for precise measurement of distorted voltages. Sensors 2019, 19, 1071.
[12] Rabcan, J.; Levashenko, V.; Zaitseva, E.; Kvassay, M.; Subbotin, S. Application of fuzzy decision tree for signal classification. IEEE Trans. Ind. Inf. 2019, 15, 5425–5434.
[13] Czapp, S. The impact of DC earth fault current shape on tripping of residual current devices. Elektronika ir Elektrotechnika 2008, 4, 9–12.
[14] Yao, W.; Kui, L.; Can, L.; Zhitao, G.; Feng, N.; Zhang, X. Study on modeling and simulation of AC/DC sensitive residual current transformer. In Proceedings of the 1st International Conference on Electric Power Equipment-Switching Technology, Xi'an, China, 23–27 October 2011. [
[15] Czapp, S.; Dobrzynski, K.; Klucznik, J.; Lubosny, Z. Low-frequency tripping characteristics of residual current devices. In Proceedings of the 2017 IEEE International Conference on Environmental and Electrical Engineering & 2017 IEEE Industrial and Commercial Power Systems Europe (EEEIC /I&CPS Europe), Milan, Italy, 6–9 June 2017; pp. 298–301.
[16] Erdei, Z.; Horgos, M.; Lung, C.; Pop-Vadean, A.; Muresan, R. Frequency behavior of the residual current devices. In Proceedings of the International Conference on Applied Sciences (ICAS2016), Hunedoara, Romania, 25–27 May 2016.
[17] Freschi, F. High-frequency behavior of residual current devices. IEEE Trans. Power Deliv. 2012, 27, 1629–1635.
[18] Czaja, P. Examination of the impact of design of a residual current protective device on the release frequency range. In Proceedings of the 2017 Progress in Applied Electrical Engineering (PAEE), Koscielisko, Poland, 25–30 June 2017.
[19] Czapp, S.; Dobrzynski, K.; Klucznik, J.; Lubosny, Z.; Kowalak, R. Improving sensitivity of residual current transformers to high frequency earth fault currents. Arch. Elect. Eng. 2017, 66, 485–494.
[20] Cocina, V.; Colella, P.; Pons, E.; Tommasini, R.; Palamara, F. Indirect contacts protection for multi-frequency currents ground faults. In Proceedings of the 2016 IEEE 16th International Conference on Environment and Electrical Engineering (EEEIC), Florence, Italy, 7–10 June 2016.
[21] Lee, T.M.; Chan, T.W. The effects of harmonics on the operational characteristics of residual current circuit breakers. In Proceedings of the International Conference on Energy Management and Power Delivery, Singapore, 21–23 November 1995; pp. 715–719.
[22] Czapp, S. The effect of earth fault current harmonics on tripping of residual current devices. In Proceedings of the International School on Nonsinusoidal Currents and Compensation, IX Conference-Seminar (ISNCC), Lagow, Poland, 10–13 June 2008.
[23] Wieland, T.; Aigner, M.; Schmautzer, E.; Pasker, J.; Fickert, L. Influences on safety issues for inverter supplied grid structures. In Proceedings of the 2012 Electric Power Quality and Supply Reliability, Tartu, Estonia, 11–13 June 2012.
[24] IEC. IEC 60755:2017. General Safety Requirements for Residual Current Operated Protective Devices; International Electrotechnical Commission: Geneva, Switzerland, 2017.
[25] IEC. IEC 61008-1:2010. Residual Current Operated Circuit-Breakers Without Integral Overcurrent Protection for Household and Similar Uses (RCCB)–Part 1: General Rules; International Electrotechnical Commission: Geneva, Switzerland, 2010.
[26] IEC. IEC 62423:2009. Type F and Type B Residual Current Operated Circuit-Breakers with and Without Integral Overcurrent Protection for Household and Similar Uses; International Electrotechnical Commission: Geneva, Switzerland, 2009.
[27] Field Test of Installed Residual Current Devices in Denmark; Sikkerhedsstyrelsen: Esbjerg, Denmark, 2009.
[28] Han, Y.; Ding, C.; Shou, X. Design & implementation of an A-type residual current circuit breaker IC. In Proceedings of the 2012 IEEE International Symposium on Industrial Electronics, Hangzhou, China, 28–31 May

2012.

[29] Danfoss Handbook on VLT Frequency Converters, "Facts Worth Knowing about Frequency Converters", PE-MSMBM Publications, 2014

[30] Gruhn, T.; Glenney, J.; Savostianik, M. Type B ground-fault protection on adjustable frequency drives. IEEE Trans. Ind. Appl. 2018, 54, 934–939.

[31] Czapp, S.; Guzinski, J. Electric shock hazard in circuits with variable-speed drives. Bull. Pol. Ac. Tech. 2018, 66, 361–372.

[32] Czaja, P.; Pluta, W. Magnetic properties of additive transformer cores of residual current devices. Przeglad Elektrotechniczny 2019, 12, 109–112

[33] Horgos, M.; Erdei, Z.; Barz, C.; Birsan, I.; Ilia, M. Contributions to testing residual current devices at different frequency values. In Proceedings of the 6th International Conference on Modern Power Systems (MPS), Cluj-Napoca, Romania, 18–21 May 2015.

[34] Eaton. Residual Current Devices. In *Application Guide*; Eaton: Vienna, Austria, 2017.

[35] Panda, R.K.; Veeramalla, J. Behavior modeling of a type B RCD. In Proceedings of the 1st IEEE International Conference on Power Electronics, Intelligent Control and Energy Systems (ICPEICES), Delhi, India, 4–6 July 2016.

IOP Conf. Series: Materials Science and Engineering **906** (2020) 012019 doi:10.1088/1757-899X/906/1/012019

Implementation of on-demand low rate wireless parameter monitoring system for industrial machineries.

Riya Shrivastava[1], Jigyasa Kodnani[2], Samridhi Gupta[3], Dakshta Mehta[4], R. Raja Singh[5], S. Meikandasivam[6]

[1,2,3,4]Student, [5]Associate Professor, [6]Professor

[1-6]School of Electrical Engineering, Vellore Institute of Technology, Vellore, India

Corresponding Author's Email: riya.shrivastava2017@vitstudent.ac.in

Abstract- Systematic and rigorous monitoring of critical equipment is the cornerstone of any successful preventive management system in the industrial environment. Recent developments in cost-effective wireless sensor network (WSN) technologies promising a wider and more sustainable deployment of distributed sensing and processing infrastructure. In this research work, it is mainly focused on the electrical monitoring parameter and enable remote switching devices for optimal power management systems using ZigBee. This paper proposes a digital system for electrical system parameters such as voltage and current using ZigBee-based wireless sensor networks (WSNs). The main feature of ZigBee protocol, to act as a communicable medium between transmitter and receiver modules, which is suitable for the industrial environment.

Keywords - Industrial Wireless Sensor Network, Power Monitoring System, ZigBee XB, Low Cost

1. INTRODUCTION

The process of rationalization and upgrading of industrial structure has a particular phase impact on economic development, and the relationship between industrialization and the rationalization of economic development is a strong constant [1,2]. Industrialization observed as the method of change, in which various countries follow various methods that depends on their primary circumstances and the time of entrance in the race [3]. Modern monitoring and control require to collect all relevant information, statistics and data relating to the various industrial processes, motors, machinery and equipment operating on the industrial premises. It aims to produce results of controlled use, improved productivity and higher quality of industrial products [4–6]. Although conventional wired communication protocols have an important place in industrial intensive care and control grids in the past, they prove to be derisory to satisfy the largely unstable and rigorous requires of recent industrial technology, mainly because of rigorous use of wired infrastructure. Wireless technology has the potential of generating a revolution among the industries because of its applications, not only by reducing the challenges regarding wired networks, but also by generating new class advancements and their uses[7–10]. Wireless protocols play an important part in industrial monitoring and control schemes, because it has certain benefits when compared to conventional wired technology [11,12]. Along with reduction of large scale and setting up costs, the topology's modest use of wired solutions allows it to be easily deployed in unsupervised areas (for example, in moving parts) [13,14]. Changes of the mesh technology can be effortlessly done by avoiding the not required cost of wiring [15].Telecommunication, electronics, and computer engineers have made a tremendous determination to advance several topologies of digital communication, with the goal of bringing performance closer to full scale [16,17].

During late nineteenth century, digital wireless technology was put into force to increase the efficiency of satellite communication systems for the International Satellite Organization in a rapidly growing situation. Later it was used in mobile phones and gradually increased the importance of the cellular business. Currently, this wireless communication technology is used in many appliances and networks. ZigBee consumes less power and the data rate is quite low [18–21].This paper presents an implementation of on-demand low rate wireless parameter monitoring system for industrial machineries using ZigBee. The contribution can be summarized as follows Section 2 discusses the comparative study of Wireless technologies, Bluetooth and Zigbee. Section 3 discusses the methodology of achieving the wireless communication between modules. Section 4 discusses the algorithm of the proposed model. Section 5 discusses the implementation and the components used. Section 6 discusses the final results and output waveform. Finally, Section 7 concludes that Zigbee is the most suitable module for industry purpose and the makes the proposed system cost-effective. XBee, takes the place of existing Infrared technology-based devices which saves the cost of instalment. Zigbee uses lithium battery which lasts long. The network set up is very simple and easy. Therefore, the proposed implementation is both cost effective and easy to implement.

2. *COMPARATIVE STUDY ON WIRELESS TECHNOLOGIES*

Table I describes the major differences between the two wireless technologies, Bluetooth and Zigbee. Each technology is grounded on the IEEE standard. Generally, Bluetooth and ZigBee are used for wireless communication about 10 m. However, in some applications ZigBee can even range upto 100 meters. In this section a detailed comparative study is carried out to adapt the suitable technologies for the industrial machineries.

Table 1 Comparison of Bluetooth and Zigbee wireless technologies

Specification	**Bluetooth Module**	**Zigbee Module**
IEEE Standard	802.15.1	802.15.4
Range of Frequency	2.54 GHz	866/915 MHz
Maximum Signal Range	1 Mbps	250 Kbps
Range of Communication	10 m	10-100 m
Bandwidth of Channel	1 MHz	2 MHz
Type of Modulation	GFSK	BPSK, QPQSK
Basic cell Extension	Scannet	Cluster Tree, Mesh
Spreading	FHSS	DSSS
Primitives	151	35
Maximum cell nodes	8	More than 65000
HCL Events	37	30

2.1 Power consumed by chipset

Table 2 presents the comparison of power consumption and the detailed features of the specific chipset for each topology. Figure 1 depicts the power consumed (mW) for each topology. It is observed that power consumption is minimum in Bluetooth and Zigbee.

Table 2 Features of chipset for each topology

Protocols	Chipset	VDD(Volt)	I_{TX}(mA)	I_{RX}(mA)	Bit Rate(Mb/s)	Power(TX)	Power(RX)
Bluetooth	BlueCore2	1.8	57	47	0.72	57.0	48.6
Zigbee	CC2430	3.0	24.7	27	0.25	24.5	28.4

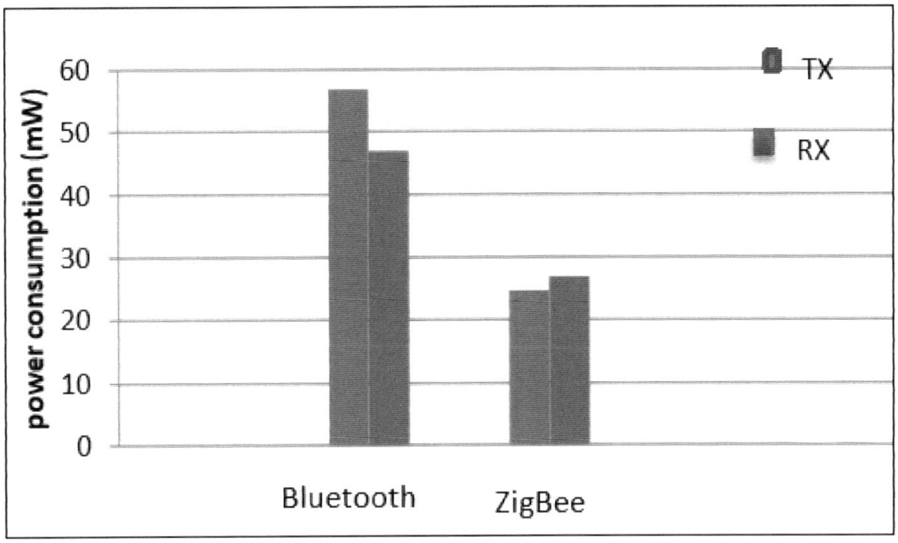

Figure 1 Power consumed in mW

2.2 Complexity of topology

The complication of each technology is distinguished on the basis of the number of primitives and events. Table 1 shows the count of primitive and Host Controller Interface (HCI) instances for Bluetooth and the PHY/MAC primitive count for ZigBee, Wi-Fi and UWB topologies. "In PHY/MAC films, Bluetooth primitives comprise HCI SAP, Client Service Access Point (SAP), Synchronous Connection-Oriented (SCO) SAP, and Logical Link Control and Optimization Protocol (L2 CAP) primitives." Bluetooth 188 is very complex topology with primitive and total events. While, "ZigBee is simpler with 48 priorities defined in 802.15.4. This is only one-fourth of the number of primitives and events defined in Bluetooth." Compared to Bluetooth, effortlessness makes ZigBee more appropriate for sensor interacting modules due to its restricted memory and computing power.

3. Methodology

Shown in Figure 2 is the proposed system that works on the fundamental of Wireless Communication. Two XBEEs are used for communicating. It is a plug-in device which consists of 2 plug-in arduinos (Transmitter and receiver). The Transmitter has sensors attached to them which takes data from the machines and transmit it and receives the data and provides the signal to the meter for various interpretations.

For achieving the wireless communication between the XBees it is required to configure the XBee to obtain a link between them. In this model, XCTU software has been used for configuring XBee.

One of the XBee is configured as coordinator node and the other XBee is configured as the router node. Table 4 shows the parameters that have to be set for enabling the XBees to transmit data. The source address of one XBee should match the destination address of another XBee. By making these changes the XBees will be able to transmit the data among themselves through the channel that they will establish with the help of same address. The frequency that is being used to communicate is defined by CH and as same frequency is required, the Channel for both the XBee has to be kept the same. The PAN ID defines the network that the XBee will communicate to. DH and DL defines the low and high destination address. DH and DL should be the serial number high and serial number low (SH and SL) respectively of the transmitter's XBee. MY enables the response of 16-bit address packets. Whenever the value changes the XBee Node2 sends the data value from the button to the XBee Node1.

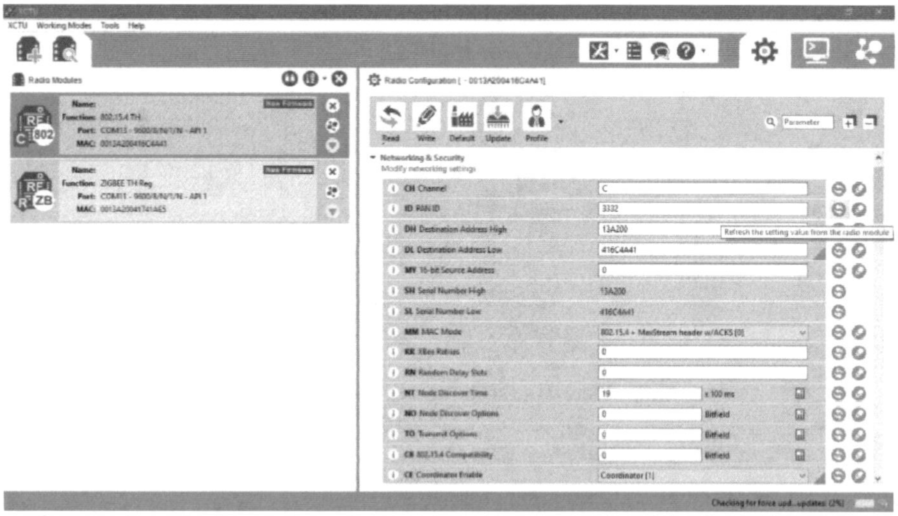

Figure 2 Selected coordinator module

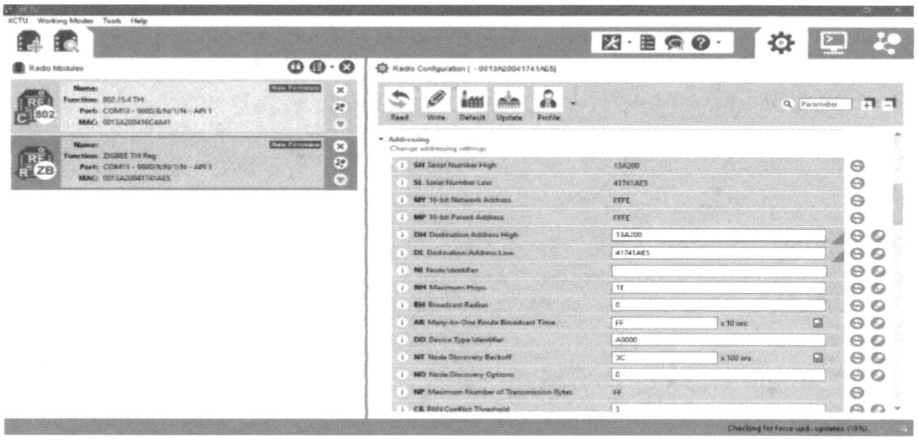

Figure 3 Selected receiver module

XCTU is a software used mainly by developers to manage radio modules such as XBee. It is a free multi-platform application which uses simple-to-use graphical interface. This application makes it easy to configure and test XBee modules as it includes all the required embedded tools. Figure 2 and 3 shows the XCTU Software and each of the coordinator and receiver modules configurations.

Table 3 Parameters for XBee Configuration

Fields to Set	Short-Form	XBee Node1	XBee Node2
Channel	CH	C	C
PAN ID	ID	3322	13A200
Source Address	MY	0	1
Destination Address High	DH	0	0
Destination Address Low	DL	1	0

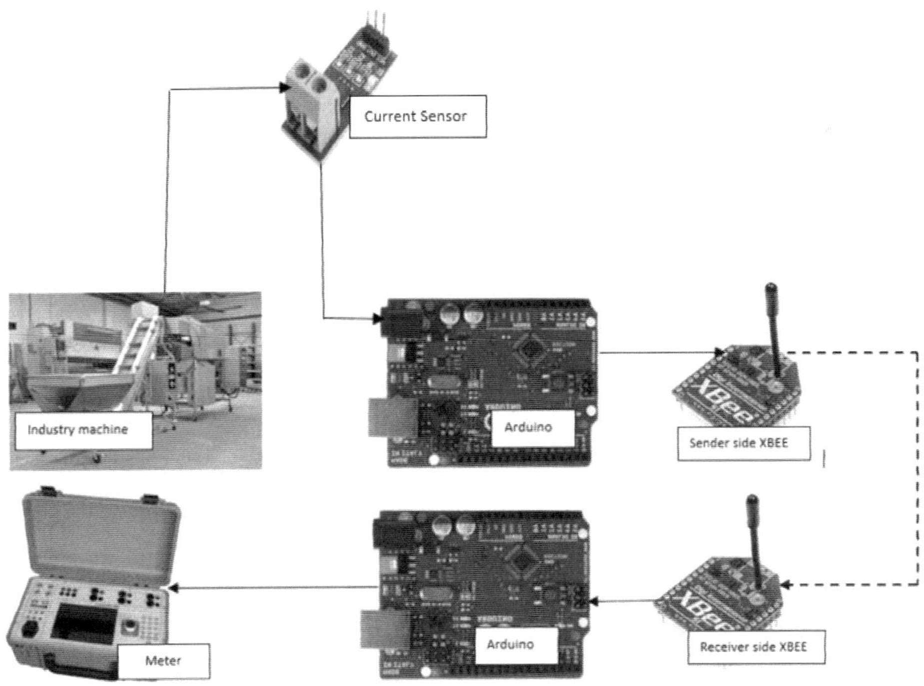

Figure 4 Schematic diagram of proposed model

Figure 4 represents the block diagram of our proposed model. It shows the flow of data from machine to the meter's monitor screen using XBee through arduino.
This model consists of S2C XBEE MODULE for communication protocol, it provides point-to-point wireless connection to various devices. At the end node or base node, it collects sensor reading from arduino and the data collected is transmitted to the monitoring junction and at the coordinator node the information sent by the base node is received. The microcontroller Arduino Uno is a development board built on the ATmega328. It helps in powering up different sensors and also establishes a connection to permit monitoring and appropriate regulation of various devices. It supports the radio communication. At the end node station or router node it helps in reading serial output and is often used as a communication station between XBee chips and sensors and at the base station or coordinator node it is used as a communication station between XBee chip and computer. LEM TRANSDUCER(LV25-P) sensor used for measuring the current. It can work upto 150khz input signal and can withstand current overloads without damage and helps monitoring the current flowing through appliance.

4. Algorithm

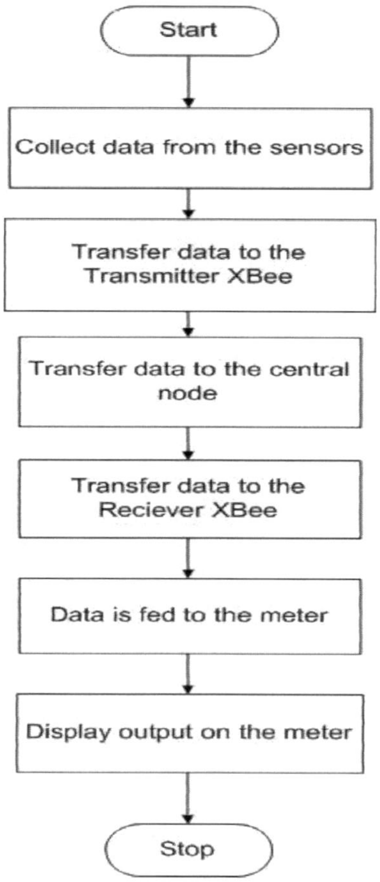

Figure 5 Flowchart of Proposed Algorithm

Figure 5 is the flowchart of the proposed system algorithm. The entire algorithm is simple and easy to implement. This model consists of current sensor which collect the current data from various machines in the industry. In order to configure the XBEEs XCTU software is used. The data collected by the sender node is then transmitted to the receiver node using XBEE wireless communication. The XBEE transfers the data received to the meter. Finally, the meter is used to display the data.

One of the plug-in devices is connected to machine, whose parameters are to be monitored. Another plugin device is to connected to the meter, that receives data which is transmitted through the wireless channel. The data is transferred in the form of small packets. Since, the plug-in device is lightweight and portable, it is easy to use and maintain. For the proposed model, there is one coordinator node and multiple router nodes. XBee can be used to achieve point to multi point communication using star topology. This paper is based on above topology.

5. IMPLEMENTATION

For proposed model, it is first required to configure the XBEE so that they can communicate among themselves for which XCTU software is used. The Arduino code that is used was divided into parts.

- 3 different modules has been imported- XBee.h , EmonLib.h for current calculation and SoftwareSerial.h for defining the ports.
- After that the sender and receiver XBee are configured.
- The next part of the code was used to set the ports and baud rate.
- After that the data was read from the sensor connected to the arduino.
- Then the created packets of data and set the communication channel to transfer the packets.
- After receiving the data on the meter side, the data is displayed on the meter.

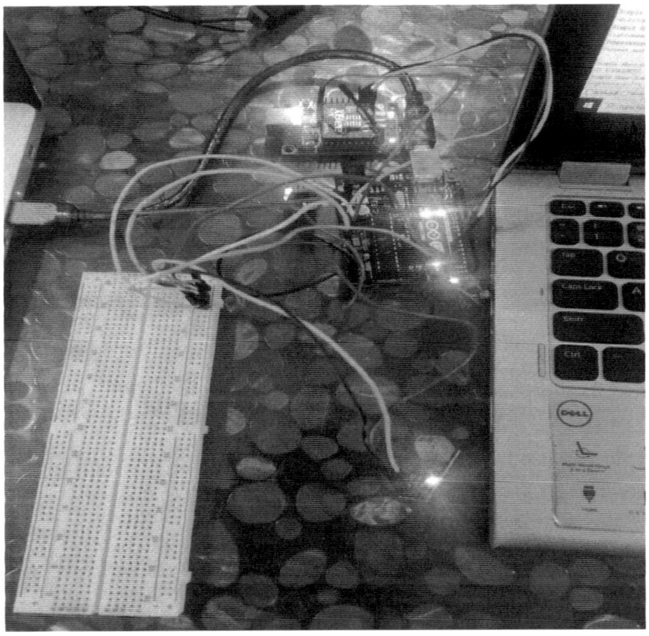

Figure 6 Proposed Model

6. RESULT AND DISCUSSION

ZigBee XBee protocol is used as the communication medium between the transmitter and receiver modules. The input signal is transmitted wirelessly to the other channel and the noise disturbance is

IOP Conf. Series: Materials Science and Engineering **906** (2020) 012019 doi:10.1088/1757-899X/906/1/012019

reduced using filter circuit. The measurement of current is carried out and a plot as shown in Figure 7 is drawn where the parameter value is on Y-axis and time is on X-axis.The above implementation can be used as a simple and cost-effective solution to transfer data to smart meters. The new ZigBee XBee standard performs well in industrial environments.

Table 4 Observed real-time data

	A	B
1	Day	Current
2	1	142.73
3	2	142.17
4	3	143.19
5	4	143.57
6	5	110.51
7	6	145.7
8	7	136.25
9	8	135.57
10	9	130.65
11	10	144.44
12	11	150.7
13	12	105.5
14	13	105.71
15	14	102.42
16	15	143.1
17	16	146.97
18	17	151.06
19	18	134.95
20	19	142.61
21	20	143.62
22	21	150.88
23	22	149.95
24	23	133.74
25	24	141.55
26	25	144.48
27	26	144.9
28	27	139.74
29	28	142.11
30	29	134.31
31	30	135.8

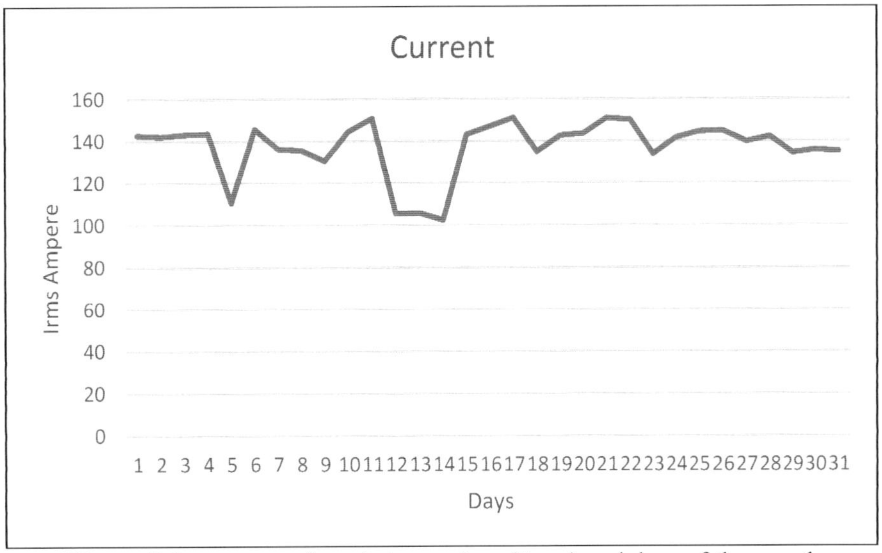

Figure 7 Output waveform between Irms(Amp) and days of the month

During the working condition, the data transmitted is wirelessly received as a small data packet on the XBee, and the XBee serial data is transmitted to the receiver pin RX of the Arduino UNO microcontroller. The required data can therefore be viewed directly on the Arduino IDE's Serial Monitor platform. The system uses a micro-SD card to store the data as a text file which can be later used to view the data on a computer. When we place XBee and Arduino in the coordinator, real time monitoring of the data received is done and it is stored as an excel file. With the help of these data the plot for variation of current is drawn.

7. Conclusion

This paper presents a prototype of a wireless communication system has been built based on XBee module. The flexibility of mesh networking Zigbee modules makes it a strong option for industrial automation. The initial stage was the collection of values obtained from various sensors. Then, using XBee communication, the data is communicated with the monitoring meter. It is a back and forth(both-way) communication between the smart plugs and the monitoring meter. Finally, the data is displayed on the meter. Wireless technology has already become an important application which usually integrated to a wide range of device and other technologies. This system uses the XBee as the wireless module. The combination of low-cost hardware, very good radio performance makes XBee a competitive choice on the wireless market. The use of XBee makes our entire system cost-effective and the plug-in devices have easy mobility and low maintenance.

8. REFERENCES

[1] Lu W 2011 Study on characteristic of emerging technology *Int. Conf. Manag. Serv. Sci. MASS 2011*
[2] Lin Y L, Lin C C and Chiu H S 2015 The development of intelligent service system for machine tool industry *Proc. 2015 1st Int. Conf. Ind. Networks Intell. Syst. INISCom 2015* 100–6
[3] Zhang J and Tan W 2010 A study on ripple effects of logistics industry in China *Proc. - 2010 Int. Conf. Optoelectron. Image Process. ICOIP 2010* **1** 164–7
[4] Zhou W, Habetler T G and Harley R G 2007 Bearing condition monitoring methods for electric machines: A general review *2007 IEEE Int. Symp. Diagnostics Electr. Mach. Power Electron. Drives, SDEMPED* 3–6
[5] Barksdale H, Smith Q and Khan M 2018 Condition Monitoring of Electrical Machines with

Internet of Things *Conf. Proc. - IEEE SOUTHEASTCON* **2018-April** 1–4

[6] Asada T and Roberts C 2011 Development of an effective condition monitoring system for AC point machines *IET Conf. Publ.* **2011**

[7] Khan Z H, Thiriet J M and Genon-Catalot D 2009 Wireless network architecture for diagnosis and monitoring applications *2009 6th IEEE Consum. Commun. Netw. Conf. CCNC 2009* 5–6

[8] Xu, Wu, Daneshmand, Liu W 2015 A data privacy protective mechanism for WBAN *Wirel. Commun. Mob. Comput.* 421–30

[9] Dzung D, Endresen J, Apneseth C and Frey J E 2005 Design and implementation of a real-time wireless sensor/actuator communication system *IEEE Int. Conf. Emerg. Technol. Fact. Autom. ETFA* **2 2 OF 2 VOL** 433–42

[10] Willig A 2003 An Architecture for Wireless Extension of PROFIBUS *IECON Proc. (Industrial Electron. Conf.* **3** 2369–75

[11] Porcino D and Hirt W 2003 Ultra-wideband radio technology: Potential and challenges ahead *IEEE Commun. Mag.* **41** 66–74

[12] Albreem M A M 2015 5G wireless communication systems: Vision and challenges *I4CT 2015 - 2015 2nd Int. Conf. Comput. Commun. Control Technol. Art Proceeding* 493–7

[13] Editorial G 2016 SPECIAL SECTION ON INTERNET OF THINGS (IoT) IN 5G WIRELESS COMMUNICATIONS Internet of Things (IoT) in 5G Wireless Communications **4** 10310–4

[14] Sachan A, Nigam S and Bajpai A 2018 An Energy Efficient Virtual-MIMO Communication for Cluster Based Cooperative Wireless Sensor Network *2018 9th Int. Conf. Comput. Commun. Netw. Technol. ICCCNT 2018* 1–6

[15] Obaid T, Rashed H, -Elnour A A, Rehan M, Muhammad Saleh M and Tarique M 2014 Zigbee Technology and its Application in Wireless Home Automation Systems: A Survey *Int. J. Comput. Networks Commun.* **6** 115–31

[16] Cuomo F, Luna S Della, Monaco U and Melodia T 2007 Routing in ZigBee: Benefits from exploiting the IEEE 802.15.4 association tree *IEEE Int. Conf. Commun.* 3271–6

[17] Wheeler A 2007 Commercial applications of wireless sensor networks using ZigBee *IEEE Commun. Mag.* **45** 70–7

[18] Zhou B, Hu C, Wang H Bin, Guo R and Meng M Q H 2007 A wireless sensor network for pervasive medical supervision *IEEE ICIT 2007 - 2007 IEEE Int. Conf. Integr. Technol.* 740–4

[19] Moron M J, Casilari E, Luque R and Gázquez J A 2005 A wireless monitoring system for pulse-oximetry sensors *Proc. - 2005 Syst. Commun. ICW 2005, Wirel. - ICHSN 2005, High Speed Networks - ICMCS 2005, Multimed. Commun. Syst. - SENET 2005, Sens. Networks* **2005** 79–84

[20] Lee J S and Wang Y M 2013 Experimental Evaluation of ZigBee-Based Wireless Networks in Indoor Environments *J. Eng. (United Kingdom)* **2013**

[21] Li C, Wang Y and Guo X 2010 The application research of wireless sensor network based on ZigBee *2010 Int. Conf. Multimed. Inf. Technol. MMIT 2010* **2** 89–92

IOP Conf. Series: Materials Science and Engineering **906** (2020) 012020 doi:10.1088/1757-899X/906/1/012020

Materials for lightweight electric motors – a review

S Darius Gnanaraj[1]*, Edison Gundabattini[1] and R Raja Singh[2]

[1]School of Mechanical Engineering (SMEC)
[2]School of Electrical Engineering (SELECT)
Vellore Institute of Technology, Vellore, Tamil Nadu, India 632014.

*Corresponding author: dariusgnanaraj.s@vit.ac.in

Abstract. Electric motors are widely used in automobiles and large electric vehicles would be introduced in market in the near future to reduce global warming. Reduction in electric motor weight also reduces the overall weight of automobiles. The use of fiber-reinforced polymer materials reduces the weight of electric motors. Lightweight electric motors promote energy savings owing to the reduction of self-weight. Researchers all over the World are trying to make all parts of electric motors using lightweight materials. This review paper presents the outcome of research carried out by researchers from Germany and UK. Fiber reinforced plastics (FRP) reduce weight and increase mechanical and electrical properties. Phenolic composite materials gave promising results due to its electrical insulation properties and corrosion resistance. Thermosets do not swell when they are exposed to chemicals unlike thermoplastics. Aluminium matrix composites (AMC) are used to improve the motor efficiency resulting in a considerable weight reduction. Electric motors have 90 percent or above, which means that a high mechanical energy is derived from electrical energy. The remaining electrical energy is converted into heat. To prevent overheating, the heat is conducted through housing to a cooling jacket having cold water. Researchers have replaced the circular wire by rectangular flat wire that can be coiled tightly on the stator. Newer cooling concepts are developed for making housings using polymer materials. Carbon fiber reinforced materials with metal pins are used for reducing weight and to increase thermal conductivity. Use of high modulus fibers increase damping properties and reduce noise produced by electric motors. Comparisons between electric motors made of lightweight materials and traditional metals in terms of weight, efficiency, cooling rate are presented.

Keywords: Lightweight materials, electric motors, Fiber reinforced plastics, Aluminium matrix composites, cooling concepts.

1. Introduction

To reduce the effects of global warming, many countries are replacing traditional vehicles by electric vehicles. Many researchers are working to design energy efficient electric vehicles by reducing overall weight of the vehicles using composite materials. Attempts are being made to reduce the weight and cost of all parts of automobiles including the power source, electric motor. We require lightweight materials with a higher strength to weight ratio, composites with high power to weight ratio and materials with high thermal resistance. Some parts of the electric motor should have high thermal conductivity or electrical conductivity whereas other parts require electrical insulation. It is a difficult task to replace conventional materials with modern materials and achieve high energy efficiency. 10 Lakhs electric motors are produced in the World every year [1]. Eco-friendly materials are preferred for easy recycling of materials after the life of the product is over. Composites help designers to combine many properties required by different parts of the electric motor into different composites. This paper reviews the progress made by a few researchers in replacing heavy parts of electric motors by lightweight materials, novel cooling methods incorporated and increase in efficiency.

2. Materials for lightweight electric motors

Fiber reinforced plastics (FRP) provide opportunities to replace parts of electric motors to reduce the overall weight. FRP materials increase the mechanical properties and electrical conductivity. Thermosetting FRP materials offer high resistance to heat and corrosion due to chemicals. Soft magnetic composites (SMC) are used for achieving high power to weight ratio at reduced manufacturing costs. Phenolic composite materials (PCM) are useful to replace many metallic parts by a single integrated part. Assembly time may decrease and maintenance also can be carried out on a faster rate. Polymer housings are easier to manufacture than housings made of aluminium. Figure 1 shows the distribution of weight of an electric motor [2].

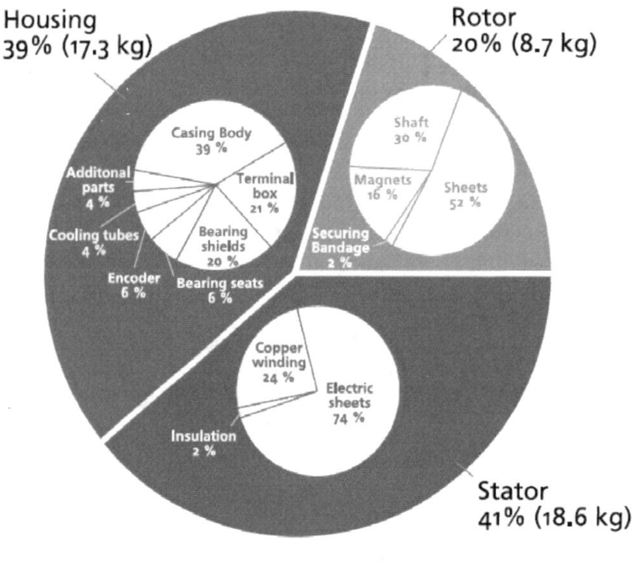

Figure 1. Distribution of weight of an electric motor [2]

It is seen that the weight of stator is 41%, housing 39% and rotor 20%. There are two types of masses: stator and housing do not rotate but only translate and rotor rotates and translates. Rotational inertia reduction is aimed for saving weight. Better dynamic behavior can be obtained by lowering the rotational inertia. While accelerating or decelerating, higher efficiency of the motor is obtained by lowering the rotational inertia [2]. Injection molded magnetic carrier and magnets mounted on a single rotor disc are shown in figure 2. This structure combines two functional requirements: torque transmission and magnetically active part. GFRP (Glass fiber reinforced plastics) is used for transmitting torque and SMC is used for the flow of magnetic field. The length of the magnet carrier was made equal to the length of the magnet. Multiple magnet carriers are assembled to accommodate required number of magnets. Splines are used for easier assembly on the rotor shaft and for a positive torque transmission between the rotor disc and the shaft [3].

IOP Conf. Series: Materials Science and Engineering **906** (2020) 012020 doi:10.1088/1757-899X/906/1/012020

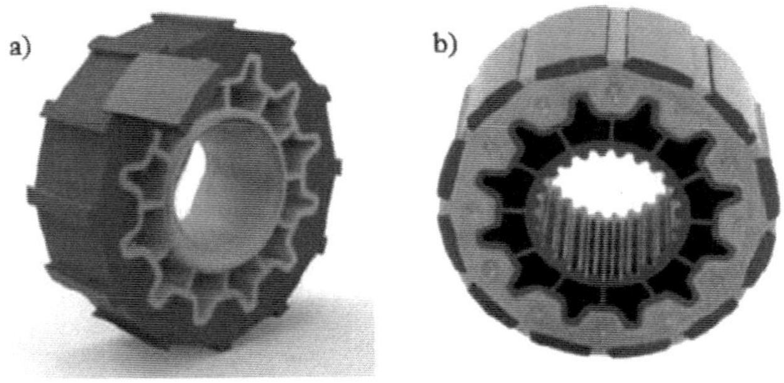

Figure 2. Injection molded magnet carrier a) Soft magnetic composite (inner) and Polyamide (outer) b) Magnets mounted on single rotor disc [3]

3. Effective cooling in reducing the weight of electric motor built with lightweight materials

Enhancing the efficiency with size, cost and weight of the motor could be succeeded by means of better materials and a better designing process. Design of efficient cooling system is of great importance in electric motors to improve performance and their applicability in various fields such as traction, wind, aviation and transport. The major challenge for designers of an electric motor is to overcome the challenges in the power to weight ratio and to accommodate within the limited available space [4-5].

In electric vehicles lighter motors are required and the working temperature of PMSMs is a big challenge to address. Losses in the electric motor reduce the transferred energy and rise the working temperature in all the applications of electric motor. Efficient cooling is paramount for any research related to a high speed and lightweight electric motors. Permanent magnet synchronous motor (PMSM) powers effectively various kinds of electric power vehicles. PMSM has useful features like high torque to weight ratio, simple structure, high power-weight ratio, high efficiency, good reliability, large power density and enormous range of speed. Because of that they are most preferred for aviation applications [6].

Major losses in PMSM are mechanical losses, iron losses and copper losses. Iron losses occur in stator such as eddy current loss and hysteresis loss. Copper losses occur due to high flow of current in armature winding. Windage loss and bearing friction losses cause Mechanical losses. Temperature during the operation of the motor leads to winding insulation failure. This may cause short circuits in the winding. High temperatures developed above critical values in rotor hot spots may demagnetize the permanent magnets in the PMSM. Hence, generated heat in the motor has to be dissipated for the safe operation of the motor [7]. Minimizing the weight of a motor and maximizing the efficiency are the real challenges of a high speed electric motor. Hence, efficient thermal management system through effective cooling influence the weight of electric motor is essential in the design process [8].

Cryogenic cooling with Liquid nitrogen reduces energy loss as resistivity of copper decrease significantly with this method. Hence, less copper utilized to carry copper compared to water cooled or air cooled machines. This improves the power to weight ratio-economically and maximum current density-efficiently, in electric motors [9]. A liquid cooling system has been proved to be good in keeping the insulation temperatures under critical thermal limit. Nitrogen cooling system was designed owing to its compactness and the light weight of the motor. The effective inlet temperatures of fluid cooling systems were higher and hence the size of the heat exchanger also reduced [10].

A novel air ventilation duct was provided as a part of cooling structure in an air cooling system without an additional fan and cooling tubes-for liquid cooling system. The rotation of the electric motor itself would provide the power to enable this method of cooling system. The enhanced convection heat transfer rate method of this cooling had brought down the weight of the motor [11]. A refrigerant cooling method was usual compared to water cooling method with respect to weight of the coolant. Refrigerant cooling method was investigated and proved to be better compared to water cooling. Refrigerant cooling method provides 60% higher torque with less weight and also, at maximum torques the motor operation time was extended by 34%. Electric motor research always aims at effective cooling and bringing down the size of the motor, to enhance the torque and power [12]. Table 1 compares the cooling methods available in literature.

Table 1. Comparison of cooling methods

	Air Cooling	Oil Cooling	Water Cooling	Refrigerant Cooling
Schematic Illustration	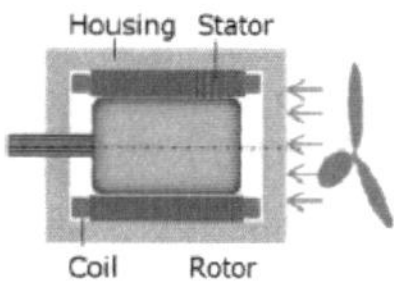	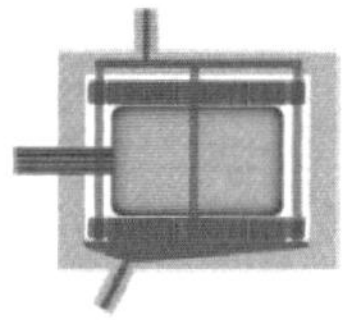	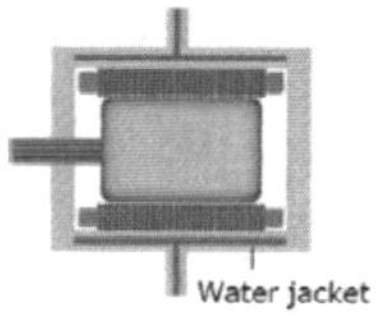	
	Heat exchange by wind	Heat exchange by oil directly flowing into heat source	Heat exchange by flowing water in a water jacket	Heat exchange by flowing refrigerant and vaporization heat
Cooling Performance	Low	High	Average	High
Cooling Uniformity	Uniform	Non-uniform	Uniform	Uniform
Weight Component	High	Average	High	Low

Indirect liquid cooling method was investigated to be applicable and appropriate for aviation purposes. Indirect cooling method was a reliable method of removing large amounts of heat with light weight [13]. In oil cooling method high output power to volume ratio was achieved using oil circulation system to cool the electric motor and its shaft bearing. Here the circulated oil is cooled by the water externally. This method of cooling is far better compared to water cooling with respect to weight of the cooling agent [14]. Zadeh et al. investigated the performance of indirect cooling method by means of advanced nano-fluids as coolants to remove the heat generated by electric motor. Aluminum-oxide nano particles are added into the base fluid to enhance the heat transfer rate. Results proved that increased flow rate and increased fraction of the nano particles to 4%, raised the heat transfer rate upto 40%, which is a phenominal improvement in performance [15].

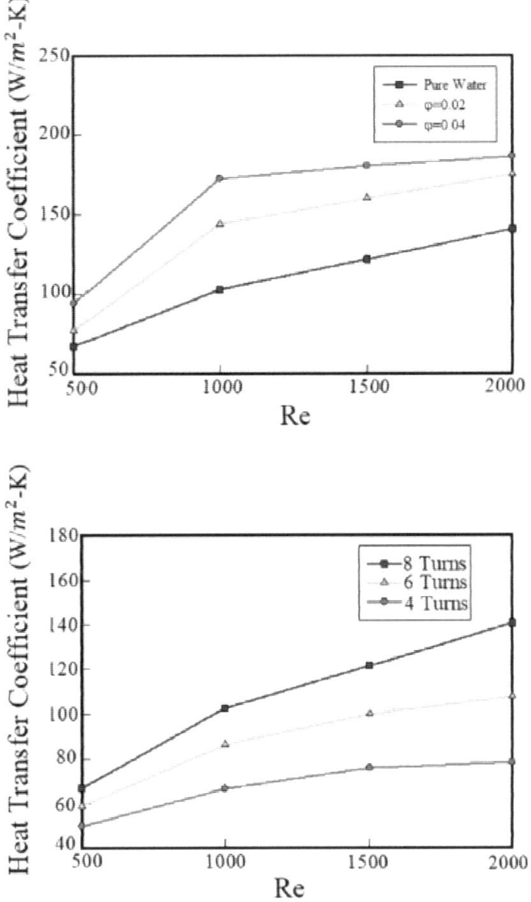

Figure 3. Effect of heat transfer with and without Aluminum-Oxide nano-particles [15]

Prius 2010 motor with V-shaped Nd magnets has cavities in the rotor is shown in figure 4. These cavities not only enhance the air flow and but also increases the effectiveness of the cooling, and reduced weight and inertia [16].

Figure 4. Prius 2010 electric motor-baseline [16]

In general weight of the motor depends upon the type of the motor. The stator core volume in case of a brushless PM motor is half of the volume of single-phase induction motor (SPIM). The conductors employed in a brushless PM motor are weighing one-third of the weight used by a SPIM [17]. Recently, heat pipes are used as efficient phase changing heat transfer devices. The low cost, light weight and high thermal conductivity heat pipe cooling method is best opted for air-cooled motors. Air cooling motors using heat pipes have good cooling effects and are used in electric vehicles [18]. High Temperature Superconducting (HTS) electric machines field windings employ HTS wires which have less weight, less volume and high power density. Temperature distributions under Axial Magnetic Motive Force (AMMF), with High Temperature Superconducting (HTS) wires of machine components are investigated and the results are shown in figure 5. Temperature distribution in the machines having copper wires and HTS wires are helpful in designing the simpler, less weight and more reliable cryogenic cooling system [19].

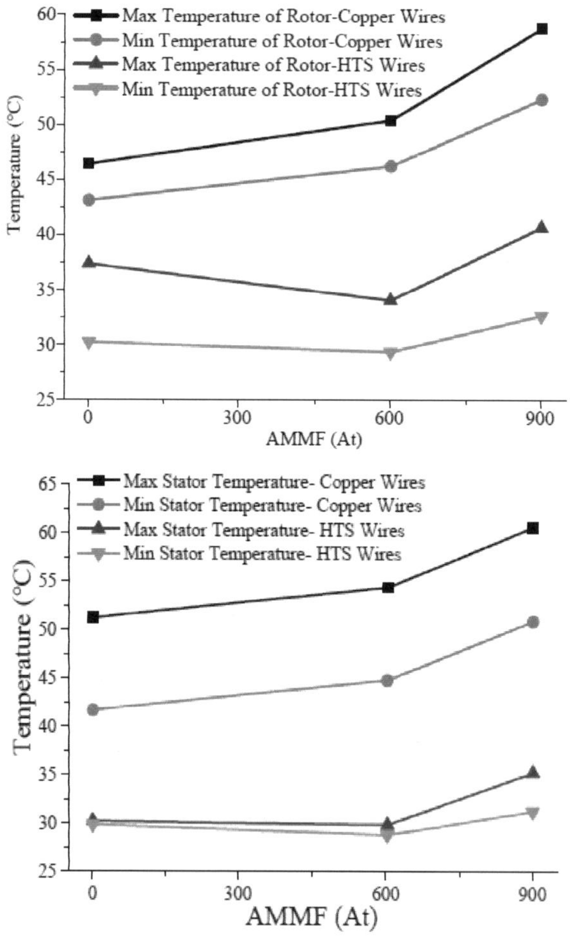

Figure 5. Temperature distribution on Rotor core and Stator core with HTS Copper wires [19]

Metal housing dissipates heat to the jacket having cold water for cooling the electric motor. Recent research replaced the circular wire used for the windings by a rectangular flat wire which could be wound more tightly on the stator, creating room for a cooling jacket adjacent to windings. Generated heat could be transmitted away from the source of heat through a short distance using this method. This motor is lighter and power-to-weight ratio is good in comparison to the existing ones [20].

4. Lightweight materials for improving performance of electric motors

Whether intended for a small or large household appliance, or used for driving hybrid and/or electric cars, or applied in industrial environments, the electric motor continues to confirm an unprecedented trend in global growth. Performance of any electric motor depends upon various factors, for an instance, efficiency is a measure of numerous mechanical and electrical imperfections within the motor. Similarly, the design of electric motors facilitates not just to reduce the cost but also to enhance the performance and the pursuit of the full potential. There are many facets of attaining the maximum performance of an electric motor. The choice of core materials, too, is among them. A task that also includes the components that makes up the core of electrical motors, from magnetic lamination to the stator case [21]. Raw materials with better properties increase the motor's performance. This seems to be doubtless from the view point of material research. Nevertheless, the achievement of the motor performance improvement goal is not adequate with the proper raw material. On a more general standpoint, in the past, magnetic losses and power losses provided the only primary architecture parameter. The losses in electric motor are distributed into various components, in three-phase motors, resistance losses (I2R losses) in stator windings and rotor bars will result in an output loss of up to 15 percent. In single phase fractional horsepower motors the I2R losses could be up to 30 percent. Similarly, losses in magnetization in the stator and rotor cores induce an output loss from around 1 percent to 7 percent. Likewise, 0.5 to 1.5 percent of losses may occur due to bearing friction and inefficient cooling fans. Losses in friction and magnetization are related only to the size and design of the motor and not dependent of motor load. The subsequent losses are called as stray load losses, which is around 2 percent of total losses. Partially loaded motors have lower efficiencies because of constant losses such as magnetizing losses, friction losses, and windage losses that will be the larger percentage of total motor power consumption. Figure 6 shows the various components of losses for a 50 hp squirrel cage induction motor as a function of motor load.

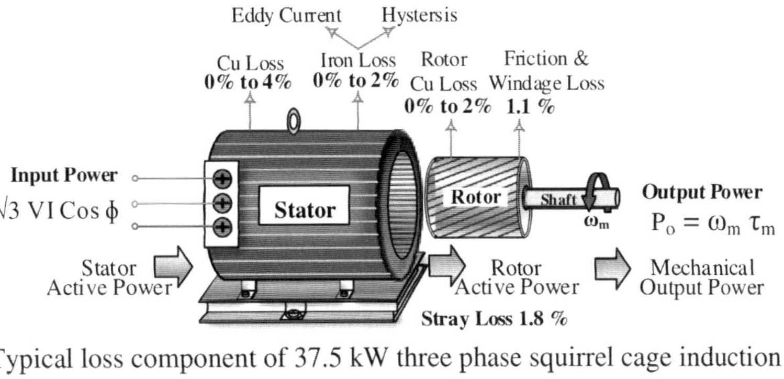

Figure 6. Typical loss component of 37.5 kW three phase squirrel cage induction motor

The overall efficiency of a motor is dictated by the construction materials, mechanical and electrical design. Power sensitive motors use high quality components and use engineered architecture to achieve higher efficiencies. Large copper wire diameter in the stator and more aluminum in the rotor minimize motor resistance losses. Enhanced rotor configuration and optimized air gap between stator and rotor results in stray loss reduction. An optimized design of the cooling fan gives ample motor cooling with a minimum loss of windage. Thinner and higher-quality rotor and stator core steel laminations allow the energy-efficient motor to work with lower magnetization losses. High quality bearings lead to reduced losses of friction [22]. The development and the architecture advancement have led in time to understand a higher importance to the dimension of the magnetic permeability. Magnetic permeability is defined as the ratio of the magnetic induction to the magnetic intensity. It is a scalar quantity and denoted by the symbol μ. Magnetic permeability helps us measure a material's resistance to the magnetic field or measure of the degree to which magnetic field can penetrate through a material. Realizing that magnetic losses, heating and power losses occur in the core magnetization process, it is also important to assess the required current intensity. Effectiveness in magnetizing the conductive material allows a lower quantity of it to be used optimally. Materials that ensure easy magnetization in isotropic manner in all directions of the lamination plan making up the ferromagnetic core for applications on rotary machines. From the viewpoint of material crystallography, it is necessary to ensure that the (crystallographic) directions of easy magnetization are distributed homogeneously in the plane. Focusing on improved crystalline texture includes maximizing magnetic properties to achieve greater efficiency of application. Magnetic lamination, i.e. the ferromagnetic content inside the core and its behavior, but also the amount of conductive content and the current strength required to produce the desired result. Thus, considering the material resistivity element, it faced the selection of materials that benefited predominantly from silicon but also from aluminum or manganese. With a limit, as the commonly used Silicon rises with values above 3.5 percent, the substance is very brittle, making their industrial development and usage very important. A further strategy then consisted of orienting the research towards improving the quality of the material surface and its cleaning. The new trend then appears to be conducting work into extremely thin materials but we should consider the potential implications right now. From mechanical parts to approaches that incorporate laser technology that are less influenced by the influence of thickness and that result in a slight deterioration of magnetic activity in the blanking / cutting regions. In the meantime, we should imagine producing super-thin laminations with a multi-layer technology that can reduce magnetic loss while facilitating their use.

5. Summary

A review on the progress in research in replacing heavy metal parts by lightweight parts made of plastics and composites are presented. The details of requirements of materials for withstanding the turning moment and their ability to conduct magnetic field are discussed. The intricacies involved with the design of composite parts to satisfy the requirements are shown. Novel cooling methods like, cryogenic cooling with liquid nitrogen, liquid cooling system, indirect cooling by nano-fluids as coolants are presented. Heat transfer rate is found to rise up to 40% while using aluminium oxide nano particles into base fluid water. High temperature superconducting (HTS) wires are found to reduce weight and increase power density. Rectangular flat wires are found to be more efficient compared to round wires in conducting heat. High quality materials and optimized design are used by energy efficient motors for achieving high efficiencies.

References

[1] Solomon D G, Greco A, Masselli C, Gundabattini E, Rassiah R S and Kuppan R 2020) *Annales de Chimie - Science des Matériaux,* **44** 1-14 https://doi.org/10.18280/acsm.440101
[2] Peter M, Fleischer J, Blanc, F S L, and Jastrzembski J P 2013 *IEEE EDPC Conf.*
[3] Koch S F, Peter M and Fleischer J 2017 *Procedia CIRP.* **66** 283-288. doi:10.1016/j.procir.2017.03.274
[4] Jaeger M, Ruf A, Hameyer K and Tongeln T G V 2018 *IEEE Transportation and Electrification Conference and Expo, ITEC 2018* pp 194–200 doi: 10.1109/ITEC.2018.8450239
[5] Jercic T, Zarco D, Martinovic M, Kovacic M, Juric J, Hanic Z and Stipetic S 2017 *IEEE Int. Electr. Mach. Drives Conf. IEMDC* pp 1–7
[6] Wang S, Li Y, Li Y Z and Xiong K 2018 *IEEE/ASME International Conference on Advanced Intelligent Mechatronics, AIM* pp 1184–89 doi: 10.1109/AIM.2018.8452450
[7] Dilshad M R, Ashok S, Vijayan V and Pathiyil P 2016 *Proceeding IEEE - 2nd Int. Conf. Adv. Electr. Electron. Information, Commun. Bio-Informatics, IEEE – AEEICB* pp 172–76 doi: 10.1109/AEEICB.2016.7538266
[8] Xu Z, Rocca A L, Pickering S J, Eastwick C, Gerada C and Bozhko S 2015 *Proc. - 2015 IEEE Int. Electr. Mach. Drives Conf. IEMDC* pp 1607–13 doi: 10.1109/IEMDC.2015.7409278

[9] Marignetti F 2010 *IEEE Trans. Ind. Appl.* **46** 984–92 doi: 10.1109/TIA.2010.2045328
[10] Deák C, Brauneiser H J and Dietel H 2011 *1st Int. Electr. Drives Prod. Conf. Proc.* pp 5–10 doi: 10.1109/EDPC.2011.6085540
[11] Akawung A F and Fujimoto Y *IEEE Int. Symp. Ind. Electron.*pp 273–78 doi: 10.1109/ISIE.2019.8781494
[12] Fujita H, Itoh A and Urano T 2019 *World Electr. Veh. J.* **10**, doi: 10.3390/wevj10020038
[13] McCluskey E P, Saadon Y, Yao Z and Camacho A 2019 *Intersoc. Conf. Therm. Thermomechanical Phenom. Electron. Syst. ITHERM* pp 1134–38 doi: 10.1109/ITHERM.2019.8757435
[14] Gieras J F *17th Int. Conf. Electr. Mach. Syst.* pp 1–16 doi: 10.1109/ICEMS.2014.7013920
[15] Zadeh A D and Monte F D *Entropy* **22** 99 doi: 10.3390/e22010099
[16] Ma Q, El-Refaie A and Lequesne B 2020 *IEEE Trans. Ind. Appl.* **56** pp 1452–63 doi: 10.1109/TIA.2020.2966458
[17] Niasar A H and Nikkhah *IEEE J. Emerg. Sel. Top. Power Electron.* **8** pp 1268–75 doi: 10.1109/JESTPE.2019.2909030
[18] Sun Y, Zhang S, Chen G, Tang Y and Liang F 2020 *Appl. Therm. Eng.* **170** p 114970 doi: 10.1016/j.applthermaleng.2020.114970
[19] Li D, Wen Y, Li W, Feng B and Cao J 2018 *Energies* **11** p 1208 doi:10.3390/en11051208
[20] Institution of Diagnostic Engineers https://institutiondiagnosticengineers.org/news/cooling-technology-allows-motors-to-be-built-from-fibre-reinforced-polymers accessed on 2nd June 2020
[21] Peet M J, Hasan H S and Bhadeshia H K D H 2011 *Int. J. Heat Mass Transf.* doi: 10.1016/j.ijheatmasstransfer.2011.01.025.\
[22] Kuehl A, Kreitlein S, Hoeft A and Franke J *IEEE Texas Power and Energy Conference (TPEC)* (College Station, TX) pp 1-5 doi: 10.1109/TPEC.2018.8312076

IOP Conf. Series: Materials Science and Engineering **906** (2020) 012021 doi:10.1088/1757-899X/906/1/012021

Harmonic elimination and dc-link voltage balancing in bipolar hybrid microgrid using Space vector modulation for Interlinking Converter (ILC)

S Vasantharaj[1*], V Indra Gandhi[2]

[1]Research Scholar, Vellore Institute of Technology, Vellore
[2]Associate Professor, Vellore Institute of Technology, Vellore
[*]Corresponding author E-mail: Vasantharaj.s@vit.ac.in

Abstract: The modernization of distribution grids developed the power systems towards the smart grid. The microgrid plays a vital role in an improving power systems area, which refers to the loads operated under an organized and synchronized way and the distributed energy resources. The microgrid, which is operated in "islanded" mode, should be connected to the main power grid, or it must be entirely off-grid. Bipolar hybrid microgrid feeds local loads with local resources because it consists of both DC and AC bus simultaneously. The converter, which is used for microgrid application, plays an important role, which is said to be Interlinking Converter (ILC). The lower order harmonics can be eliminated by these ILCs. By reducing the THD and increasing the dc-link voltage utilization SVM is used. In this paper, there will be an increase in system performance and great reliability with a fully automated system. There will be a decrease in switching losses, current ripple and THD of grid current by the use of a new switching modulation method, which is validated by the Matlab simulation results.

Keywords:
Interlinking Converter, THD, dc-link, carrier based modulation, voltage ripple, grid and hybrid microgrid.

Nomenclature:

ILC	Interlinking Converter	NPC	Neutral Point Clamped
SVM	Space Vector Modulation	V_{dcl}	DC-link Voltage
DC	Direct Current	VSC	Voltage Source Converter
AC	Alternating Current	BHM	Bipolar Hybrid Microgrid
U_{ref}	Normalized Reference Voltage Vector		
MG	Microgrid	V_{dc}	DC voltage
MLC	Multilevel converter		

1.Introduction:

The ILC performs the bidirectional flow of power without any power conversion circuit between dc and ac buses. The distributed power is transferred between utility grids to MG and vice versa [1–3] will be managed by ILCs, which are the essential device of microgrid [4–6]. The proper bidirectional ILC should be selected for the load management and the resources that are connected to the microgrid. The BHM consists of two self-governing AC and DC buses [7]. ILC can control variables of ac or dc bus that act as an interface between the buses. Hence, in an AC microgrid the operation of ILC has two conditions: (1) DC Grid Feeding and AC Grid Forming, (2) AC Grid Feeding and DC Grid Forming [8–11]. The performance of ILC will not be affected by the changes in ac bus but the dc bus voltage will be formed by the ILC. In hybrid microgrid, two level VSC act as a suitable ILC topology for BHM [12]. The proposed work contains more efficiency compared to multilevel ILCs. In various cases by having, more number of voltage levels in dc bus we can able to connected different voltage levels of loads and resources. A boost converter has to be applied to link PV sources to the dc bus, though, when lower dc-link is in range, avoids additional boosting voltage, cost of the converter and volume by joining the PV

IOP Conf. Series: Materials Science and Engineering **906** (2020) 012021 doi:10.1088/1757-899X/906/1/012021

to this lower V_{dc} [13]. Alternatively, the voltage balancing challenge will be raised by various V_{dc} levels, which is offered by multilevel converters as ILC.
During high switching frequencies MLC does not accomplish well due to switching power loses. It is well known that in medium and high power applications, multilevel converter needs extra semiconductors [14]. The 10-switch converter performs more advantages in both two level and three level converters that negotiates amongst two level VSCs and MLCs. In [15], it explains about the different types of MLCs, having combination of three-level NPC and two level VSC. In preceding works on 10-switch converter, V_{dcl} balancing problems are not discussed because it is restricted to SPWM methods [16, 18]. The Z source converter that is connected to 10-switch converter produces output voltage of low quality with unbalanced currents with the use of four-vector SVM method. Furthermore, in the switching sequence of SVM by applying two small voltage vectors there will be an increase in THD and decrease in dc-link voltage utilization.
In this proposed SVM a full scale V_{dcl} balancing approach is designed and simulated. The survey has taken for harmonizing the dc-link poles and the same is evaluated for dc-link balancing strategies with the help of this proposed 10-switch converter. The main features of this 10 switch converter is elaborated in Section 2 and explains about the elimination method for lower order harmonics with the analyses of universal control pattern for a grid-tied ILC. The simulation results of this proposed SVM for BHM under balanced and unbalanced conditions are described in Section 4. Similarly, the current THD, voltage across ILC, dc link voltages and dc voltage ripple will be evaluated [19–21]. By the use of this simulation results the benefits of V_{dcl} balancing and suggested modulation method is evaluated.

2. Diagrammatic Representation and Circuit model of ILC

The BHM contains ILC which is shown in Fig. 1. An ac output voltage of 2/3 level with three-wire bipolar dc topology is proposed by 10 switch converter. The conventional voltage source converter (VSC) added with S1A to S4A switches to obtain this proposed converter. This converter comprises of 3 main legs and an auxiliary leg contains 4 switches having three different forms of switching states (P, O, N). The switching states and their illustration is shown in table 1.

Table 1: switching states of ILC

Switching State	ON switches		Terminal voltage
P	S1A	S1 or S3 or S5	E
O	S2A S3A	S1 or S3 or S5 S2 or S4 or S6	0
N	S4A	S2 or S4 or S6	-E

The auxiliary leg of switches S2A or S3A produces a switching state O and these switches depend on former leg states. The remaining switching states are said to be understandable. eight and twenty seven switching vectors of two-level and three-level VSI is compared with the 10-switch converter which has twenty one switching vectors that contains eighteen active and three zero vectors (PPP, OOO, NNN). The twenty-one switching vectors of 10 switch converter is splited as three zero vectors which are sited midpoint of a hexagon including six large and twelve small vectors. The dc-link poles are short-circuited due to this 10-switch converter by having this medium level vectors. The SVM, which is applied for 10-switch converter, consists of six sectors and these sectors are subdivided into small parts having the shapes of trapezoid and triangle. By using this SVM there will be an increase in ac grid THD, switching configuration and NP voltage deviation.

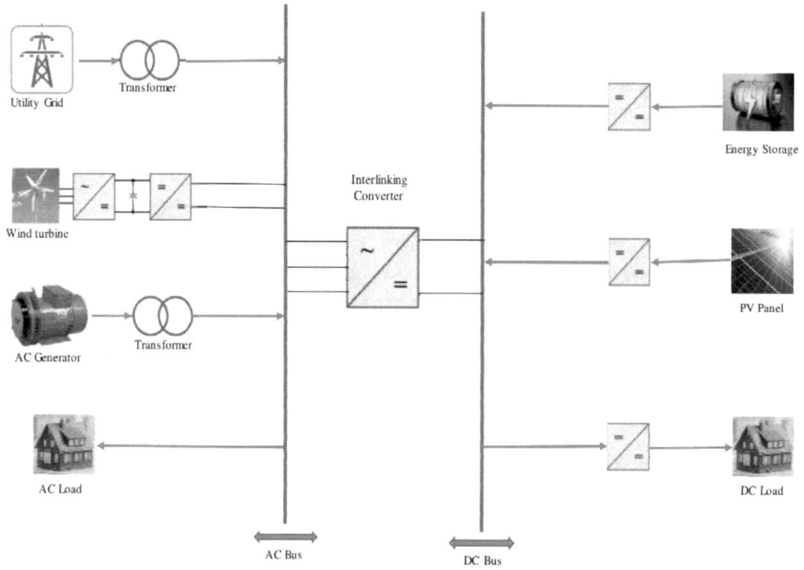

Figure 1. Diagrammatic Representation of Interlinking Converter for Bipolar Hybrid microgrid.

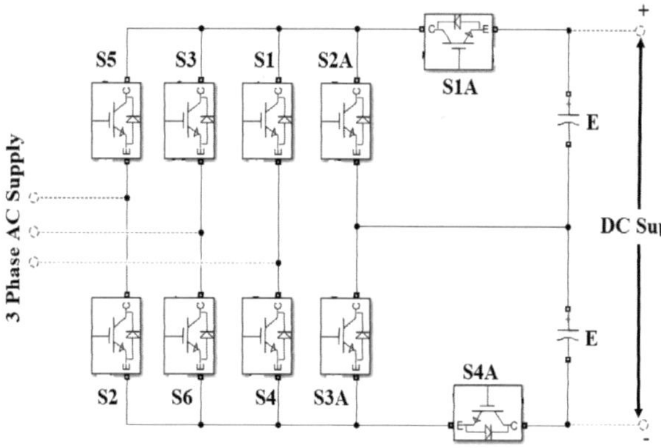

Figure 2. Circuit diagram for ILC.

Fig. 1 shows the bipolar hybrid microgrid with grid tied converter. Similarly, fig 2 represents the circuit diagram of 10 switch interlinking converter. The voltage and frequency on ac bus is controlled by the use of diesel generator, micro turbine or some other converters, that's why here the ILC is linked to an ac bus. The main purpose of this ILC is to control and stabilize the pole voltages of the DC bus and it is used to inject active and reactive power to ac bus. The THD of ac current in the MG is affected by slight variations in grid, which is produced by the low-order harmonics.

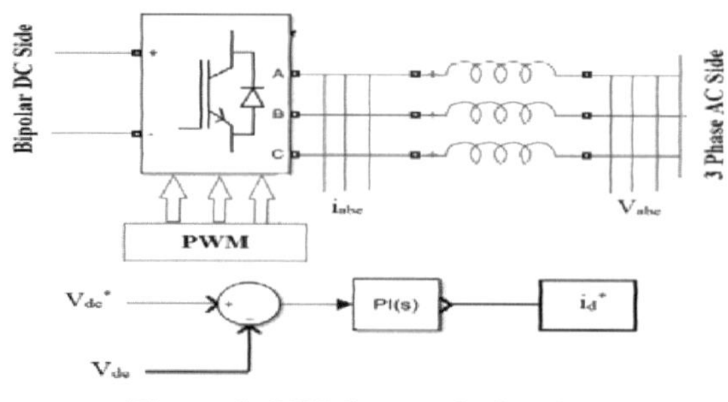

Figure 3. PWM control of ILC.

To overwhelmed these minor distortions effect it consists of two different methods, 1 — feed-forwarding grid voltage [21, 22]. 2 — by interfacing harmonic controller parallel with PI controller. To eliminate lower-order harmonics like fifth and seventh the resonant controller (Gr) is tuned to the same frequency. The basic PWM control of ILC is shown in fig. 3 is upgraded by these usual methods.

$$G_r(s) = \frac{K_r s}{S^2 + (6\omega_n)^2} \tag{1}$$

Where, ω_n is the nominal angular frequency and K_r is the gain.

3. SVM topology and its switching sequence

A V_{dcl} balancing ability is elaborated in this subdivision by the use of proposed SVM technique. The main objective of this novel SVM method is to improve dc-link voltage and to decrease the output current THD and common mode voltage. Similarly, this proposed SVM need not have any passive or active elements. The SVM operates under both balanced and unbalanced conditions that lead to decrease in extra converter expenses, volume and loss due to power.

3.1. Region determination

Initially, corresponding SVM for three level VSI, the V_{ref} is induced to measure its angle and amplitude. Then, it is to be determined that were V_{ref} is located and to identify the sector and region for this reference voltage vector. Each sector region is similar to one another and it is validated in fig 4. The region and the sector in which it is situated will be stipulated by the angle and amplitude of the V_{ref}.

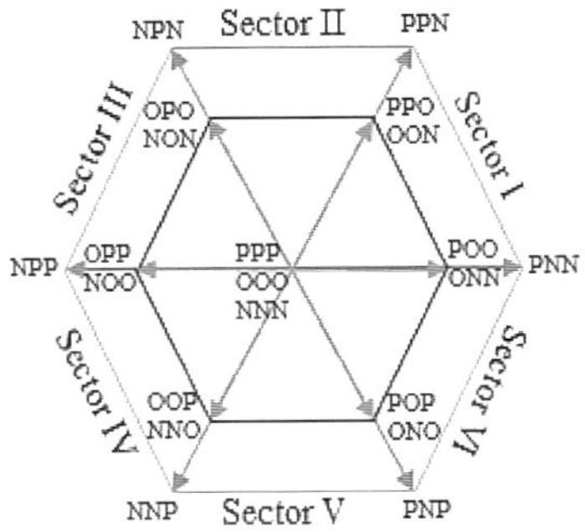

(a)

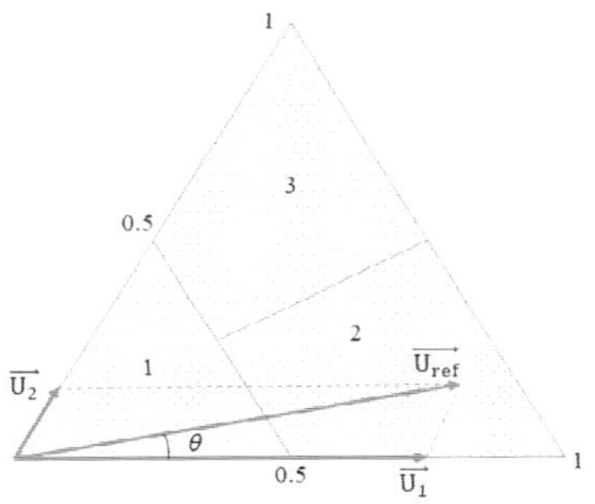

(b)

Figure 4. (a) SVM for 10 switch converter (b) Sector representation for normalized voltage vector

The U_{ref} is attained to compute and locate the region of V_{ref} for this 10-switch interlinking converter. The vectors U_1 and U_2 obtain the resultant vector of U_{ref}, which is shown in equation (2), (3) and table 2.

$$U_1 = U_{ref}\left(Cos\theta - \frac{sin\theta}{\sqrt{3}}\right) \tag{2}$$

$$U_2 = 2U_{ref}\left(\frac{sin\theta}{\sqrt{3}}\right) \tag{3}$$

To eliminate over modulation these calculations are used. The U_{ref} angle and its amplitude are used to change the region of respective sectors with the values U_1 and U_2 maximum of 1. The U_{ref} amplitude goes in the trapezoidal area, which contains two equal areas after passing region 1 area, which is shown in fig 4. The trapezoidal region changes from 2 to 3 when the reference voltage vector attains 30^0 which is equal to $\sqrt{3}/2$. The non-existence of medium voltage is reimburse with the split up trapezoidal regions of 2 and 3.

Table 2. Region identification based on U_{ref}

Case	Angle (θ)	Region
$U_1 \leq 0.5$ and $U_2 \leq 0.5$ & $U_1+U_2 \leq 0.5$	NA	1
$U_1 > 0.5$ or $U_2 > 0.5$ or	$\theta \leq 30^0$	2
$U_1 + U_2 > 0.5$	$\theta > 30^0$	3

3.2. Symmetrical switching sequence

This type of switching sequence is organized to minimize the THD, which is considered as the main objective of this proposed method.

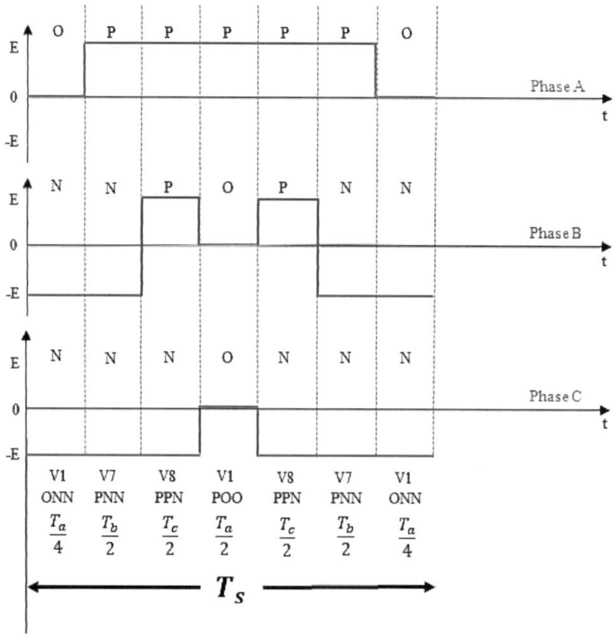

Figure 5. Switching sequence of sector 1.

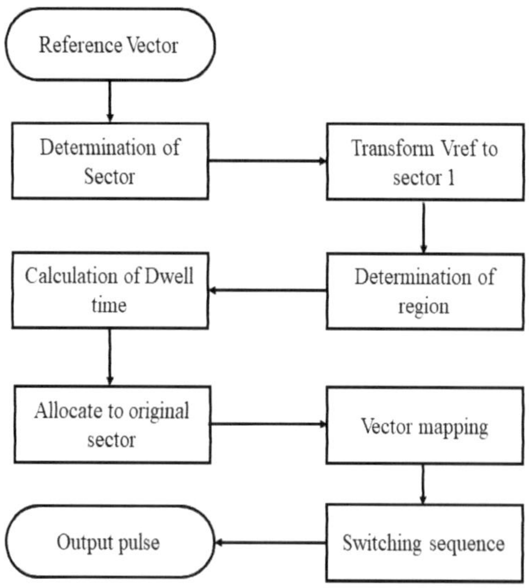

Figure 6. Flowchart for SVM calculation.

The switching time (Ts) for the sector 1 with switching sequence is shown in Fig. 5. Since fig. 4 shows identical sectors and regions, hence by having the reference of first sector, the vector selection and switching sequence of this proposed SVM for all the six sectors eighteen regions are calculated. The selected voltage vectors finalize the switching pattern. Fig. 6 shows the flowchart for the proposed SVM method.

3.3. Voltage balancing with modified switching sequence

In bipolar microgrid, by injecting different and contradictory small voltage vector for 10-switch converter the unbalanced DC-link pole voltage will be rectified. Fig 5 that shows the seven segment switching sequence selects two forms of small vector voltages similarly under balanced condition. Moreover, according to these conditions, it may have five or six segments of switching sequence. For an unbalanced condition, the switching sequence is modified and its states are shown in fig. 7.

V1	V7	V8	V1	V8	V7	V1
ONN	PNN	PPN	POO	PPN	PNN	ONN
$\frac{kT_a}{2}$	$\frac{T_b}{2}$	$\frac{T_c}{2}$	(1-k)Ta	$\frac{T_c}{2}$	$\frac{T_b}{2}$	$\frac{kT_a}{2}$

← T_s →

Figure 7. Unbalanced switching Sequence in sector 1.

There will be slow reduction in dc-link pole voltage and escalation in another pole by extracting unequal power from those poles. By using modest and forthright procedure, a coefficient k is determined to find the unbalanced voltages [23]. The interlinking converter that is operated as both rectifier or inverter mode has the value of K regarding the direction of power flow. The 10-switch converter acts as a rectifier mode has the following equations (4) and (5).

$$V_{dc1} - V_{dc2} = V_{Balancing} \tag{4}$$

$$V_{Balancing} > V_{ripple} \Rightarrow K = 1;$$

$$-V_{ripple} \leq V_{Balancing} \leq V_{ripple} \Rightarrow K = 0.5;$$

$$V_{Balancing} < V_{ripple} \Rightarrow K = 0; \tag{5}$$

The dc-link pole voltage ripple is obtained by difference in two dc-link pole voltages and it is denoted as V_{ripple}. The coefficient K becomes 0.5 when the converter works under balanced condition with the absolute value of $V_{balancing}$ is lesser than the V_{ripple}. The coefficient K equals 1 when the converter works under unbalanced condition with the absolute value of $V_{balancing}$ is larger than the V_{ripple}. Hence, by using equations (8) and (9), the coefficient K value can be determined.

4. Simulation Results:

The MATLAB simulated output of the 10 switch-interlinking converter with SVM method for BHM is discussed in this section. The proposed SVM is used to eliminate the harmonics, which is shown, by using FFT spectrum of grid converter output current and voltage. Similarly, controlled dc-link voltage for balanced and unbalanced conditions is examined. Fig 8 shows the MATLAB simulated block diagram for this BHM.

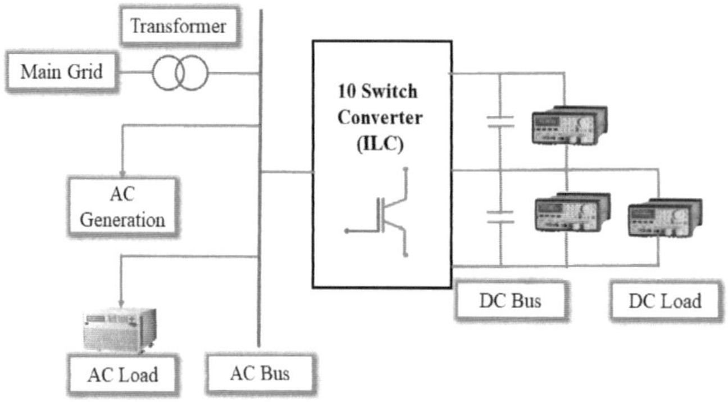

Figure 8. Block Diagram of MATLAB Simulation.

By the use of Matlab Simulink 2014a the ILC with BHM is simulated and the results are discussed below. The Simulink model of MG is simulated under balanced and unbalanced condition for the grid side. Fig. 9 shows the simulated result of PV array and wind energy conversion system.

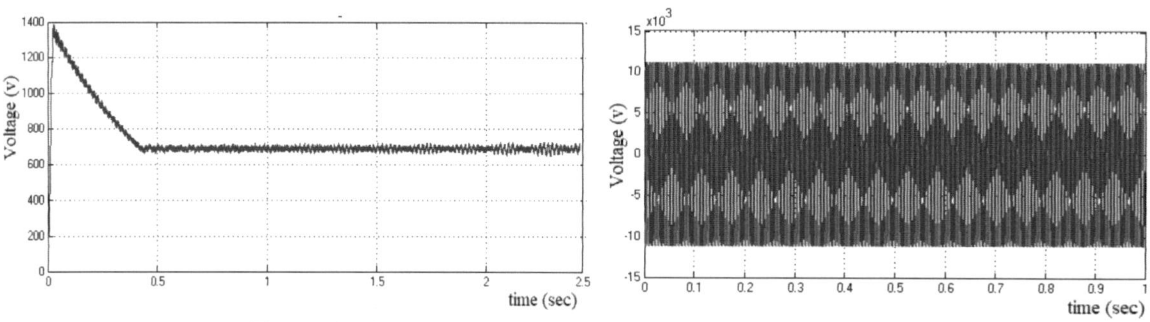

Figure 9. Simulated output of solar PV and Wind energy system

The battery voltage of bipolar hybrid microgrid has a value of 230 V and utility grid that supplies a voltage of 9 kW, which is shown in fig 10. The ac grid under unbalanced condition having a simulated output voltage of 500 V and a current of 5 A which is shown in fig. 11.

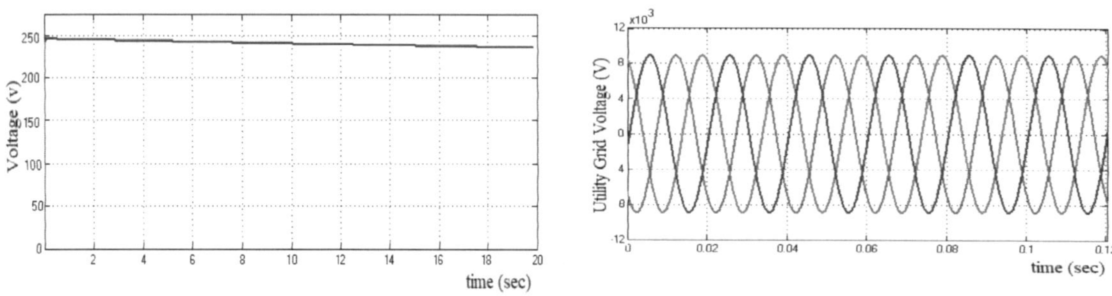

Figure 10. Simulated output of battery and ac utility grid.

The THD analysis for a proposed interlinking converter for an unbalanced condition as a FFT spectrum ac output voltage and current is displayed in fig 12. The output voltage and current has a THD of 20.51% and 12.71%.

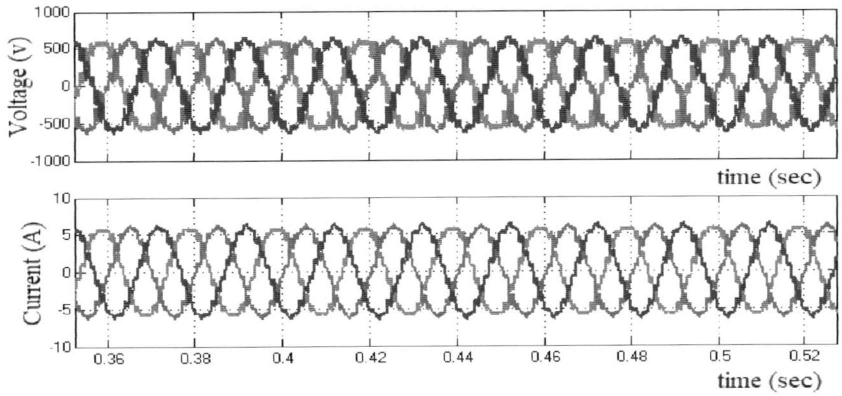

Figure 11. Simulated ac grid side output of Voltage and current for unbalanced condition.

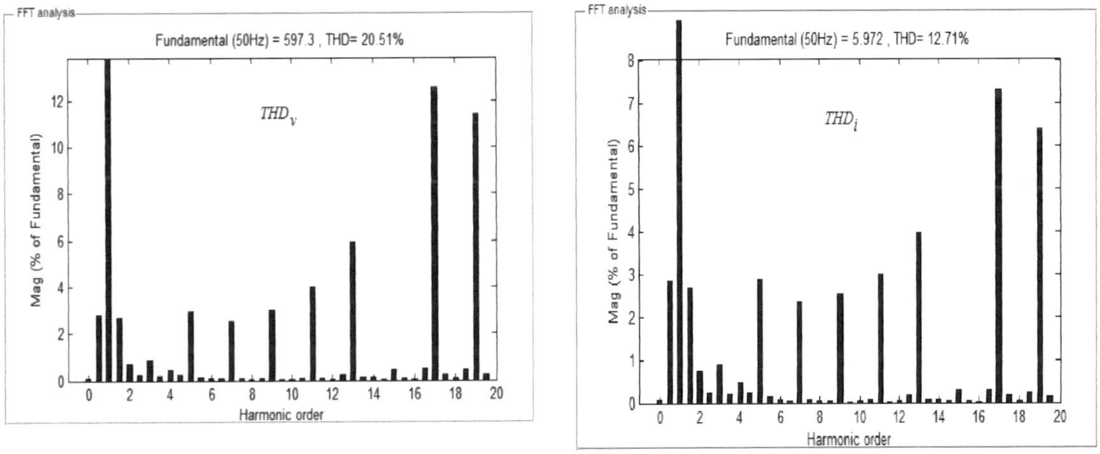

Figure 12. FFT Analysis for grid side output voltage and current for unbalanced condition.

The balanced output voltage and current waveform of grid-connected system is shown in fig. 13. Similarly, fig. 14 shows the balanced condition of grid for this proposed system the THD analysis and FFT spectrum of ac output voltage and current, and their THD are decreased to 12.97% and 4.73%.

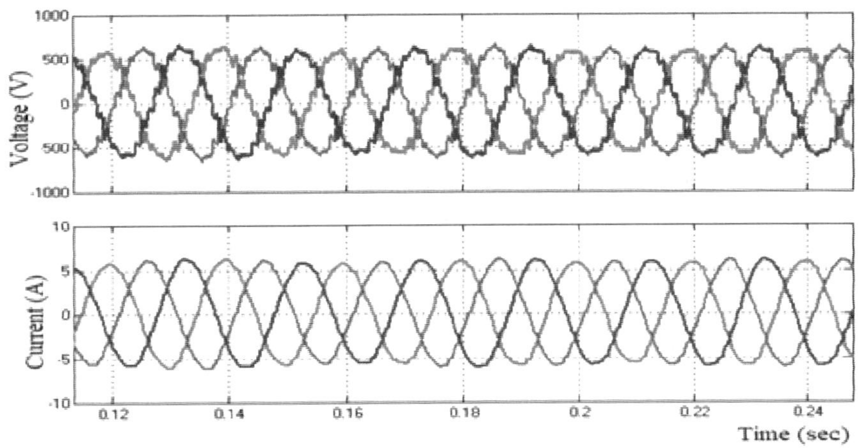

Figure 13. Simulated ac grid side output of Voltage and current for balanced condition.

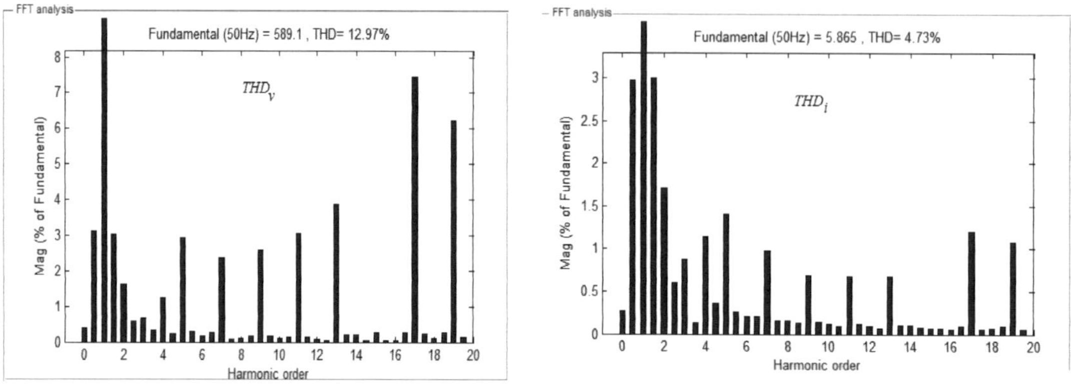

Figure 14. FFT Analysis for grid side output voltage and current for balanced condition.

5. Conclusion:

This paper proposes the working of MG with two-way power flow with the help of ac/dc bidirectional converter and SVM technique. In this proposed SVM method, the grid obtains constant power flow by the use of bidirectional converter under unbalanced condition. This will enrich the consistency of the bidirectional ac/dc converter in weak MG. The efficiency of the MG with bidirectional converter is simulated for balanced and unbalanced grid voltages under dissimilar control schemes. In addition, the renewable energy sources like solar PV and wind energy system features are simulated and the results were discussed. By this proposed SVM technique, the lower order harmonics in the ac grid side are eliminated. The objective of this SVM comprises reduction of current THD in ac grid and increase in V_{dcl} consumption. Still, the proposed SVM technique is very tough and unites the pole V_{dcl} in severe unbalance circumstances. The power quality issues and the elimination of lower order harmonics in BHM with 10-switch converter evaluated by the use of proposed SVM is simulated with the help of Matlab Simulink.

Reference:

[1] Dileep G, A survey on smart grid technologies and applications, Renewable Energy 146 (2020) 2589-2625.

[2] Saurabh Kumar, Vijayakumar K, Simulation and experimental comparative analysis of the DC-DC converter topologies for wind driven SEIG fed DC Nano grid, Electric Power Systems Research 181 (2020) 106196.

[3] Elizabeth Noghreian, Hamid Reza Koofigar, Power Control of hybrid energy systems with renewable sources (wind-photovoltaic) using switched systems strategy, Sustainable Energy, Grids and Networks 21 (2020) 100280.

[4] F. Nejabatkhah, Y.W. Li, Overview of power management strategies of hybrid AC/DC microgrid, IEEE Trans. Power Electron. 30 (2015) 7072–7089.

[5] C. Yin, H. Wu, F. Locment, M. Sechilariu, Energy management of DC microgrid based on photovoltaic combined with diesel generator and super capacitor, Energy Converters. Manag. 132 (2017) 14–27.

[6] E. Hossain, E. Kabalci, R. Bayindir, R. Perez, Microgrid testbeds around the world: State of art, Energy Converters. Manag. 86 (2014) 132–153.

[7] P.C. Loh, D. Li, Y.K. Chai, F. Blaabjerg, Autonomous control of interlinking converter with energy storage in hybrid AC-DC microgrid, IEEE Trans. Ind. Appl. 49 (2013) 1374–1382.

[8] J.Y. Kim, J.H. Jeon, S.K. Kim, C. Cho, J.H. Park, H.M. Kim, K.Y. Nam, Cooperative control strategy of energy storage system and micro sources for stabilizing the microgrid during islanded operation, IEEE Trans. Power Electron. 25 (2010) 3037–3048.

[9] J. Rocabert, G.M.S. Azevedo, A. Luna, J.M. Guerrero, J.I. Candela, P. Rodrguez, Intelligent connection agent for three-phase grid-connected microgrids, IEEE Trans. Power Electron. 26 (2011) 2993–3005.

[10] J. Rocabert, A. Luna, F. Blaabjerg, P. Rodríguez, Control of power converters in AC microgrids, IEEE Trans. Power Electron. 27 (2012) 4734–4749.

[11] X. Lu, J.M. Guerrero, K. Sun, J.C. Vasquez, R. Teodorescu, L. Huang, Hierarchical control of parallel AC-DC converter interfaces for hybrid microgrids, IEEE Trans. Smart Grid 5 (2014) 683–692.

[12] Amir Reza Hassani Ahangar, Gevork B. Gharehpetian, Hamid Reza Baghaee, A review on intentional controlled islanding in smart power systems and generalized framework for ICI in microgrids, Electrical Power and Energy Systems 118 (2020) 105709.

[13] Mohsen Eskandari, Li Li, Mohammad. H. Moradi, Improving power sharing in islanded networked microgrids using fuzzy-based consensus control, Sustainable Energy, Grids and Networks 16 (2018) 259–269.

[14] H. Abu-Rub, J. Holtz, J. Rodriguez, G. Baoming, Medium-voltage multilevel converters State of the art, challenges, and requirements in Industrial applications, IEEE Trans. Ind. Electron. 57 (2010) 2581–2596.

[15] R. Rahimi, G.R. Moradi, E. Afshari, B. Farhangi, S. Farhangi, Study of leakage current phenomena in hybrid 2/3-level three-phase transformerless photovoltaic grid-connected inverters, in: 1st Int. Conf. New Res. Achiev. Electr. Comput. Eng., 2016.

[16] Hammad Armghan, Ming Yang, Ammar Armghan, Naghmash Ali, M. Q. Wang, Iftikhar Ahmad, Design of integral terminal sliding mode controller for the hybrid AC/DC microgrids involving renewables and energy storage systems, Electrical Power and Energy Systems 119 (2020) 105857.

[17] J.H.G. Muniz, E.R.C. Da Silva, R.B. Da Nobrega, E.C. Dos Santos, An improved pulse-width-modulation for the modified hybrid 2/3-level converter, in: 2013 Brazilian Power Electron. Conf. COBEP 2013 - Proc. 2013, pp. 248–253.

[18] Reza Ghanizadeh, Gevork B. Gharehpetian, Distributed hierarchical control structure for voltage harmonic compensation and harmonic current sharing in isolated microgrids, Sustainable Energy, Grids and Networks 16 (2018) 55 – 69.

[19] Mohamed R. Amer, Osama A. Mahgou, New switching technique for current control of grid converters for wind power systems, Sustainable Energy, Grids and Networks 9 (2017) 1–12.

[20] T. Abeyasekera, C.M. Johnson, D.J. Atkinson, M. Armstrong, Suppression of line voltage related distortion in current controlled grid connected inverters, IEEE Trans. Power Electron. 20 (2005) 1393–1401.

[21] X. Wang, X. Ruan, S. Liu, C.K. Tse, Full feedforward of grid voltage for grid connected inverter with LCL filter to suppress current distortion due to grid voltage harmonics, IEEE Trans. Power Electron. 25 (2010) 3119–3127.

[22] Sunghyok Kim, Songchol Hyon, Chonung Kim, Distributed virtual negative – sequence control for accurate imbalance power sharing in islanded microgrids, Sustainable Energy, Grids and Networks 16 (2018) 28 – 36.

IOP Conf. Series: Materials Science and Engineering **906** (2020) 012022 doi:10.1088/1757-899X/906/1/012022

SMART WATER GRID: AN IoT FRAMEWORK

D.Saravanan[1], Aryan Agrawal[2], R. Raja singh[2],V.Indragandhi[2]

[1]Trane Technologies, Thermoking,ETC-Bangalore,Karnataka,India.
[2]School of Electrical Engineering, VIT Vellore,Tamil Nadu,India.

Corresponding author's E-mail ID:saravanan13051119@gmail.com

Abstract. Water scarcity became most inexorable problem these days. In urban areas most of the people live in flats or apartments where, they need to get water from the corporation and then they ought to supply for each flat and apartment. Supplying water in a normal way arises many technical issues. In this paper, a working prototype is implemented to solve these problems. This model contains one Main water tank and two small sub water tanks. Each node (tank) can function automatically based on the tank and ground water levels or for other problems and also can be controlled manually from the water level through android app anywhere in the world with the help of IOT. As the Indian populace keeps on developing, India's water infrastructure should be prepared to take care of an expanding water demand while keeping up great resource condition to guarantee good quality of water is provided to customers around the clock.

1. Introduction

A Smart Water Grid framework information and communications technologies (ICT) into the classical administration of the water conveyance framework as shown in figure 1. Sensors, meters, advanced controls and analytic tools are utilized to automate, monitor and control the transmission and appropriation of water, ensuring that water is effectively with great quality. Today, numerous ventures are making utilization of ICT to enhance their activities and procedures, especially through digitization and robotization.

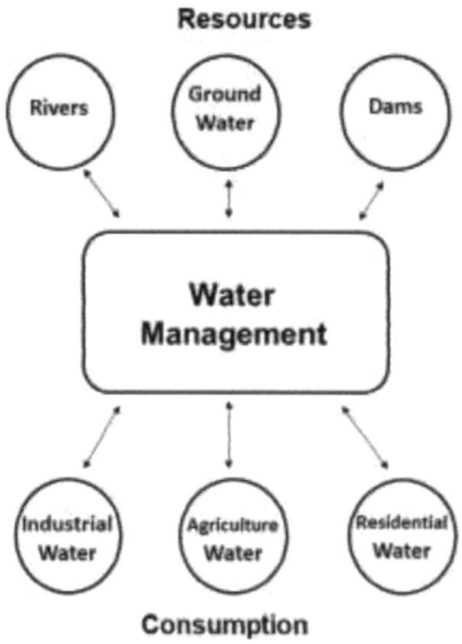

Figure 1: Smart Water Grid – A system overview

IOP Conf. Series: Materials Science and Engineering **906** (2020) 012022 doi:10.1088/1757-899X/906/1/012022

Digitization and mechanised gathering of information at site, and remote transmission to a focal framework for checking and investigation. Late progressions in ICT have been to a great extent driven by the rise of Internet of Things, cloud computing and enormous information investigation. At the point when connected to the water business, these advances in ICT empower to catch and store tremendous pools of huge information and perform capable examination and prescient investigation to contribute towards more prominent productivity and viability in water asset administration.

There is an expanding need to use on these inventive advances so as to guarantee the water infrastructure is feasible and strong. Wireless sensor networks (WSN), considered to be with lot of the operational procedures, for example, water spillage checks, water quality checks and meter readings, are as yet done physically. Against the scenery of a contracting and progressively instructed work drive, the usage of the Smart Water Grid will likewise streamline labour necessities and overhaul work forms via mechanizing modest undertakings. Decision support apparatuses, for example, predictive hydraulic modelling, valve activity recreation, request expectation demonstrating and pipeline disappointment examination, help to enhance arrange evaluations and arranging.

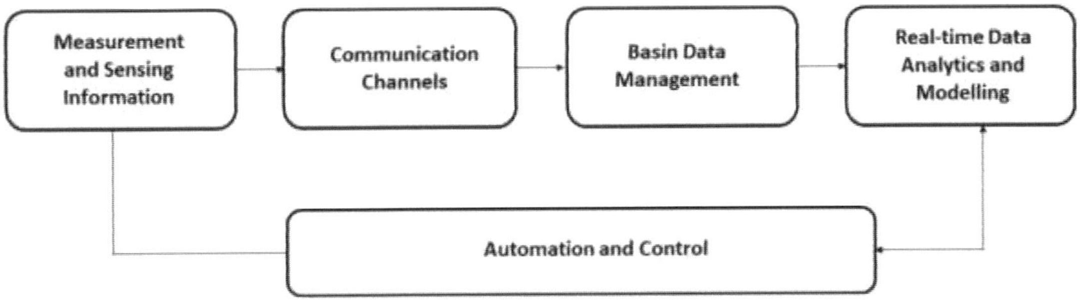

Fig. 2: Flow of modern water grid system

Saving water is essential idea of Smart water Grid. Flow of modern water grid shown in figure 2 which involves ultrasonic sensors to monitor the levels of water in a users' tank, and fill it up if the water is low. The filling up also stops if the tank is full, allowing no wastage of water. Use of IoT to monitor the ultrasonic sensors and thus provide real-time feed about the ongoing process. The tanks have been fitted with sensors that are connected to Wi-Fi modules. These Wi-Fi modules collect data from sensors, like water level, whether motor is ON or OFF and indicates when it is low. This helps to monitor the water levels and be ahead of our needs and conserve water through real-time monitoring.

2. Literature survey

2.1. *Internet of Things enabled real time water quality monitoring system*

In this paper the authors describes a new concept at are being developed in the smart water monitoring which gives some easy solutions to monitor the in-pipe water quality using the Internet of Things Technique with an efficient power usage. By developing a model where the sample water and then the data which was collected are uploaded to the internet using the IoT (Internet of Things). Also analysing the data using this IoT. Additionally there is an added remote to the model which is used to alert the user. This work predefined the standard values for the quality of water in the system, if there is any difference in the water quality compared to the standard values then the remote will alert the user. Finally this paper provided a water quality monitoring system which has a less manufacturing cost.

2.2. *Smart Water Grid: The future water management platform:*

In this particular paper the authors explains the method called 'how to use Smart Water Grid in the water scarcity area's'. Communication technology and information technology plays a big role in the world. The Smart Water Grid project is connecting these two things in a water management system. Also explaining about that Smart Water Grid is the good technology for solving the water storage and water monitoring related problems. This paper deals with the important research related topics that should be integrated in the Smart Water Grid. It should have a good platform configuration, intelligent control, better water infrastructure management and most importantly the efficiency of energy. Finally concluded this paper with, how to integrate these things to smart water grid and make the best water management platform for future.

2.3. *Researches on Smart Water Grid*

In this paper the authors explains the necessity of Smart Water Grid research. How it combines the two technologies, which are, communication and information in a single system and how it efficiently monitors the water supply and how it also used to detect the climate change. All of these things are explained in this paper.

2.4. *Case Study: A Smart Water Grid in Singapore:*

In this paper, a system called Water wise. Because the old water distribution systems are not efficient nowadays because of the increased frequency of usage. So there is a need for a system that has an integrated system which also monitors the water flow, the available percentage of water and quality of water. Because a Smart Water Grid means it should monitor the leakage of water and control the leakage, and it should optimize the operation. Water Wise is a perfect platform which is having an integrated and real-time monitoring. So by this way, monitoring the Water Grid in Singapore in a Smart and an efficient way.

2.5. Smart Water Grid for smart cities: A sustainable prototype demonstrator:

In this paper, a prototype of smart water grid for the smart cities. Because the main aim of smart cities is that all things should work in a smart and efficient way. Also these smart cities have the best communication technology and good network protocols. In this paper, demonstrated a prototype that has water monitoring technology, network sensor for the water height measurements and a smart meter which monitors the total usage of water. By using a Wireless metering bus network at a frequency between 120 Hz to 200 Hz. Finally given the results by simulations and some other experiments.

2.6. Need of Smart water systems in India

In this paper, in India due to the shortage of the water and how we can give better solution for these water shortage problems using the smart water system technology. The reports say that half of the population in the world will go under shortage of water. Especially counties like India, Bangladesh are developing countries. So these countries will definitely face water scarcity. In this paper, a low cost smart water management system to reduce the water scarcity in India. Because within 30 to 40 years India will go through water scarcity and half of the people in India will leave the Indian cities.

3. Analysis and Proposed Technology

Idea of the project mainly tells by using IOT (Internet of Things) as the platform to transmit the information which can be accessed from any place at any point of time. An app called blynk in Google play store with its own cloud. NODE MCU-ESP8266 module is used to transfer water level values to the cloud from where it can be accessed through internet.

This module enables microcontrollers to interface with a Wi-Fi system. It mainly helps to monitor the level of water in centimetres. Once uploading the values into the cloud is done, for the question of getting the values from cloud is done with the help of ultrasonic sensor. An ultrasonic sensor is a gadget that can quantify the separation by utilizing sound waves. It allots a separate by sending a sound wave at a particular recurrence and tuning in for that sound wave to reflect back. By recording the time passed between the sound waves being created and the sound wave reflect back, it is conceivable to figure the separation between the sonar sensor and the water level.The last thing in this project is to monitor and switch on the motor through the Blynk app. The level monitor is set as an analog meter and a switch is placed in the app. The novelty of this project is planned to next level by

IOP Conf. Series: Materials Science and Engineering **906** (2020) 012022 doi:10.1088/1757-899X/906/1/012022

designing a water smart grid project. Finally, we designed our project in such a way to measure the level of water in three sub tanks, a main tank and ground water level from where water is pumped. At first level the sub tanks are measured when it shows the solenoid valve of respective tank opens and water flows from the main tank to the sub tanks. Then the water level in the main tank is measured and when it is less than a threshold value, the motor is switched on with the help of a switch in the app. Whether the motor switches off or on depends on the level of water in both the tanks and the ground water level at that region.

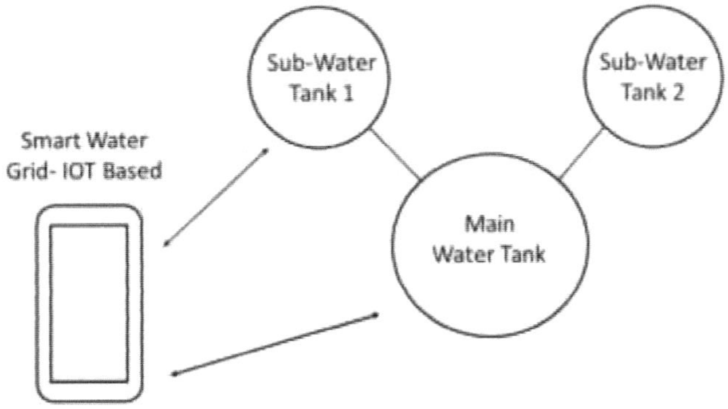

Figure 3: A architecture of IoT based smart water grid system

As shown in figure 3, system works automatically to refill the main and the sub tanks when the water level goes below a threshold value. And on filling the water back, once the water reaches the brim of the tank it stops flowing into the tanks because of another limiting value is set so as to stop the supply of water once the tank gets loaded in order to avoid overflow of water from the tanks.
If in case, there is a shortage or no presence of water even in the underground source, the motor wouldn't even start to supply waters so as prevent the motor from malfunctioning and causing any harm. Only if there is a sufficient amount of water in the Underground source will the motor start working in order to supply the water to the main tank and thereby to the other two sub tanks.

This is mainly controlled through an internet host which connects both the sensor modules and the reception through a mobile or any other internet accessible device. By sensing the level of ground water through the ultrasonic sensor module HC-SR04 the pump will be switched ON provided the ground water condition HIGH and main water tank LOW. This main water tank gets fill only once the two sub tanks gets condition FULL. Pump remains OFF in cases like both ground water and main water tank is LOW & also only when the ground water is LOW. A NODE MCU module has been set in each sub tank & the main tank so as to send the signals to the control system i.e to the host and the operational module which in turn starts recognizing either the shortage or surplus of water which would be prevented from happening.

All such controls and operations can be monitored and also controlled from anywhere through the one thing which would be always with us which is our handheld devices or our mobile phones or laptops through which we can access the internet and by that be able to see these operations. Algorithm of operating conditions has been shown in figure 4.

IOP Conf. Series: Materials Science and Engineering **906** (2020) 012022 doi:10.1088/1757-899X/906/1/012022

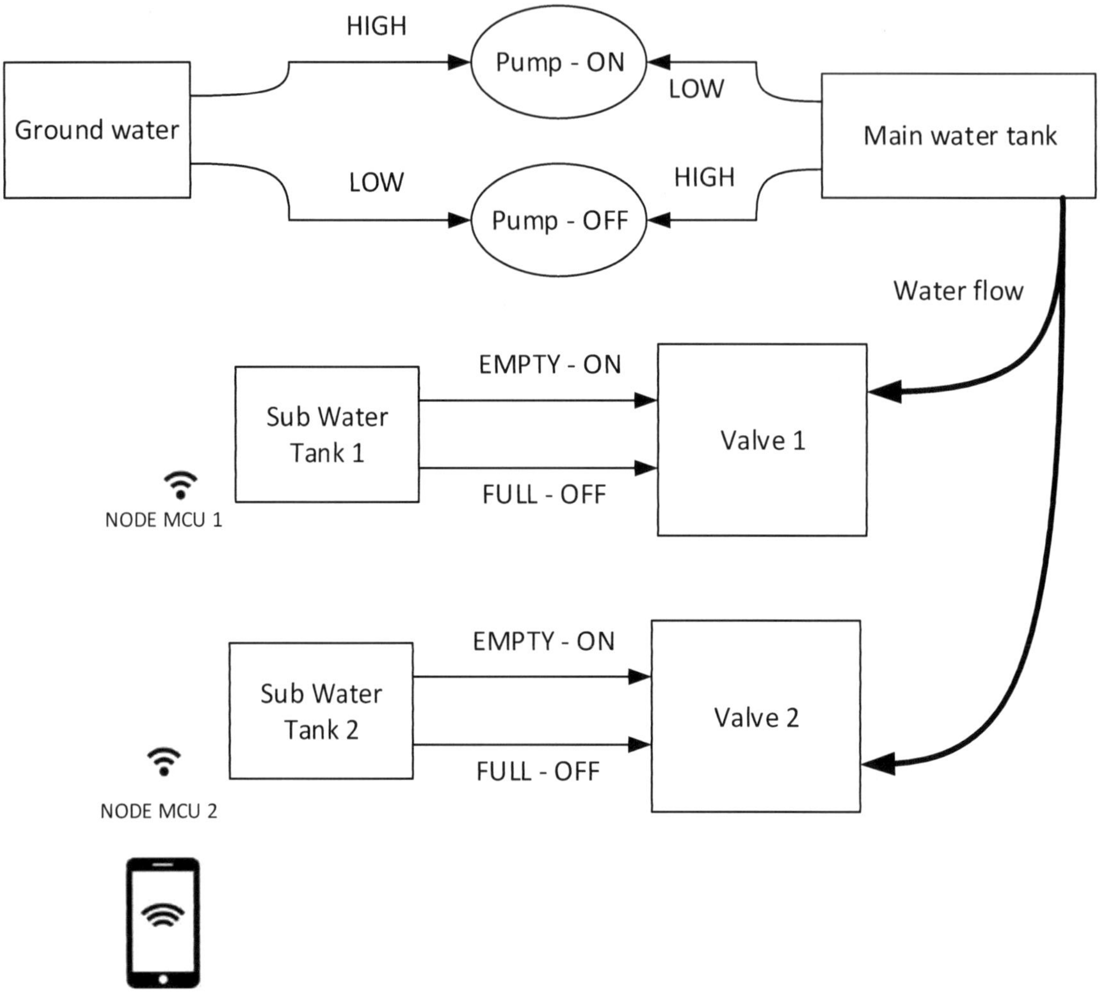

Figure 4: Algorithm for smart water grid technique

4. Design of smart water Grid system

Relay module with optocoupler used in this design is about 250V AC,10A relay contact current capacity where this can be used for controlling various appliances & other equipment with large current ratings. Arduino water level sensor also called as soil hygrometer module which used to measure the ground water level where the two stripes plated with copper of it will be start conducting by forward biasing the NPN resistor when it touches water. It has 3 pins consists of 5V supply. ground and an analog output pins.

Main-tank is used to feed water into the sub-tanks around it. To feed water its water level is to be checked beforehand so that the water is available for sub tanks. Water to the main tank is pumped with the help of ground water pump. As precautionary measure ground water level is also measured so that the water pump does not get switched on. Ground water level is measured with the help of soil hygrometer module. When the water level in the main tank is measured with the help ultrasonic sensor and it is low water ground level is checked and when this is measured to be low nothing changes but if it is measured high water pump is switched on as shown in control circuit diagram figure 5.

IOP Conf. Series: Materials Science and Engineering **906** (2020) 012022 doi:10.1088/1757-899X/906/1/012022

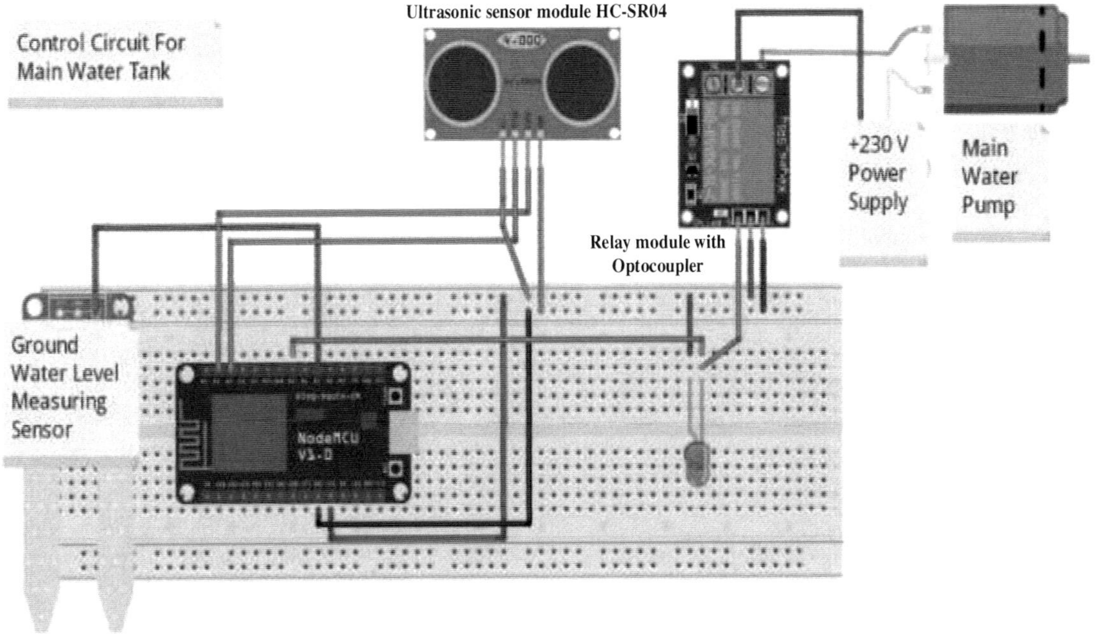

Fig 5: Main Water tank Circuit Diagram

Main Tank is connected to the Sub Tank through a pipe with an electrically operated solenoid valve. Water is initially stored in main tank. From figure 6, water level in the sub tank is measured using ultrasonic sensor and if it is found to be low the solenoid valve is electrically opened .When the water level is high the solenoid valve is electrically switched off. The Diode D2(1N4007) shown in Figure 6 is connected parallel to solenoid to dissipate the energy where it tries to continue forcing the current downward when the solenoid gets off.

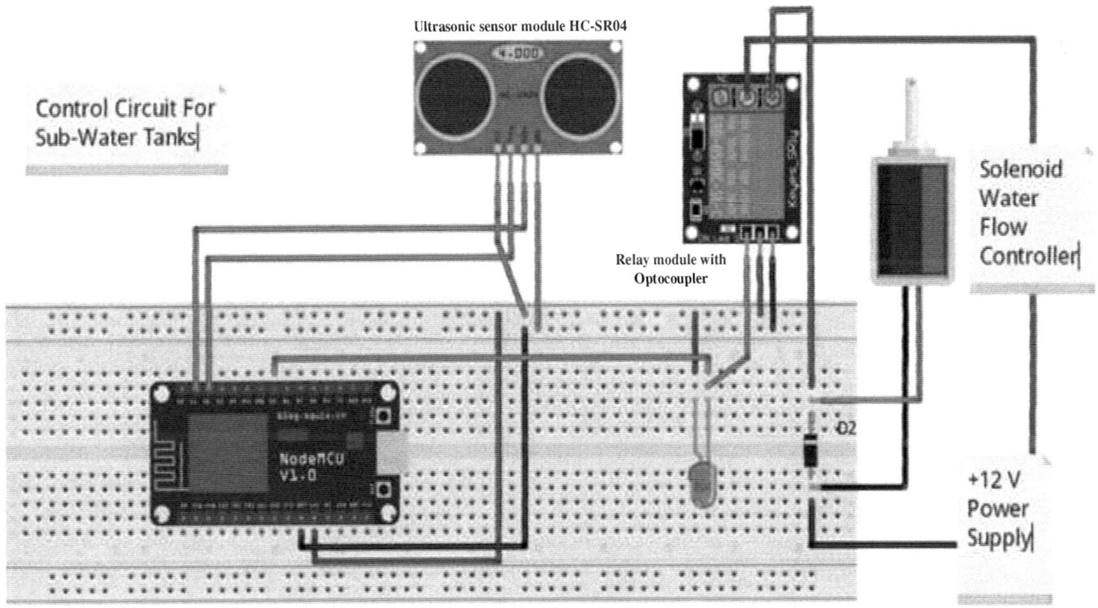

Fig 6: Sub Water tank Circuit Diagram

5. Experimental implementation & Investigation

The problem of Water Scarcity can be limited by realizing this project in either a small scale or even for a large scale area or locality. The main maxim of our project was to reduce water consumption by monitoring the amount of water utilized in each area and with reference to the data obtained shall regulate just the required amount of water to that area or sector. Neither more nor less. And by bringing about this smart water grid to the real world, to our metropolitan cities and towns shall make the life of people more convenient by letting them control and see the water level at their home or office or any other place from anywhere and anytime. They can also control the motor even through the internet because the project is mainly based on IOT. The user would be able to see the amount of water available or left in real time without any delay by having an interface with android app as shown in figure7.

Figure 7: Android app Interface

We have thus made a prototype for this 'SMART WATER GRID' in a small scale level as for this project. It consists of an Underground Water Source, a main tank and 2 sub tanks attached to it.When there is a paucity of water in the sub tanks or the tank, which can be found by the ultrasonic sensor attached to all the three tanks and also to the underground water source, a signal is sent from the system indicating that there is a deficit of water. And hence, thereby the system works automatically to refill the main and the sub tanks when the water level goes down below a threshold value. And on filling back the water, once the water reaches the brim of the tank it stops flowing into the tanks because another limiting value is set for so as to stop the supply of water once the tank gets loaded in order to avoid overflow of water from the tanks.

If in case, there is a shortage or no presence of water even in the underground source, the motor wouldn't even start to supply waters so as prevent the motor from malfunctioning and causing any harm. Only if there is a sufficient amount of water in the underground source will the motor start working in order to supply the water to the main tank and thereby to the other two sub tanks.

On working this prototype, it has been found that all the systems are working perfectly and the desired output is obtained, which is the live relay of the amount of water available and also the controllability of the each sub tanks and tanks valves so as to supply water even to the individual tanks with just a touch of the finger from anywhere. So if at all, a person thinks that it's going to be a very hot summer and there is an imminent danger for water, people can choose so as to accumulate the desired amount of water for their use with the help of Smart Water Grid. And by implementing this project as shown in figure 8 (a) & (b) to vast metropolitan cities, it is possible to control water shortage to a certain level and in the process save water up to a certain extent. Which in itself would serve as a result and success of this imperative project for the generations to come.

Figure 8(a): Smart water grid & IOT installation **(b):** Working prototype model

6. Conclusion:

This developed project can be used to build a water efficient society. It is very helpful in saving water, electricity and more importantly man power. To make it possible sub-tanks are installed for a smaller region, main-tanks are installed in a place with high ground water level so that it may be used to feed water to these sub-tanks. The beauty of this system is that it can be operated manually and also automatically. Man may make mistakes sometimes so it is programmed to work automatically but still the water levels in the tanks can be monitored from any place and can be controlled if wanted. Future scope is to make an advancement in the system by adding water quantity measuring meter, which will help people to identify the leakages during its transmission. Adding many sensors like PPM sensor and PH value sensors in the main tank will help to identify the quality of water and by analysing all the data such as quality of water, water consumed, water replenished will help the system to take feasibility solutions for the real time water enhancement.

References

1. Y. Lalle, L. C. Fourati, M. Fourati and J. P. Barraca, "A Comparative Study of LoRaWAN, SigFox, and NB-IoT for Smart Water Grid," 2019 Global Information Infrastructure and Networking Symposium (GIIS), Paris, France, 2019, pp. 1-6.
2. M. Fatima, S. Jain, A. Chikara and M. Luthra, "Review on Implementing Smart Water Grid for Smart Cities in India: Challenges and Solutions," 2019 5th International Conference on Advanced Computing & Communication Systems (ICACCS), Coimbatore, India, 2019, pp. 216-219.
3. M. Saravanan, A. Das and V. Iyer, "Smart water grid management using LPWAN IoT technology," 2017 Global Internet of Things Summit (GIoTS), Geneva, 2017, pp. 1-6.
4. A. M. Manoharan and V. Rathinasabapathy, "Smart Water Quality Monitoring and Metering Using Lora for Smart Villages," 2018 2nd International Conference on Smart Grid and Smart Cities (ICSGSC), Kuala Lumpur, 2018, pp. 57-61.
5. Michael Allen, Ami Preis, Mudasser Iqbal and Andrew J. Whittle, "Case Study: a Smart Water Grid in Singapore", Water Practice and Technology, vol. 7, no. 4, November 2012.
6. Aditya Gupta, Sudhir Mishra, Neeraj Bokde and Kishore Kalat, "Need of Smart Water Systems in India", International Journal of Applied Engineering Research, vol. 11, no. 4, pp. 2216-2223, 2016, ISSN 0973-4562.

7. J. Pacheco, D. Ibarra, A. Vijay and S. Hariri, "IoT Security Framework for Smart Water System", IEEE/ACS 14th International Conference on Computer Systems and Applications (AICCSA), Oct2017, ISSN 1285-1292.
8. G. Gosavi, G. Gawde and G. Gosavi, "Smart water flow monitoring and forecasting system", 2nd IEEE International Conference on Recent Trends in Electronics.
9. T. Malche and P. Maheshwary, "Internet of Things (IoT) Based Water Level Monitoring System for Smart Village", Advances in Intelligent Systems and Computing, vol. 508.
10. R Chandra, K.A Nishadh and P.A Azeez, "Monitoring water quality of Coimbatore wetlands Tamil Nadu India", *Environ Monit Assess*, vol. 169, pp. 671, 2010, [online] Available: https://doi.org/10.1007/s10661-009-1206-0.

Secure File Sharing System in Cloud Using AES and Time Stamping Algorithms

R.Sheeja[1], Chidambaranathan Bibin[2], Priya R Krishnan[3], R.Nishanth[3], S.Gopinath[2] and K.G.Ashok[4]

[1]Saveetha School of Engineering, Thandalam, Chennai, Tamil Nadu.

[2]RMK College of Engineering and Technology, Chennai, Tamil Nadu.

[3]Cochin University College of Engineering Kuttanad, Kerala.

[4]Easwari Engineering College, Chennai, Tamil Nadu.

Email: cjabbn@gmail.com

Abstract. In recent days, the growth of digital data has been increased explosively due to data-intensive applications like healthcare, education and electronic commerce. Due to the huge volume of data created and the utilization of data in emerging applications, a local storage device has to be deployed. Cloud storage services allow both the individuals and commercial users for the outsourcing of data to the cloud server and to access the data remotely with the help of the internet, so the demand for cloud storage has been raised. However, one of the critical problems of cloud data storage is security. Some researchers have used third party auditor for verifying the data stored in the cloud. This paper introduces a novel method for cloud storage and for protecting the organizations' data. The proposed system strengthens the level of authentication with the help of AES and time-stamping algorithms. The experimental results demonstrate the efficiency of the proposed method when auditing the shared data integrity

Keywords:

Cloud computing, AES, Timestamping, KDM, Data owner

1. Introduction

One of the first features of distributed computing is the utilization of the cloud [1]. Additionally, the cloud offers limitless storage for the customer to store their information.

The distributed storage gives the most effective approach to remote information reinforcement, with the goal that client can able to recover the data whenever utilizing the cloud administrations. Cloud likewise lessens the money related over-burden of ventures and associations keep up their information. There are more contextual investigations that are identified with distributed storage for remote information reinforcement [2]. Additionally, people can store their individual information to the cloud utilizing Dropbox and Google Drive and then forth [3, 4]. Nowadays progressively the number of individuals groups is utilizing apparatus like Dropbox to store their information in the cloud. Nonetheless, we've to think about the safety worries in putting away the fragile information in the cloud which is maintained by third-party cloud administrations. In our defined work, two of the main security issues are considered especially. Within the first place, we've to ensure that just approved gatherings have access to the re-appropriated information in cloud through proficient key dissemination component and access strategy. Second, to confirm secure information we've to execute cryptography plans for giving security when clients transfer or download information from cloud administrations. In this paper, we utilized RSA and AES calculation for accomplishing the proposed issues and some cryptographic key activities to shield the data which is obtained from the cloud Wang et al. in [5]. Consider the dynamic storage of data in a distributed scenario. They proposed a protocol that would be capable of determining the accuracy of the data.

A few investigations [6] are identified with the insurance of re-appropriated information utilizing cryptographic strategies. S.Pawar al.[7] proposed an examining framework that empowers the clients to confirm the respectability of redistributed information. Pawar et al. [8] thought of a safe redistributed information get to the system with access rights. W.Huang [9] talked about the uprightness and security on a redistributed information utilizing hash-based component. The key offers are put away in the Hash Table in an appropriated way. RSA is the most well-known open key calculation. Rivest, Shamir and Adleman [10] imagined this calculation where both open and private key is utilized for encryption and unscrambling. All the messages are scrambled utilizing the general population key and it is sent to the beneficiary. The recipient utilizes the private key to unscramble the message. B.chor et.al [12] proposed a technique to anchor information in the cloud utilizing RSA. Cramer and JB Nielsen [11] designed AES a symmetric calculation. AES utilizes a similar key for both encryption and of messages. Sheren A. El-Booz, Gamal Attiya and Nawal El-Fishawy proposed a secure cloud storage system combining time-based one-time password and automatic blocker protocol [13].

2. Proposed work

Here the first user should register with all personal details they need to fill while registering time that individual user should select one particular group because here, we are going to divide into four groups the user should select one particular group. After selecting that, the details are stored in the database. After storing that the details of each of them are sent to admin then if admin accepts then only that individual user can log in. After accepting the user request then the user can log in and can upload the file. While uploading that file the private key will be generated for that file. Like that, if any user comes to registration after login that user can view that files in user inbox if the user needs that file they need to send an invitation to admin if that user is expounded to the identical group of that file then only that request will allow passing admin. Then if the second-hand needs the file, they send a letter of invitation to admin, In admin box admin can view what are the files available they will view and people who send request they will be viewed in the admin page. If admin accepts the request then that user can get that file with the private key while sending that file admin will allocate 30 mins time if that user wants that file within 30 means they need to enter that non-public key to induce individual file that point should be taken from our current desktop timings.

2.1 The Cloud

Data redistributing and Data reinforcement are completed in the cloud by the knowledge proprietor. To shield the knowledge from the unapproved client, information is put away in an exceeding sort of encryption in the cloud. Classification is accomplished by putting away the knowledge in an exceedingly scrambled frame. The cryptographic activity and the transfer and download document activity are finished utilizing our proposed strategy. So, there's not much association of cloud explicit activity in our work.

2.2 Key Distribution Manager

The Key Distribution Manager (KDM) acts as confided in outsider where all the cryptographic related activities are done here. ACL is additionally maintained in KDM for the arrangement of individual records. At whatever point the client must transfer or download the document in the cloud, first, the client must enlist with KDM and KDM confirms for validation. KDM will be maintained by the association itself to make a trust for the client who is getting to the information.

2.3 Data Owner

Data proprietor will transfer the scrambled information to the cloud with the help of KDM. Information Owner sends the Access Approach (AP_i) which is expounded with the document alongside the rundown of clients who can get to the present record to KDM. This arrangement of access approaches is maintained in ACL document which could be a piece of a KDM. All entrance approaches are related with each document and also the approved clients are permitted to urge to the documents

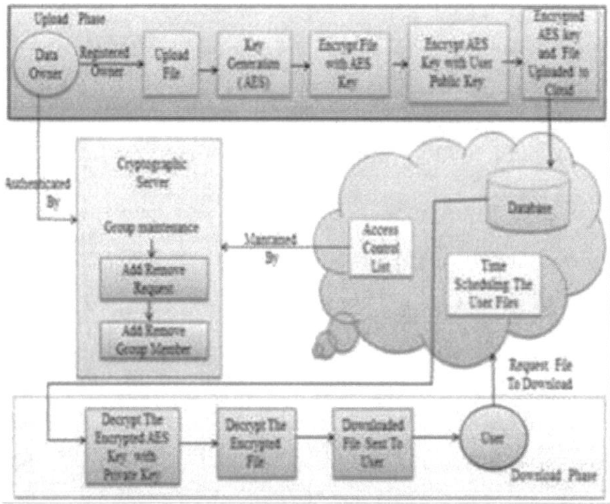

Figure 1. Proposed System Architecture

2.4 User Registration

Information Owner and every other individual client must first enrol with KDM. KDM monitors rundown of the client and their Gathering ID (IDi) which is given by the gathering individuals while enlistment. Amid the procedure of encryption and unscrambling, KDM confirms and verify the rundown of clients who wish to induce to its document.

2.5. Key Generation

The KDM is to blame of RSA key age of our strategy which has both Symmetric and Asymmetric calculations. In our proposed work, we use RSA calculation to form the key combination which is open and personal key.

1. Two prime number a and b chosen randomly.
2. Compute c = a * b

3. Calculate the function f(c) i.e.,

$$f(c) = (a-1) * (b-1)$$

4. variety e is chosen randomly such the range as

$$1 < e < f(c) \text{ and also GCD } (e, f(c) = 1.$$

5. With these parameters private key (d) is generated using the equation

$$d=e-1 \bmod f(c) \tag{1}$$

2.6 Encryption

Data Owner sends the document to KDM for encryption. Encryption is conveyed our utilizing the Symmetric calculation AES using below equation.

$$D1 = EAES\ (File, Rk) \tag{2}$$

The KDM creates the irregular key (Rk) of the length required for record encryption. Presently the document is an encoded frame. The knowledge is put away within the cloud by Data proprietor after receiving from KDM. After the procedure of record encryption, the AES key created is scrambled utilizing people normally key of RSA.

$$D2 = ERSA\ (Rk, KPu) \tag{3}$$

2.7 Decryption

At the purpose when the gathering part must get to the record, either the client can download the scrambled document from the cloud or the demand will be made to KDM for recovering the scrambled document. The KDM sends the scrambled record and therefore the private key combine (KPr) to the information Owner. After confirming the legitimacy of the client, the KDM sends the Encrypted record to the gathering individuals. When the record is acquired by the client, at that time a requirement is sent to the data proprietor for AES key to unscramble the record. The information owner utilizes the private key (KPr) and D2 esteem get from KDM to recover the AES key (Rk) using the equation.

$$KAES = DRSA(D2, KPr) \tag{4}$$

As per the request provided by the user and also with relevance to user access policy, the generated AES key (KAES) is shipped to the group of users. The group members after receiving the AES key will ready to decrypt the file using the equation (5)

IOP Conf. Series: Materials Science and Engineering **906** (2020) 012023 doi:10.1088/1757-899X/906/1/012023

$$\text{File} = \text{DAES}(\text{D1}, \text{KAES}) \quad (5)$$

2.8 Policy-Based Data Access

Each document and also the clients are related to single access arrangement. Amid the record transfer, the data proprietor sends the record alongside the arrangement related with the document and rundown of clients who can get to the document. Each client's necessities to induce to the record is with the constant or distinctive arrangement because it is chosen by the data proprietor. The KDM validates the clients amid the record download activity. The arrangement can shift with diverse applications. Assume when some representatives are working in an association and each one engaged with the distinctive venture, we can relate a client-based strategy will be the documents for the client. At whatever point the representative leaves the association the KDM will create diverse RSA key match furthermore, will re-scramble the AES key utilizing the new open key. AES secrets likewise recovered before encryption by RSA open key if important. At that time KDM will send the new private key to the knowledge proprietor. The knowledge proprietor will send the recently created private key to the gathering individuals who approach strategy to urge thereto record. The individual from gathering who left won't have any entrance thereto information and also the arrangement related with the clients may be additionally erased from ACL which is a piece of KDM. The new client first registers with the KDM. While joining the client presents his personality for confirmation. At the purpose when the client has to join the present gathering, he sends a requirement to KDM by giving the group ID for confirmation. The new part has been sent by the knowledge proprietor, at that time KDM adds the part to the gathering which is referenced by gathering Id. Likewise, ACL refresh the new client data and therefore the strategy is expounded to the client with the record that has to be caught. At the purpose when the knowledge proprietor has to recharge with the new approach, sends a requirement to KDM to refresh the approach related to the record. The KDM refreshed the ACL list as indicated by the new approach and if necessary, it yet again recovers the new match of RSA open and a personal key. At the purpose when a client is expounded with various document and since each record is connected with a solitary arrangement, KDM will make utilization of the Boolean mixture of approaches for that client.

3. Performance Evaluation

This segment assesses the execution of our proposed framework. The proposed strategy is tested and actualized in java. For our investigation, we utilized Google App Motor as distributed storage. We utilized Google Cloud SQL as a database to store and recover information whenever and at anyplace. In our proposed work, we utilize both symmetric and kilter key for look after security Utilizing RSA key pair created, AES key is encoded. The testing has been done with various key lengths. Fig.2 and Fig.3 demonstrate that the time taken to create the private key and open key increments with admiration to the expansion in the key length. Our evaluation is based on the time taken to upload and download the file.

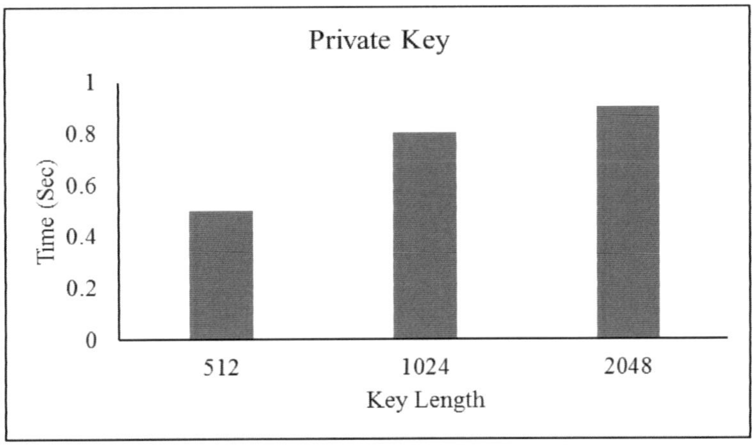

Figure 2. Time taken for private key generation

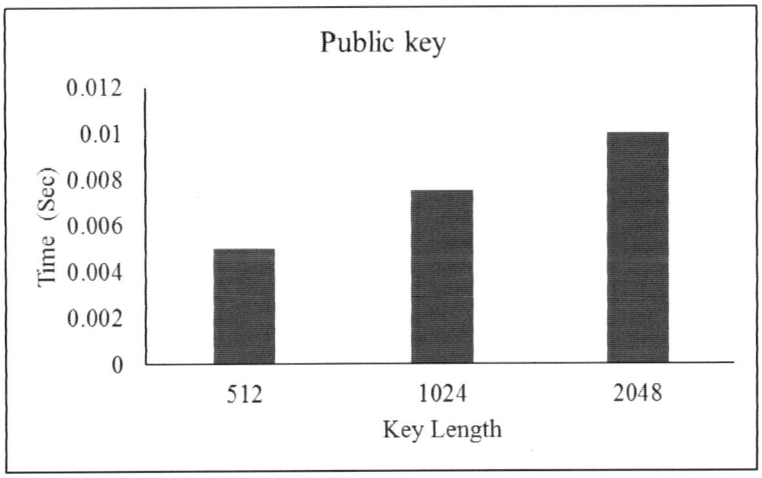

Figure 3. Time taken for public key generation

4. Future Enhancements

Based on our analysis, categorization of dynamical networks prevention mechanisms and we identify some possible directions. Future works embrace investigations of cyber-security management for systems with state and input delays. In real applications, due to completely different environmental influences or enforced equipment's, it's conjointly vital and sensible to consider the heterogeneous case wherever A, B, C and performance D are completely different for the followers. Moreover, since input saturation nonlinearities are omnipresent in physical and engineering systems. It is necessary to explore the implications of input saturation in these Publications/privacy security.pdf systems in presence of cyber-attacks on their communication channels. Distributed intrusion detection and resilient distributed estimation are fascinating topics for complicated cyber-physical networks.

5. Conclusion

This paper dealt with the protection of privacy while maintaining public auditing. A new secure cloud storage system is proposed to ensure that the details of companies are secure from all data operator and third-party inspector, respectively by protected sharing of information utilizing RSA and AES calculation to keep up security inside the cloud server. KDM will be in charge of all key age and key appropriation process in our proposed plan. The execution is assessed what's more; the outcomes are acquired dependent on RSA key age and AES encryption process. From the outcome, it is seen that our proposed strategy will be pertinent for sharing information in the cloud safely. We use approaches-based access instrument to give security with the information in the cloud and to give validation. In future, we can utilize different KDM to deal with the information with various access strategies to maintain a strategic distance from insider attacks.

References

[1] A. G. Dimakis, P .B. Godfrey, Y. Wu, M. J. Wainwright, and K. Ramchandran, "Network coding for distributed storage systems," *IEEE Transactions on Information Theory*, **vol.56, no. 9**, Sep. 2010.

[2] J. A. Garay, R. Gennaro, C. Jutla, and T. Rabin, "Secure distributed storage and retrieval," *Theoretical Computer Science*, **vol. 243, no. 1-2**, pp. 363-389, Jul. 2000.

[3] A. Parakh and S. Kak, "Space efficient secret sharing for implicit data security," *Information Sciences*, **vol. 181, no. 2**, pp. 335-341, Jan. 2011.

[4] N. B. Shah, K. V. Rashmi, K. Ramchandran, and P. Kumar, "Privacy-preserving and secure distributed storage codes," http://www.eecs.berkeley.edu/ retrieval," *Journal of Computer and Systems Sciences*, **vol. 71, no. 2**, pp. 213-247, Aug. 2005.

[5] C Wang, Q Wang, K Ren, W Lou, Privacy preserving public auditing for secure cloud storage. *IEEE Transactions on Computers* 62(2), 362–375 (2011)

[6] N. B. Shah, K. V. Rashmi, and P. Kumar, "*Information-theoretically secure regenerating codes for distributed storage*," in Proc. IEEE Global Communications Conference (GLOBECOM), Houston, TX, Dec. 2011.

[7] S. Pawar, S. Rouayheb, and K. Ramchandran, "Securing dynamic distributed storage systems against eavesdropping and adversarial attacks," *IEEE Transactions on Information Theory*, **vol. 57, no. 10**, pp. 6734-6753, Oct. 2011.

[8] O. O. Koyluoglu, A. S. Rawat, and S. Vishwanath, "Secure cooperative regenerating codes for distributed storage systems," *IEEE Transactions on Information Theory*, **vol. 60, no. 9**, pp. 5228-5244, Sep. 2014.

[9] W. Huang and J. Bruck, "*Generic secure repair for distributed storage*," arxiv preprint arXiv: 1706.00500, Jun. 2017.

[10] A.Shamir, "How to share a secret," *Communications of the ACM*, **vol. 22, no. 11**, Nov. 1979.

[11] R. Cramer, I. B. Damgard, and J. B. Nielsen, *Secure Multiparty Computation and Secret Sharing*, Cambridge University Press, 2015.

[12] B. Chor, O. Goldreich, E. Kushilevitz, and M. Sudan, "*Private information retrieval*," in Proc. the 36th Annual Symposium on Foundations of Computer Science, Wisconsin, USA, pp. 41-50, Oct. 1995.

[13] Sheren A. El-Booz*, Gamal Attiya and Nawal El-Fishawy," A secure cloud storage system combining time-based one-time password and automatic blocker protocol", *EURASIP Journal on Information Security*, 2016.

IOP Conf. Series: Materials Science and Engineering **906** (2020) 012024 doi:10.1088/1757-899X/906/1/012024

Design and Analysis of Series Loaded Resonant Converter fed by Standalone/PV Source

Sri Lakshmi Tejaswi Pallavarjhula[1], **V. Sree Vidhya**[2], **Sai Suma Mahanthi**[3]
[1,3] B.Tech. Final Year , Department of EEE, G.V.P. College of Engineering for Women
[2] Assistant Professor, Department of EEE, G.V.P.College of Engineering for Women

E-mail: sri.vydhyanadhan@gvpcew.ac.in

Abstract. The deficiency of power is an enormous issue due to higher growth demands among the utility customers which led the focus on renewable energy sources for electricity generation. Solar energy has gained high popularity due to highly abundant, clean, economical and sustainable when compared to other alternative energy sources. The DC-DC converter topologies are used in solar photovoltaic (PV) applications for power conversion due to less cost, highly reliable, highly efficient, and robust operation. These converters incorporate a wide variety of voltage variations and protection against overvoltage. In the recent times, solar power converters using various soft switching techniques due to high switching frequencies which reduce the size, cost, weight and loss of the converter. The series resonant power converters are one such soft switching converter which is widely employed due to their high - efficiency with the benefit of low switching losses. The paper confers on the analysis of Series Loaded Resonant Converter with low loss switching by applying solar PV as an input source. The converter behaviour is observed in various operating modes concerning the resonant frequency. The design parameters of the proposed converter are validated in MATLAB/SIMULINK environment.

1. Introduction

The demand for energy is increasing at a considerable rate in every country. Higher demand for electrical energy and depletion fossil fuels necessitates the use of renewable energy [10],[11]. This environmental aspects put restrictions on the use of conventional energy sources. Various types of power converters are used in such applications. For applications involving PV and battery, DC-DC converters are commonly used. In certain instants, there may be need to step-up and step-down the dc input supply. In some applications, there is need for bidirectional power transfer. They are used to match the loads of the power supply. Because of this ease, today the DC-DC converters are increasingly found in applications such as regulated power supply, domestic, automobile, battery charging, etc. These converters are also useful as power sources with multiple outputs for control circuits. [1].

In the last few decades, the attempts were redirected towards the use of resonant converters. The idea was to include a resonant tank in the converter to create sinusoidal waveforms of current and/or voltage using lossless/soft switching [4]. These low loss switching such as zero-voltage switching (ZVS) or zero-current switching (ZCS) can be shaped for power switches for achieving transient free waveform and also allows the switches to work at higher frequencies and in turn reduces electromagnetic interference (EMI). The performance of SLRC will undergo low loss switching, when the switching frequency is higher or lower than the resonant frequency, [7]-[9]. The majority of works have covered the operation of half bridge resonant converter. Few converters with half and full-bridge configurations have been discussed for medium and high voltage levels and evolved in many industry applications [7]. This study presents the design and analysis of series loaded resonant converter , working at various operating modes to achieve soft switching across all power switches .The paper is organised as follows: Brief discussion of series loaded resonant converter with low loss switching in section 2, circuit description and analysis of series resonant converter in section 3, modes of operation of SRC in section 4, discussion of simulation and validation of series resonant DC-DC converter using MATLAB/SIMULINK software in section 5.

2. Series Loaded Resonant Converter with Low Loss Switching

Resonant Power Converters are named after the concept resonance. This happens when the reactance of inductor and capacitor cancel each other in LC resonant circuit. Resonant power converter operates in such a manner that switching losses are minimized [12]. The devices are switched on and off when the voltage across the device is zero or current through switch is zero. This results in zero voltage or zero current switching respectively. The switching losses are proportional to the product of voltage across and current through the switch. The low loss switching maintains one of the parameters that is either voltage across or current through the switch close to zero during switching. This makes their product close to zero and switching losses becomes almost zero. In turn, this enhances the efficiency of power conversion and also higher switching frequencies can be used for power conversion. Other benefit of this is reduction in size of inductor, capacitor and transformer. This reduces overall weight and volume of the converter making it compact. In these converters, the shapes of voltages and currents being close to sinusoidal, dv/dt and di/dt are reduced. This reduces EMI effects [13]. Depending on low loss switching, the behaviour of ZVS and ZCS are carried out in four possible ways:

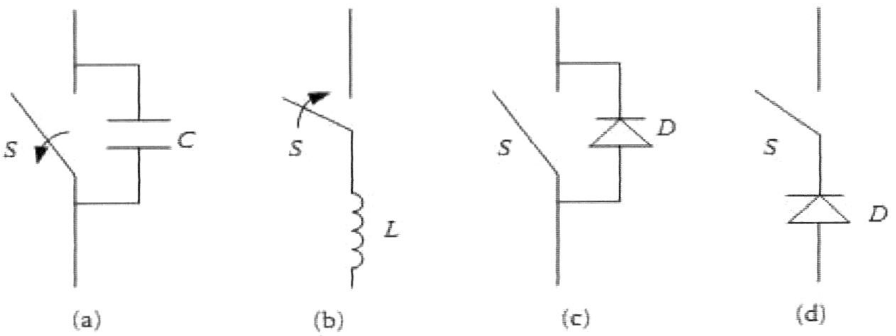

Figure 1. Possible switching configurations of active/main switch.

1. A capacitor when connected in parallel with the main switch the circuit acts for ZVS OFF condition. During turn-on, the voltage across the capacitor is zero, and it cannot change the state of the switch to OFF condition at the same time. So, in switch OFF period, the voltage rests close to zero, which is true in the case of modern power switches as they operate rapidly during switching instants and therefore quasi resonant condition that is ZVS OFF is reached. Nevertheless, during turn-on, special consideration is to be ensured for the voltage to be zero across the switch, otherwise short circuit occurs which may lead to the damage of main switch. A capacitor when connected in parallel, the current flowing through the switch is determined and hence, ZCS ON and OFF conditions may not be attained.

2. An inductor when connected in series with the main switch, ZCS ON condition is achieved. This takes place as the current through the inductor will be zero when the switch is OFF, and when it is turned ON, the change cannot take place instantaneously. So during switching-off period the current remains close to

zero, and at this instant quasi ZCS ON state is reached. Hence, care must be taken to make sure that zero current flows in the inductor during the switch turn -on state.

3. The diode when connected in parallel with the main switch approaches to attain ZVS ON/OFF during the conduction period of the diode. This happens when the diode is in forward bias, it's voltage is clipped, so quasi ZVS ON and OFF of the main switch can be reached. Some switches, such as BJT's, cannot conduct for negative current through it as the current flowing through the switch is zero but in switches like MOSFETs a value sharing with the diode takes place if the switch has bidirectional conductivity.

4. In this configuration, the diode when connected in series with the main switch, it operates in discontinuous mode. In this condition, the flow of current in the diode is blocked. Therefore, no current flows through the diode and ZCS ON and OFF can be achieved. Hence it is be noted that the above analysis is based on ideal operation of active/main switch. These switching configurations are depicted in the Figure 1.

3. Circuit Description and Analysis of SLRC

The operational circuit of the SLRC is illustrated in Figure 2. Each part of the converter carries out a specific operation. The input DC source fed to the switching network of SLRC is supplied by the PV array at standard irradiance and temperature. The switches of full bridge inverter undergo commutation according to the switching frequency, to generate square wave output voltage. To reduce EMI and harmonic distortion, a resonant tank containing inductor (L) and Capacitor (C) components are connected in series to generate sinusoidal voltage and/or current signals [2],[4]. It acts as an energy buffer between output and input. The output of resonant tank is fed as input to rectifier network and filtered using capacitor component to produce the desired DC output voltage.

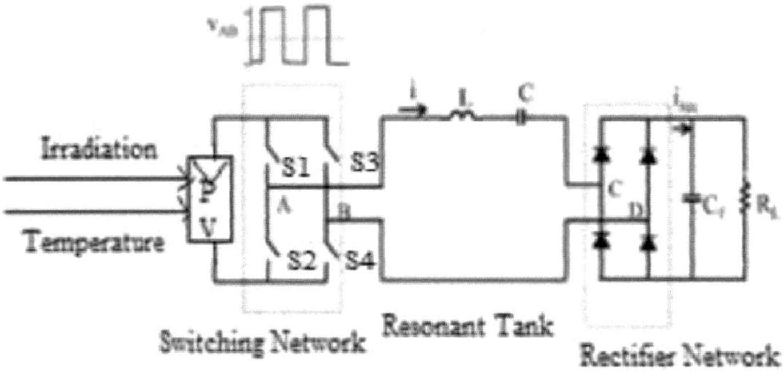

Figure 2. Schematic diagram of SLRC

In SLRC circuit, the tank circuit is connected in series with the load which acts as a voltage divider. So, the input voltage is shared between the impedance of the tank circuit and the effective resistance. The impedance of resonant tank varies with the frequency of driving voltage, as represented in Figure 3.

In this paper, we have considered the resonant tank impedance to be 53.4 ohms and at resonant frequency of 85.1 KHz. At this point, the current from the tank circuit flows to the output. The parameters of resonant are estimated and determined for the resonant frequency and the tank impedance using the below equations:

Determination of Design Parameters:

$$R_o = \sqrt{\frac{L}{C}}$$

Multiply & divide by $\sqrt{L}$

$$R_o = \frac{L}{\sqrt{LC}} = \frac{L}{1/\omega_o}$$

$$=> L_r = \frac{R_o}{\omega_o} \quad (1)$$

On substituting values R_o, ω_o in eqns.(1), we get

$$L_r = \frac{R_o}{\omega_o} = \frac{53.45}{2*\pi*85.1KHz} = 100\mu H. \quad (2)$$

$$\omega_o = \frac{1}{\sqrt{LC}}$$

Squaring on both sides

$$\omega_o{}^2 = \frac{1}{LC}$$

$$C_r = [\frac{1}{\omega_o\sqrt{L}}]^2 \quad (3)$$

On substituting values of ω_0, L_r in eqn. (3) we get

$$C_r = 1/L_r\omega_o^2 = \frac{1}{(2*\pi*85.1KHz)^2*100\mu H} = 35nF \quad (4)$$

$$\text{Quality factor } Q = \frac{R_o}{R_L} \quad (5)$$

$$R_e = \frac{V_{R1}(t)}{I_R(t)} = \frac{8}{\pi^2}\frac{V}{I} \quad (6)$$

$$R_L = \frac{\pi^2}{8}R_e = 1.23R_e \quad (7)$$

The relation between switching frequency and resonant frequency is given by

$$F = \pi\frac{f_s}{f_r} \quad (8)$$

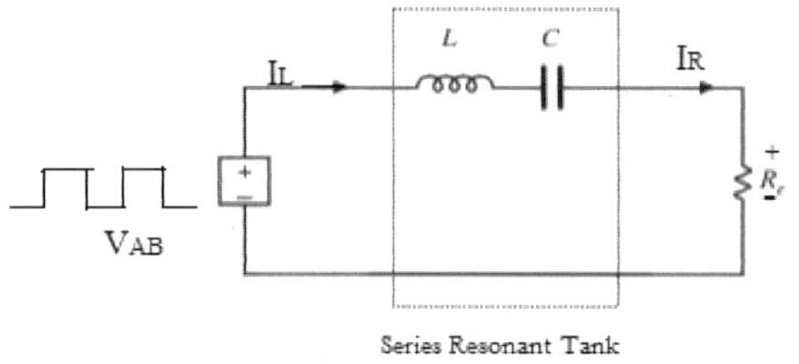

Figure3. Equivalent circuit model of SLRC.

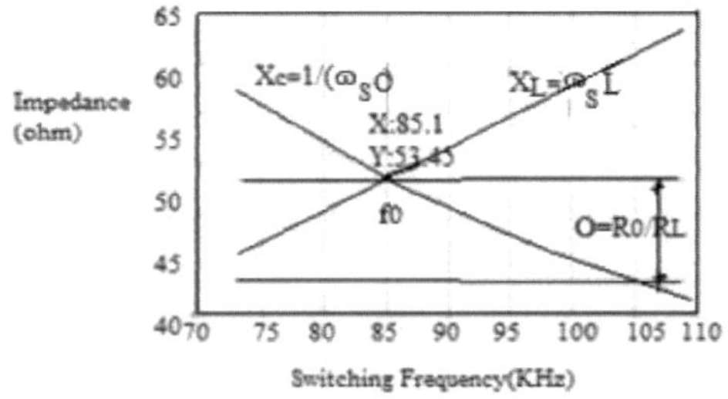

Figure4. Relation between impedance and switching frequency.

Here X_L and X_C are frequency dependent. X_L is directly proportional to frequency so we have a straight line. On the other hand X_C is inversely proportional to frequency. So, we have parabolic curve and resonance takes place when both of are equal. This is been depicted in Figure 4.

4. Modes of operation of SLRC

The operation of SLRC is mainly categorised into three parts based on mode of conduction, switching frequency, and soft switching technique. According to the load parameters and control for the limits of F>0, Q>0 the main switches are to be operated in various modes [7, 9]. Based on the ranges of the switching frequency, the modes of operation are classified as illustrated in Table 1.

TABLE 1. Switching Modes of SLRC

Conduction Modes	Frequency Period	Switching Condition	ZVS
Discontinuous Conduction Mode	0.5 $\leq$ F $\geq$ 1	Turn -On Turn - Off	No Yes
Continuous Conduction Mode	0.5 $\leq$ F $\geq$ 1	Turn -On Turn - Off	No Yes
Continuous Conduction Model	F $\geq$1	Turn -On Turn - Off	Yes No

In steady state, the SLRC above the resonant frequency is operated in four modes. The end point of the previous mode defines the initial values of current and voltage in the present mode. The operational waveforms for different modes are depicted in Figure5.

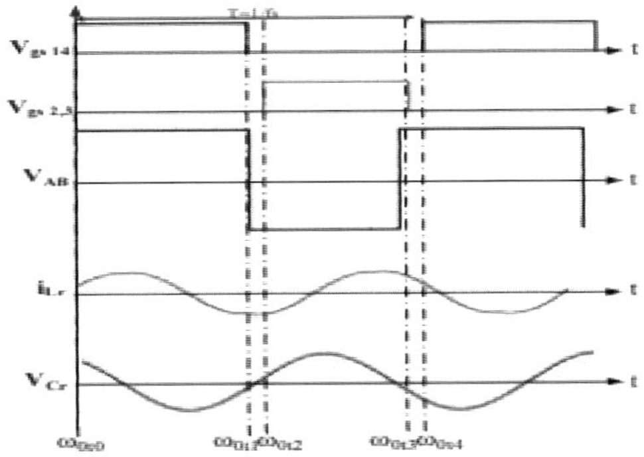

Figure5. Operational waveforms of SLRC

MODE 1: In this mode, the current flows through the diodes of the switches S1 and S4, forming the tank. In this mode, the load is supplied energy and the remaining reverts to the source. Thus the current I_{LR} (current in the resonant inductor) falls to zero at $\omega_o t_o$. This deviate the current from its original diode path and shifts to the power switch path. Thus, the switches S1 and S4 are turned on using ZVS. The expressions for the resonant voltage and current are given by

$$I_{Lr}(t) = \frac{1}{Z_o}[V_{in} - V_o - V_{cr}(0)] \sin \omega_o t \tag{9}$$

$$V_{cr}(t) = V_{in} - U_0 - [V_{in} - V_o - V_{cr}(0)] \sin \omega_o t \tag{10}$$

MODE 2: In this mode, switches S1 and S4 are forcibly turned off at $\omega_o t_1$. This happens due to zero crossing of the voltage across the resonant capacitor Vcr. This result in hard switching, however, can be eliminated with small parallel capacitors connected to the switches. Further, this current starts flowing through the diodes of switches S2 and S3 resulting in the power stage. Thus, the source supplies energy to the resonant tank, thereby resulting in zero current by the end of this mode, due to the tank's negative voltage. The expressions for voltage and current are given by:

$$I_{Lr}(t) = \frac{1}{Z_o}[V_{in} - V_o - V_{cr}(1)] \sin \omega_o (t - t_o) + I_{Lr}(1) \cos \omega_o (t - t_o) \tag{11}$$

$$V_{cr}(t) = V_{cr}(1) + \frac{I_{Lr}(1)}{\omega_o} \sin \omega_o (t-t_o) + [-V_{in} - V_0 - V_{cr}(1)][1-\cos \omega_o (t - t_o)] \tag{12}$$

MODE 3: This mode starts at $\omega_o t_2$, with diodes of switches S2 and S3 in on state, resulting in resonance between the tank elements. The current flow through the diodes of switches S2 and S3 turns on the switches at ZVS. In this mode, the load is supplied with the energy available in the tank; hence with the resonant current the polarity of the tank reverses. Also, the voltage of the resonant capacitor is shifted and reaches the maximum value when the current reaches zero. The resonant current and the voltage are expressed as

$$I_{Lr}(t) = \frac{1}{Z_o}[-V_{in} + V_o - V_{cr}(2)] \sin \omega_o (t - t_2) \tag{13}$$

$V_{cr}(t) = -V_{in} + V_o - [-V_{in} + V_o - V_{cr}(2)] \cos \omega_o (t - t_2)$ (14)

MODE 4: In this mode at $\omega_o t_3$, due to hard switching off of the switches S2 and S3 the current flow deviates from the diodes of switches S1 and S4, and slowly the resonant current reaches the value zero at $\omega_o t_4$. The expressions for the resonant current and voltage are given by

$I_{Lr}(t) = \frac{1}{Z_o}[V_{in} - V_o - V_{cr}(3)] \sin \omega_o (t - t_3) + I_{Lr}(3) \cos \omega_o (t - t_3)$ (15)

$V_{cr}(t) = V_{in} + V_o + I_{Lr}(3) Z_0 \sin \omega_o (t - t_3) - [V_{in} + V_o - V_{cr}(3)] \cos \omega_o (t - t_3)$ (16)

5. Simulation Results and Discussion:

In this paper, the design parameters of series loaded resonant converter are analyzed and simulated using MATLAB/SIMULINK software. In, this section, the converter waveforms and the conduction modes are discussed based on switching frequency period. The design parameters used for the simulation are tabulated.

TABLE 2. Design Parameters of SLRC

PARAMETERS	VALUES
DC input value	40V
Switching frequency	20 KHz≤f_s≤100 KHz
Resonant frequency	85.1 KHz
Resonant inductor	100μH
Resonant capacitor	35nF
R Load	23 Ω
Impedance	53.4 Ω

In, this section, the converter waveforms and the conduction modes are discussed based on switching frequencies between 20KHz to 100KHz.. The design parameters of SLRC are tabulated and simulated using MATLAB/SIMULINK software. The input to the full bridge switching network consists of four switches is fed from a PV source and the switches are triggered alternately. From the instant, t=0 to t=Ts/2 the switches S1 and S3 are triggered for 0.5 duty ratio for the first half cycle. On the other hand from t=Ts/2 to t=Ts switching instant, the switches S2 and S4 are triggered for the second half cycle to produce a square wave voltage and is fed to the resonant tank. The figure 6 represents the driving voltage across the switches and the total inverter output voltage. This paper, the design parameters of series loaded resonant converter are analyzed .

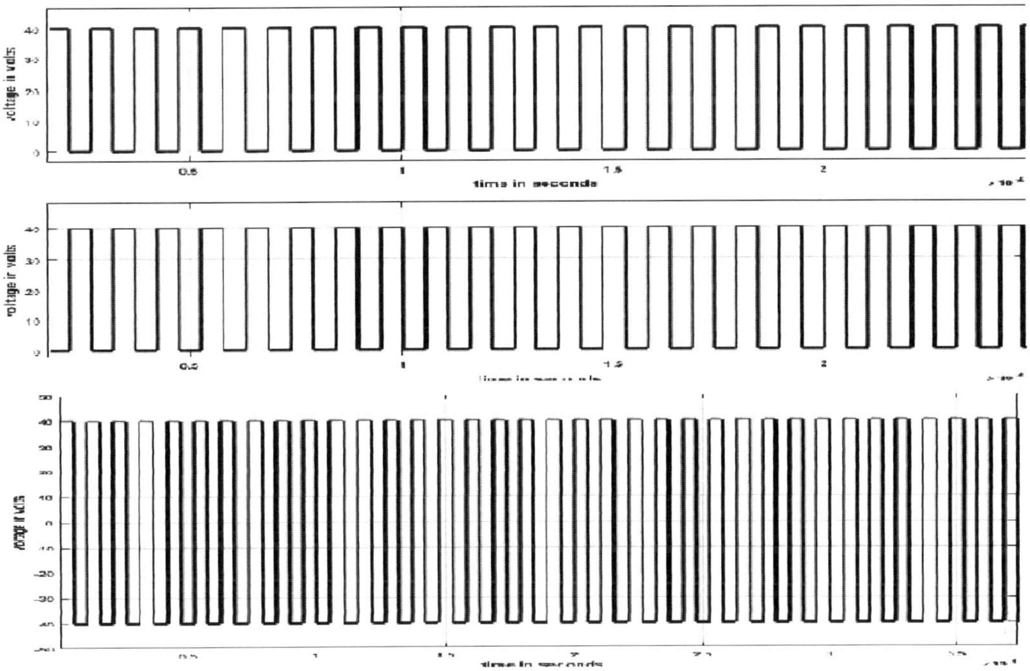

Figure 6: The voltage across the switches S1, S3 & S2, S4 and inverter output voltage

In order to investigate the SLRC behaviour based on the three conduction modes, we have considered the frequency periods at various switching frequencies between 20 KHz and 100 KHz.

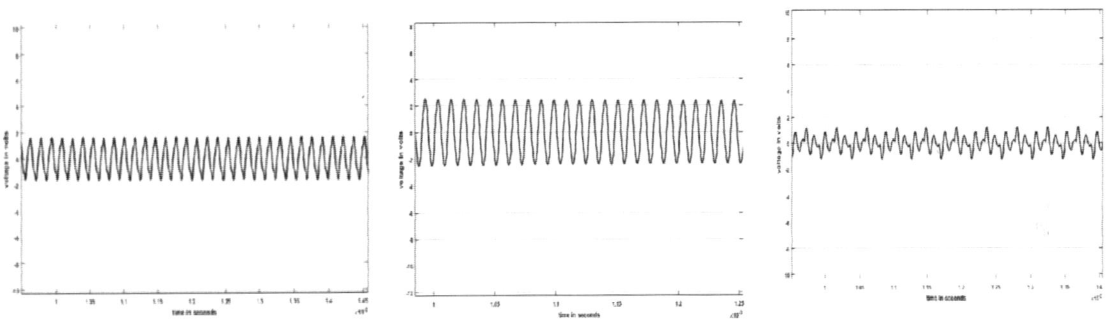

(a). Continuous Conduction Mode-1 (F $\geq$1) **(b).** Continuous Conduction Mode (0.5 $\leq$ F $\geq$ 1) **(c).** Discontinuous Conduction Mode (0.5 $\leq$ F $\geq$ 1)

Figure7. The figures7(a), (b) and (c) represents the tank current in various modes.

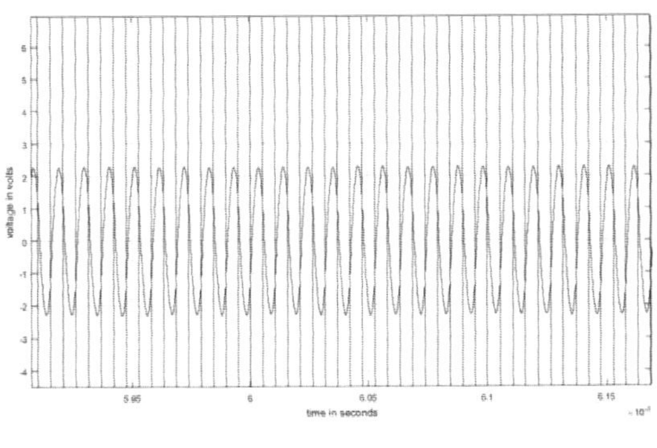

Figure 8. Zero Voltage Switching (ZVS) Output

To achieve ZVS operation, the converter is switched above resonant frequency of the tank circuit. This results in cancellation of capacitive reactance by a part of inductive reactance. And still some value of inductive reactance is left over in the resonant circuit. Hence, resonant tank is mainly inductive. this makes the tank current to lag the inverter output voltage. It can be observed that the tank current lags the square wave voltage. This feature helps in achieving ZVS operation.

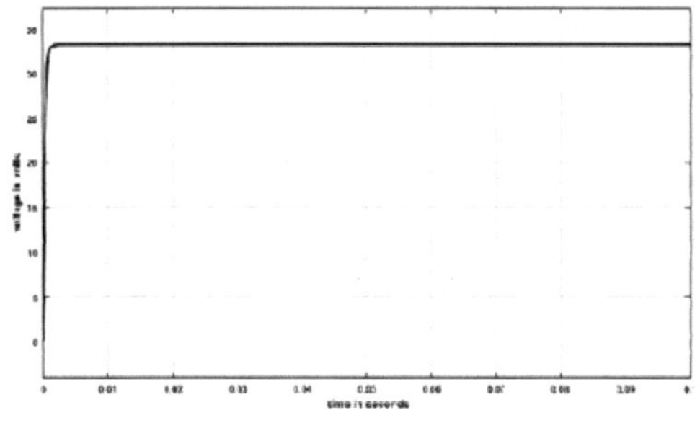

Figure 9. The filtered DC output voltage of the rectifier. The tank current which is close to the sine wave is rectified by the rectifier and filtered by output filter capacitance to produce DC output voltage. It is shown in figure 9.

6. Conclusion

The above discussed resonant converter finds are wide applications in renewable systems, LED drivers, battery chargers, regulated power supplies and bidirectional power conversion systems. They offer advantages such as high efficiency, compactness, low EMI, etc. The low loss switching is achieved by ZVS or ZCS operation of the switching devices.

Based on the soft switching techniques described, the devices behaviour, the devices in the inverter are turned on and off by ZVS. The diodes in the rectifier are operated with ZCS. Thus, the switching losses on the switching devices are reduced. In this paper, the operation states analysis of a SLRC above resonant frequency is described. The condition for operating the proposed converter in above resonance mode is achieved by selecting the characteristic impedance of the resonant tank has been discussed. This converter can be used for high frequency DC-DC power conversion with Photovoltaic (PV) or any other DC source. The validity of the work under any condition of frequency ranges has been presented using MATLAB/SIMULINK software.

References:

[1] Mouncif Arazi, Alireza Payman, Mamadou Bailo Camara,Brayima Dakyo, Study of different topologies of DC-DC resonant converters for renewable energy applications, *Thirteenth International Conference on Ecological Vehicles and Renewable Energies (EVER)*, IEEE 2018.

[2] Surja Sekhar Chakraborty, Saranya Patnala, Shekhar Bhawal, Kamalesh Hatua, Selection procedure of resonant tank parameters for an SiC MOSFET based DC/DC Series resonant converter, *IEEE International Conference on Power Electronics, Drives and Energy Systems (PEDES)* 2018.

[3] B.-R. Lin and C.-H. Liu, "Analysis, design and implementation of an interleaved three-level PWM DC/DC ZVS converter," *International Journal of Electronics*, vol. 103, pp. 322-341, 2016

[4] M. T. Outeiro, G. Buja, and D. Czarkowski, "Resonant Power Converters: An Overview with Multiple Elements in the Resonant Tank Network," *IEEE Industrial Electronics Magazine*, vol. 10, pp. 21-45, 2016.

[5] J. Javidan, "Design of a ZVS PWM DC–DC converter for high gain applications," *International Journal of Circuit Theory and Applications*, 2015.

[6] Zaka Ullah Zahid, Zakariya Dalala and Jih-Sheng (Jason) Lai, "Design and Control of Bidirectional Resonant Converter for Vehicle-to-Grid (V2G) Applications", *IECON 2014–40th Annual Conference of the IEEE Industrial Electronics Society*, pp. 1370-1376, 2014.

[7] B.-R. Lin and C.-H. Chao, "Soft-switching converter with two series half-bridge legs to reduce voltage stress of active switches," *IEEE Transactions on Industrial Electronics*, vol. 60, pp. 2214-2224, 2013.

[8] N. Mohan and T. M. Undeland, Power electronics: converters, applications, and design: *John Wiley & Sons*, 2007.

[9] R. W. Erickson and D. Maksimovic, Fundamentals of power electronics, *Springer Science & Business Media*, 2007.

[10] Nor Hanisah Baharudin, Tunku Muhammad Nizar Tunku Mansur, Fairuz Abdul Hamid, Rasnazri Ali, Muhammad Irwanto Misrun, Topologies of DC-DC converter in solar PV Applications, *Indonesian Journal of Electrical Engineering and Computer Science* Vol. 8, No. 2, November 2017.
[11] M. H. Taghvaee, M. A. M. Radzi, S. M. Moosavain, H. Hizam, M. Hamiruce Marhaban. A current and future study on non-isolated DC-DC converters for photovoltaic applications. *Renew. Sustain. Energy Rev.* 2013; 17: 216–227.
[12] Mohamed Salem, Awang Jusoh, N. Rumzi N.Idris, Tole sutikno, Iftikhar Abid, ZVS Full-bridge Series resonant Boost converter with series connected transformer, *International Journal of Power Electronics and Drive System (IJPEDS)* Vol. 8, No. 2, June 2017, pp. 812~825.
[13]] M. Salem, A. Jusoh, N. R. N. Idris and I. Alhamrouni, "Performance study of series resonant converter using zero voltage switching", *Proc. IEEE Conf. Energy Conversion (CENCON) , IEEE -2014.*

Robot based Smart Water Pipeline Monitoring System

Sujatha.R[1], Patel Parixit Hariprasad[2]

[1] Associate Professor, Dept. of Embedded Technology, Vellore Institute of Technology, Vellore, Tamil Nadu, India.

[2] Student, Dept. of Embedded Technology, Vellore Institute of Technology, Vellore, Tamil Nadu, India.

E-mail: sujatha.r@vit.ac.in, parixit.vit@gmail.com

Abstract. Water scarcity is a serious challenge posed on us and the cost of providing fresh potable will provoke over time. The adoption of robots for the water transportation is inevitable for the productive rehabilitation. Real-time water pipelines monitoring is one of the potential application where robotics play a major role. Monitoring the huge long water pipelines is the biggest challenge. Thus a robust and reliable artificial water pipe monitoring and cleaning robot is proposed to monitor the pipelines of various diameters that interconnect the distributed network for leaks and clogs and to remove the debris. The automated system can be made smart by providing online water leakage or block audit is feasible through a web Socket based application program interface in the cloud platform. An early warning dashboard will be enabled for the water supply management team. This prototype when implemented on a large scale will have potential applications in multipurpose industrial huge oil or gas pipelines along with water pipelines.

1. Introduction

Water is a precious asset to our Country to be conserved. A report says that approximately 163 million Indians do not have access to clean drinking water and 21 percent of the communicable diseases are associated with unsafe drinking water. Every day children die because of diarrhea in India due to the consumption of unclean water. At present, 50 percent of potable water supplied to a city is going waste due to the blocks or leaks in huge long water pipelines. Water is a precious resource and using the water supplies wisely is our prime duty. Pipelines are part of our infrastructures since ages that provides a more comfortable mode of water transportation. The health of the pipelines to be monitored regularly without human intervention. We have to preserve the water supply Infrastructure from innumerable number of problems in the delivery pipelines to meet the global consumption. Leakage, mechanical damage, corrosion and blocks are the major problems encountered by a water distribution network. Velocity and pressure of the flow may vary because of these problems encountered. More underwater pipeline systems are utilized for various applications. In the near future, around the world more pipeline projects are planned where advanced health monitoring of pipelines, both underground and above the ground should be carried out without human intervention.

2. Robots for water pipeline Monitoring

In-pipe flexible, mobile robot is a cutting-edge innovative technology for water pipe lines monitoring application. Implementing, operating and maintenance costs for underwater pipelines are very high. There are more risks of accidental failures or damages. In this situation, robots can provide a robust solution for long-distance water pipeline monitoring systems, to detect the leakage and its exact location. Most of the pipelines are lying underground and most of them are huge and cannot be monitored frequently by the concerned people. To keep the pipeline systems functional on a daily basis and prevent forthcoming disaster. Robot prototype developed, carrying all sensors will move inside the pipeline with the help of the wheels designed and powered by a DC to identify the blocks or leakages before they create a disaster, infect the freshwater and creating health hazards. The shapes of the industry pipelines are not common and sometimes due to corrosion chemicals present in the iron pipes coating may cause the release of poisonous gas. Robots are a suitable choice to clean those pipes. Metallic and non-metallic materials are used for pipelines. Of these most commonly used for water supply are galvanised iron, unplasticized polyvinylchloride (PVC), polyethylene (PE) and chlorinated polyvinylchloride (CPVC). Without some form of control in installation, damages to the pipes or blocks may arise which can ultimately pollute the potable water or lead to water loss. An in-pipe handy mobile service robot is the best solution to facilitate for water pipelines inspection for blocks or cracks and regular cleaning.

3. Related Work

Significant research has been carried out in recent years by researchers to overcome the long distance water pipeline issues faced by the PWD. Most of the researchers worked on sensors, data acquisition systems, mathematical formulas and data processing algorithms. Norbert Elkmann et al(2006) [1] developed prototype for motion through the sewer and achieved a new inspection system to audit the damage level. For the Emscher sewer system, an automatic pipe inspection and cleaning system was developed. Changlong Ye et al (2010) [2] developed a robot for a central air-condition named PCV-1 for the pipe environment adaption. They verified the movement by conducting a few experiments. A pipe cleaning vehicle was designed that can move both in the upward and downward direction. The robot completed a variety of pipeline movements successfully. Nguyen Truong-Thinh et al (2011) [3] proposed a new cutting method for cleaning and underwater inspection. The authors discussed a communication method for inspection and cleaning and evaluated their performance through computer simulations. Yoon-Gu Kim et al (2011) [4] focussed on the design of implementation for a pipe navigation system and to overcome the situations inside steel pipes. Authors focussed on navigation ability without slipping and various control tests were performed for the driving test. The robot proposed by them offers adaptability to different pipe diameters. Anthony Harris et al (2012) [5] proposed a concept for pipe traversing robot and its end product. The document describes the design and fabrication of a robot traversing dryer ventilation to clean it. Speed, weight, effectiveness and cost are the parameters considered in the design. Mohammed S. BenSaleh et al (2013) [6] presented a wireless sensor networks based technology for water pipeline monitoring scenarios. Even though manual sweepers were ruled out in 2013, hundreds of people die due to underground water pipelines cleaning work. To provide a solution for this tedious problem automated sewer robot "Bhrutiartan" was used.

4. Proposed Model

The proposed design involves an in-pipe mobile flexible robot that audits the supply lines for leaks and clogs. The robot has both a hardware and software interface. It is being tested for a PVC domestic pipe of 300mm inner diameter. The battery is the heaviest part in the housing and it needs a strong and stable housing. The software interface is created for the robot movement section. The entire housing is done with water- proof material to avoid shocking. A fusion of sensors is mounted on the robot body with the camera. Flow sensors and obstacle detecting sensors will detect the leak and clog. The robot with four wheels is driven by the DC motor.

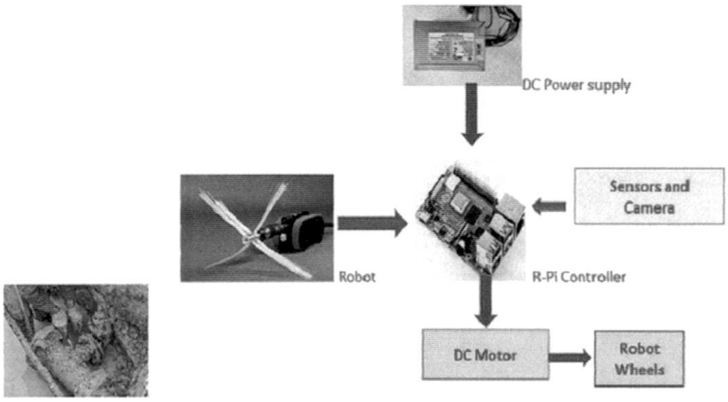

Figure 1. Block diagram of full prototype.

The figure illustrates the block diagram of the model proposed, Raspberry pi acts as the gateway that takes data from the camera and it's controlling actions such as Trach, power screw and clean mechanism are given through the mobile app. The body of the robot accommodates sensors, camera, DC motor to drive the wheels and the power supply. The robot is designed to move inside a pipe of diameter 8 inches using PVC material to make the whole body feel lightweight. The four wheels powered by DC motor will touch the ground for increased traction and grips. For a stable and strong attachment, the wheels are screwed using aluminum clamps.

The robot takes a peristaltic motion inside the pipe multiple gripping and contraction units. With the help of the DC motor, the gripping unit moves in a radial direction and generates friction with the pipe tightly. A brush is attached at the front end of the robot, as the robot moves forward, it will scrape off the block on the inner surface of the pipe and a camera is mounted at the top, for the remote monitoring of the cleaning process. The speed of the hard pipe is determined by the number of movements in one cycle of the robot per second.

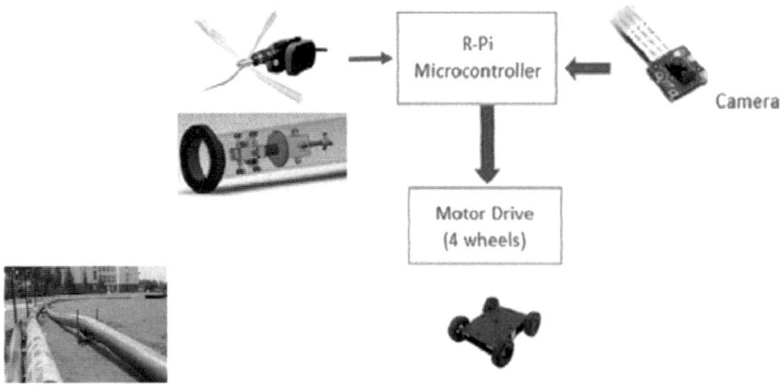

Figure 2. Robot motion control mechanism.

4.1. Analysis
The prototype is tested on two cases (i) No pipe leak (ii) with pipe leak (1.5 mm). Based on Bernoulli's principle water pressure is related to flow rate. If there is a leakage, there will be more water pressure and less flow value. Water pressure and vibration are inversely proportional.

4.2. Data Flow Graph (DFG) for the Robot movement
The data flow graph DFG (figure 3) is created based on the cleaning process and it is a software interface for robot movement using Arduino programming. Real-time monitoring of the robot movement is viewed in the cloud environment using ESP8266 Node MCU. If the test movement is a success, then through Node MCU mobile dashboard is connected. If there is any error, the process will be repeated again.

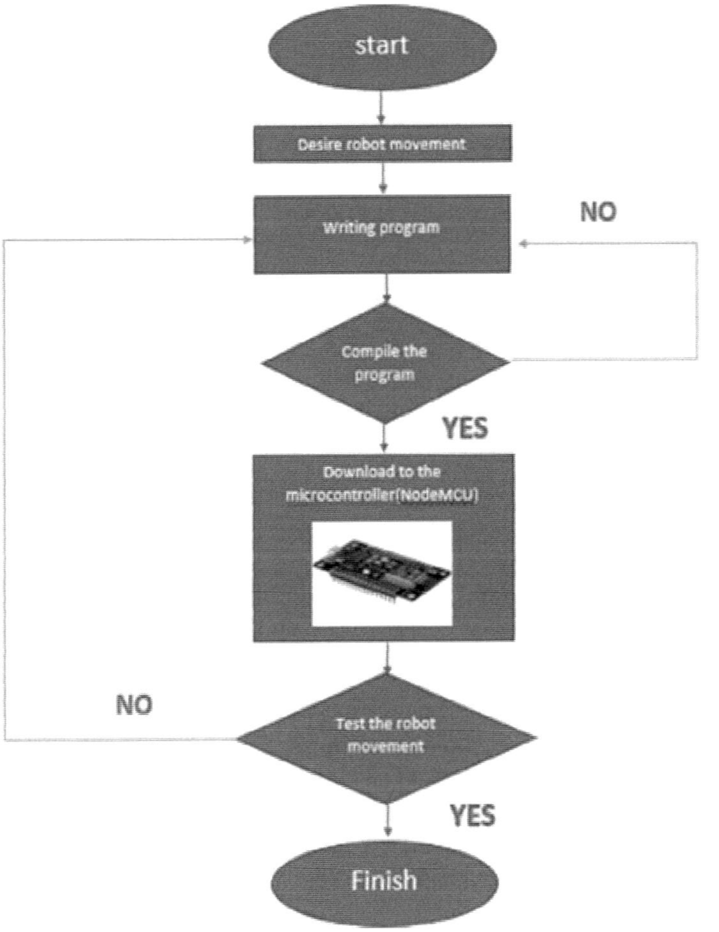

Figure 3. Data Flow Graph(DFG) for the Robot movement.

4.3. Mobile App for Robot movement control
Figure 4 shows the dashboard of the mobile app for the Robot movement. When the power button is "ON" the robot can be moved forward or reverse and the pipe cleaning is done through a motor sliding approach.

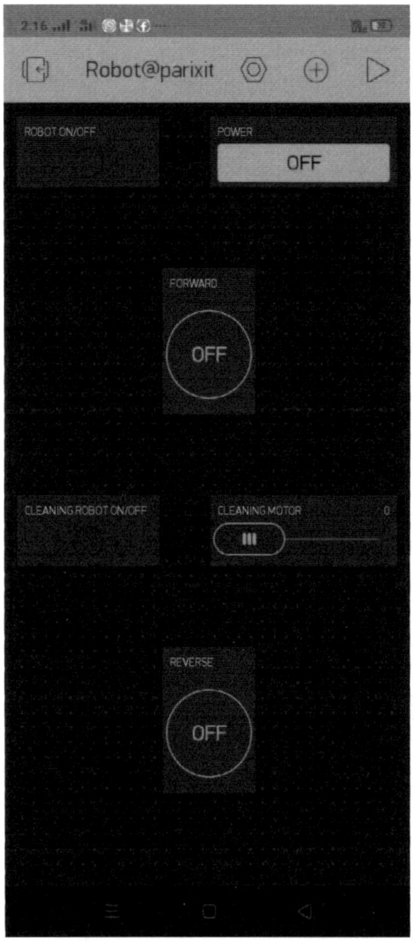

Figure 4. Dashboard for Robot movement control.

5. Conclusion

Water is an essential natural resource to be protected to meet out the global increasing consumption.To provide water security, we have to preserve the health of water Infrastructures through continuous real-time monitoring without human intervention. A robust, effective, customizable, scalable and reliable smart robot prototype is built and tested. Pipeline Infrastructures are used in transporting oil, gas and other liquids apart from water, where the leakage is risky. On a large scale, the proposed robot prototype can be implemented for the oil and gas industry to monitor the leakage. Overall, the proposed robot-based water leakage monitoring along with a mobile app provides a great advantage by avoiding human involvement. Thus the service robot has enabled the pipelines to work efficiently with minimum cost. In the future, we can use LoRa sensors deployed on the surface of the mobile robot to reduce overall power requirement and for wider coverage.

References

[1] N Elkmann, S Kutzner, J Saenz, B Reimann, F Schultke and H Althoff 2006 Fully Automatic Inspection Systems for Large Underground Concrete Pipes Partially Filled with Wastewater *IEEE/RSJ International Conference on Intelligent Robots and Systems* pp. 4234-4238 Available: 10.1109/IROS.2006.281949.

[2] C Ye, S Ma, H Li and J Yang 2010 Development of a pipe cleaning robot for air conditioning system *IEEE International Conference on Robotics and Biomimetics* pp. 1525-1529 Available: 10.1109/ROBIO.2010.5723556.

[3] N Truong-Thinh, N Ngoc-Phuong and T Phuoc-Tho 2011 A study of pipe-cleaning and inspection robot *IEEE International Conference on Robotics and Biomimetics* pp. 2593-2598 Available: 10.1109/ROBIO.2011.6181695.

[4] Yoon-Gu Kim, Dong-Hwan Shin, Jeon-Il Moon and Jinung An 2011 Design and implementation of an optimal in-pipe navigation mechanism for a steel pipe cleaning robot *8th International Conference on Ubiquitous Robots and Ambient Intelligence (URAI)* pp. 772-773 Available: 10.1109/URAI.2011.6146010.

[5] A Harris, J Davis, J Meeks, T Monfort and J Conrad 2012 Concept for a pipe-traversing robot *Proceedings of IEEE Southeastcon* pp. 1-5 Available: 10.1109/SECon.2012.6196927.

[6] M BenSaleh, S Qasim, A Obeid and A Garcia-Ortiz 2013 A review on wireless sensor network for water pipeline monitoring applications *International Conference on Collaboration Technologies and Systems (CTS)* pp. 128-131 Available: 10.1109/CTS.2013.6567217.

[7] S K Subramaniam, I Y Panessai, R A Ramlee and R Sujatha, N Rajagopal 2019 Tcp Performance and Throughput Fairness Optimization in a Multi-Hop Pipeline Network, *International Journal of Recent Technology and Engineering* **8(3S2)** pp. 499-505 Available: 10.35940/ijrte.c1114.1083s219.

[8] T Gujrati 2017 Nipm robot: Nuclear industry pipes maintenance robot *IEEE International Conference on Power, Control, Signals and Instrumentation Engineering (ICPCSI)* pp. 2286-2291 Available: 10.1109/ICPCSI.2017.8392124.

[9] B Bordel, D Martin, R Acarrial and T Robles 2019 A Blockchain-based Water Control System for the Automatic Management of Irrigation Communities *IEEE International Conference on Consumer Electronics (ICCE)* pp. 1-2 Available: 10.1109/ICCE.2019.8661940.

[10] P Shah, A Patil and S Ingleshwar 2017 IoT based smart water tank with Android application *International Conference on I-SMAC (IoT in Social, Mobile, Analytics and Cloud) (I-SMAC)* pp. 600-603 Available: 10.1109/I-SMAC.2017.8058250.

[11] A Predescu, M Mocanu, C Lupu and A Bercovici 2019 A real-time architecture for collaborative IoT applications in urban water management *23rd International Conference on System Theory, Control and Computing (ICSTCC)* pp. 839-844 Available: 10.1109/ICSTCC.2019.8885509.

[12] N Le and T Ngo 2019 Proposal of a Sewerage Cleaning Robot to Collect Garbage Applying for Ho Chi Minh City *IEEE/SICE International Symposium on System Integration (SII)* pp. 216-221 Available: 10.1109/SII.2019.8700386.

[13] S. R, S. V.P, S. E and J. B. 2017 Real Time Water Monitoring System using IoT *IJARCCE* **6(3)** pp.378-380, Available: 10.17148/ijarcce.2017.6386.

[14] J Stankovic 2014 Research Directions for the Internet of Things *IEEE Internet of Things Journal* **1(1)** pp.3-9 Available: 10.1109/JIOT.2014.2312291.

[15] Shanzhi Chen, Hui Xu, Dake Liu, Bo Hu and Hucheng Wang 2014 A Vision of IoT: Applications, Challenges, and Opportunities With China Perspective *IEEE Internet of Things Journal* **1(4)** pp.349-359 Available: 10.1109/JIOT.2014.2337336.

[16] S Wadekar, V Vakare, R Prajapati, S Yadav and V Yadav 2016. Smart water management using IOT *5th International Conference on Wireless Networks and Embedded Systems (WECON)* pp. 1-4 Available: 10.1109/WECON.2016.7993425.

[17] W Heinzelman, J Kulik and H Balakrishnan 1999 Adaptive protocols for information dissemination in wireless sensor networks *Proceedings of the 5th annual ACM/IEEE international conference on Mobile computing and networking - MobiCom '99* pp. 174-185 Available: 10.1145/313451.313529.

[18] F Tedesco, C Ocampo-Martinez, A Casavola and V Puig 2018 Centralized and Distributed Command Governor Approaches for Water Supply Systems Management *IEEE Transactions on Systems, Man, and Cybernetics: Systems* **48(4)** pp.586-595 Available: 10.1109/tsmc.2016.2612361.

[19] H Kuntze and H Haffner 1998 Experiences with the development of a robot for smart multisensoric pipe inspection *IEEE International Conference on Robotics and Automation (Cat. No.98CH36146)* **2** pp. 1773-1778 Available: 10.1109/ROBOT.1998.677423.

[20] S Poozesh, M Mehrandezh, H Najjaran and R Paranjape 2007 Design and Development of a Smart Vehicle for Inspection of In-Service Water Mains *Canadian Conference on Electrical and Computer Engineering* pp. 1203-1206 Available: 10.1109/CCECE.2007.307.

[21] W Adams, S Sridar, C Thalman, B Copenhaver, H Elsaad and P Polygerinos 2018 Water pipe robot utilizing soft inflatable actuators *IEEE International Conference on Soft Robotics (RoboSoft)* pp. 321-326 Available: 10.1109/ROBOSOFT.2018.8404939.

[22] A Jamaluddin, D Harjunowibowo, D Rahardjo, E Adhitama and S Hadi 2016 Wireless water flow monitoring based on Android smartphone *2nd International Conference of Industrial, Mechanical, Electrical, and Chemical Engineering (ICIMECE)* pp. 243-247 Available: 10.1109/ICIMECE.2016.7910449.

[23] Z Shi, M Wang, H Lin, Z Gao and L Huang 2019 NB-IOT Pipeline Water Leakage Automatic Monitoring System Based on Cloud Platform *IEEE 13th International Conference on Anti-counterfeiting, Security, and*

Identification (ASID) pp. 272-276 Available: 10.1109/ICASID.2019.8925228.
[24] A Salunke, S Ramani, S Shirodkar, O Vas and K Acharya 2019 Pipe Cleaning Robot *International Conference on Nascent Technologies in Engineering (ICNTE)* pp. 1-6 Available: 10.1109/ICNTE44896.2019.8945930.
[25] S Deb, J Rokky, T Mallick and J Shetara 2017 Design and construction of an underwater robot. *4th International Conference on Advances in Electrical Engineering (ICAEE)* pp. 281-284 Available: 10.1109/ICAEE.2017.8255367.
[26] F Trojan and R Marcal 2007 Study and Implement of a Specialized Monitoring System (SMS) to Orient Decisions and Reduce Losses in the Public Water Supply Systems *Second International Conference on Innovative Computing, Informatio and Control (ICICIC 2007)* pp. 451-451 Available: 10.1109/ICICIC.2007.545.

IOP Conf. Series: Materials Science and Engineering **906** (2020) 012026 doi:10.1088/1757-899X/906/1/012026

Implementation of FOC based speed control for an E-Rickshaw brushless DC drive

T. J.Prem Prasanna[1], Devavrat Shyam Anikhindi[1], S. Afzal Ahamed[1], B. Ashok[2], R. Raja Singh[3]

[1]Department of Automotive Engineering, Vellore Institute of Technology, Vellore, India.
[2]School of Mechanical Engineering, Vellore Institute of Technology, Vellore, India.
[3]Advanced Drives Laboratory, Department of Energy & Power Electronics, Vellore Institute of Technology, Vellore, India.
Coorospondance: premprasanna1997@gmail.com

Abstract. This paper describes an approach to implement Field Oriented Control (FOC) using PID Controller for a brushless direct control (BLDC) motor. The research proposes reduction in total harmonic distortion of the supply current using space vector pulse width modulation technique. Minimization of total harmonic distortion becomes mandatory for a smooth operation in electric vehicle applications. As the harmonics are generated in the supply current, caused by non-linear load in E-Rickshaw. The lower, the value of total harmonic distortion givers better performance of the motor when operated under light load. Here the different strategies employed for controlling BLDC motor speed is analyzed and compared. It also proved that the space vector pulse width modulation technique improves the reliability due to the balanced switching frequency achieved from the solid-state switches in the inverter. Through simulating, the proposed control strategy is analyzed and compared with the existing techniques results in smooth change in current during sudden loading, which improves system performance and as a result the torque ripple gets minimized. The comparative study was made in the Matlab/Simulink 2019b, by designing the mathematical model of motor, inverter and the switching sequence, and the max-min algorithm for the SVPWM.

1. Introduction

Recent years have seen a huge spike in the development of high-performance motor drives. These drives have proven immense importance in a plethora of applications like automotive, computer, steel rolling mills, electric trains, robotics etc.[1], [2]. Common methods for designing industrial speed controllers are experimental, root locus, frequency and optimization methods [3]. A lot of research has been done on speed control of a BLDC motor using a proportional-integral-differential (PID) controller. However, BLDC motors might give rise to large overshoots, slow response times and large steady state error. For these reasons, the gain values of the PID controller need to be tuned to obtain satisfactory output from the motor under varying load and speed conditions.The ziegler-nichols method which is widely used for tuning the PID controller with systems whose plant model can be single input single output systems. But for complex systems like the BLDC motor modeling it cannot be applicable. And as the PID controllers have the following advantages like easier to implement, uses lesser resources, more robust to tuning mismatches and better response to unmeasured disturbances. And the recent methodology for tuning the gains is using fuzzy logic controllers in addition to PID [4], [5], [6]. In this method, the PID controller sends the main control signals in the form of pulse width

modulated (PWM) waves. As a feedback, speed error and change in error are taken into consideration and fuzzified. Then based on pre-set lookup tables, the values of the gains are obtained by the process of defuzzification. The next advancement on tuning the PID controller is by sing adaptive fuzzy controller [7],[8],[9],10]. In this method, in addition to the fuzzy PID control process, an adaptive process of normalizing the inputs and outputs of the fuzzy controller is employed [11]. This puts a cap on the error obtained and provides slightly finer results under certain operating conditions.
There are different types of PWM techniques, among them the SVPWM and SPWM are popularly employed modulation techniques for industrial applications, especially with semiconductor switch converters. Due to the robust nature of the mode of operation at various speeds, the sinusoidal PWM technique is being used [12], [13]. The semiconductor switches that are bring used along with the driver circuit application are Metal-Oxide Semiconductor Field-Effect Transistor (MOSFET), Insulated Gate Bipolar Transistor (IGBT) and Bipolar Junction Transistor (BJT). These switches exhibit continuous drain current property, i.e., generally 80 A at 25°C and it decreases to 65 A at 100°C. Due to this decrease in the continuous drain current property of the switches multiple switches are connected in parallel, this makes the circuit more robust to operate at any operating temperature and is more safe. So, for the motor driver applications generally 6 switches are required to build the convertor. But due to the property of continuous drain current, 12 switches are used. And the pulse wave form from the controller triggers the gates of the switches. When the gate is triggered in a sequence of operation, so that combined operating behavior of all the switches gives the 3 phase AC output to the Motor windings [14]. The BLDC motor and the Permanent Magnet Synchronous Motor (PMSM) both have the permanent magnet rotor and there are 3 stator windings connected in star pattern. But the only difference between the PMSM and the BLDC is interconnection of coils in the stator windings. The PMSM have distributed stator windings (coils spanning more than one tooth) and BLDC have concentrated stator windings (coils wound around each tooth). This makes the back EMF of the PMSM as sinusoidal and for BLDC as Trapezoidal[15]. And while designing the mathematical model of the motor, the back EMF is created using the lookup table based on the position of the motor (degrees of rotation). The BLDC motor stator is made up of stacked steel laminations wherein windings are placed in groups that are axially cut along their inner periphery[16]. Mostly, all the BLDC motor have 3 start connected stator windings. Each winding is constructed with numerous interconnected coils, with one or more coils placed in the stator slot. These windings along with the stator periphery form an even number of poles. This type of winding arrangement in the BLDC motor makes the torque ripple to be higher than the PMSM motor. In our present work, the PID controller, helps to minimize the torque ripples in the motor.

The BLDC motor rotor is made of permanent magnet and can vary from 2 to 8 pole pairs with alternate North (N) and South (S) poles. The magnetic material is decided based on magnetic field density. Ferrite magnets, rare earth materials like Neodymium (Nd), Samarium Cobalt (SmCo) and the alloy of Neodymium, Ferrite and Boron (NdFeB) are some examples of rare earth alloy magnets are used to make the permanent magnet pole pieces[17],[18]. The position of the motor while rotating can be determined with the help of hall effect sensor and quadratic axis encoder also called as rotary encoder. For low cost application the hall sensor is used, which can be found inside the stator on the driving end. Generally, 3 hall effect sensors are mounted at 120° angle apart from each other. Whenever the magnetic poles pass the hall sensors give out a signal, high indicating North and low indicating South. Based on the combination of these 3 hall effect sensors, the position of motor is determined. Thereby the speed of the motor is being monitored and controlled by using closed loop feedback control mode of operation. The working principle of brushless DC motors is the same as regular DC motor, the Lorentz law, which states that a current carrying conductor placed in a magnetic field experiences a force. This, in turn generates an equally opposite reaction force on the magnet. In BLDC motor, the current carrying conductor stays stationary while the permanent magnet rotates. In practical operation of the 3phase BLDC motor, it operates in a 2-phase active fashion, i.e. energizing the two phases producing the highest torque while keeping the third phase inactive [19]. Energization of the two phases depends on the rotor position i.e. each sequence has current flowing into one winding (positively energized), current flowing out of one winding (negatively energized) and one winding non-energized. This principle of operation is due to the star winding of the stator present to energize the coils. And to complete one electrical cycle, it generally takes 6 steps of operation. However, electrical cycle is not equal to mechanical revolution. The rotor pole pairs sets base for the amount of electrical cycles to be repeated in order to complete a mechanical rotation. For each rotor pole pairs, one electrical cycle is completed. So, the mechanical revolution (rpm) is one half of the electrical revolution times the number of poles.

Based on the previous work of research work related to the speed control of the BLDC motor, a comparative table of research work conclude that the PID controller is suitable for sinusoidal PWM technique, as described in table 1. The motor speed is being controlled by mechanical revolution. Based on the desired input from the driver to run at the rpm, the PID controller controls the PWM signals. And the controlled signals are being sent to the inverter for the switching the switches. The inverter converts DC power from the battery into the AC power with the helpof these switches. And the sequences of operation of each switch are based rotor position.

Table 1. Comparison of Control Strategies for BLDC motor.

Control Logic and Methodology	Outcome	Ref. No.
PID controller with sing Adaptive Fuzzy controller	In this method, in addition to the Fuzzy PID control process, an adaptive process of normalizing the inputs and outputs of the fuzzy controller is employed. This puts a cap on the error obtained and provides slightly finer results under certain operating conditions.	[3]
PI with SPWM and SVPWM	With SVPWM technique, the system performance is enhanced, motor requires lesser current, speed response is smoothened and torque ripple is pretty much flattened.	[13]
PI and Fuzzy	PI with Fuzzy is better over conventional PI with faster steady state speed and lesser torque ripple	[20]
PI with Sensorless FOC	Phase locked loop and back EMF used to estimate position with minute angular position error and peak to peak estimation error.	[21]
PI with SPWM and SVPWM	Under normal operating conditions, both techniques are on par with each other. SVPWM however, gives a higher power output thus providing a bit better performance.	[22]
FOC with dither signal	Dither signal injection quench the limit cycles appearing in the system caused by non-linearity, thereby reducing the overall torque ripple and improving the response of FOC	[23]
PI with FOC, SFOC & Hysteresis	SFOC shows the least torque ripple out of the three methods	[24]

2. BLDC Motor Control

BLDC motor control is carried out with mathematical model of the BLDC motor, and then implementation of controller [25] [26],[27]. The mathematical model is developed with the idea of state space system design wherein using the state variables, state-space models describe a system through a set of differential or difference equations [28]. And the model is developed using continuous-time representation. The controller is used for controlling this developed model of BLDC motor [29],[30]. The controller regulates the actual speed with regards to the reference speed, thereby reducing the error percentage in overshoot, undershoot, and torque ripple. These problems in the controller can be minimized by using the optimum control strategy. Based on our literature review of various control strategies, the PID control with SVPWM is the best method, through which the system performance is enhanced, motor requires lesser current, speed response is smoothened and torque ripple is pretty much flattened.

IOP Conf. Series: Materials Science and Engineering **906** (2020) 012026 doi:10.1088/1757-899X/906/1/012026

2.1. Mathematical Model of the 3 – Phase BLDC Motor

The modelling of the BLDC motor is based on assumption:

- Balanced three phase connections
- The loss in the variable load like stray and copper losses are assumed to be negligible and loss in the constant load like hysteresis and eddy current losses are also neglected while modelling
- There is uniform air gap is maintained between the stator and the rotor
- The switches used in the inverter are considered to be operated at ideal condition, neglecting losses inside them.
- The internal resistances of the three phase stator winding are considered to be equal.
- The value of inductance are assumed to be constant, where self and mutual inductance are considered equal

The motor parameters that are used for modeling are explained in the appendix in the later section.[31][32] The Mathematical model of the armature winding of the BLDC motor is expressed in the form of equation (1) to equation (7).

$$V_{an} = Ri_a + \frac{d(\lambda_a)}{dt} + e_{an} \tag{1}$$

$$V_{bn} = Ri_b + \frac{d(\lambda_b)}{dt} + e_{bn} \tag{2}$$

$$V_{cn} = Ri_c + \frac{d(\lambda_c)}{dt} + e_{cn} \tag{3}$$

$$\lambda_a = L_s i_a - M(i_b + i_c) = L_s.i_a + M(i_a) = (L_s + M).i_a \tag{4}$$

$$\lambda_b = L_s i_b - M(i_a + i_c) = L_s.i_b + M(i_b) = (L_s + M).i_b \tag{5}$$

$$\lambda_c = L_s i_c - M(i_a + i_b) = L_s.i_c + M(i_c) = (L_s + M).i_c \tag{6}$$

Thus these equations can be represented in this form of matrix, where p is $\frac{d}{dt}$.

$$\begin{bmatrix} V_{an} \\ V_{bn} \\ V_{cn} \end{bmatrix} = \begin{bmatrix} R + p(L+M) & 0 & 0 \\ 0 & R + p(L+M) & 0 \\ 0 & 0 & R + p(L+M) \end{bmatrix} \begin{bmatrix} i_a \\ i_b \\ i_c \end{bmatrix} + \begin{bmatrix} e_{an} \\ e_{bn} \\ e_{cn} \end{bmatrix} \tag{7}$$

As the BLDC motor rotates, each winding generates a voltage known as back EMF, which opposes the supply voltage, according to Lenz's law. The polarity of the back EMF is opposite to the supply voltage and is dependent on the rotor position. The systematical order of equations that are being used in the BLDC are explained below. As the system is a multi input multi output, the traditional state space form of representation cannot be used, as the differential term is used. Thus using these equations, the complete mathematical model of BLDC motor is modeled using Matlab / Simulink, as shown in figure 1.

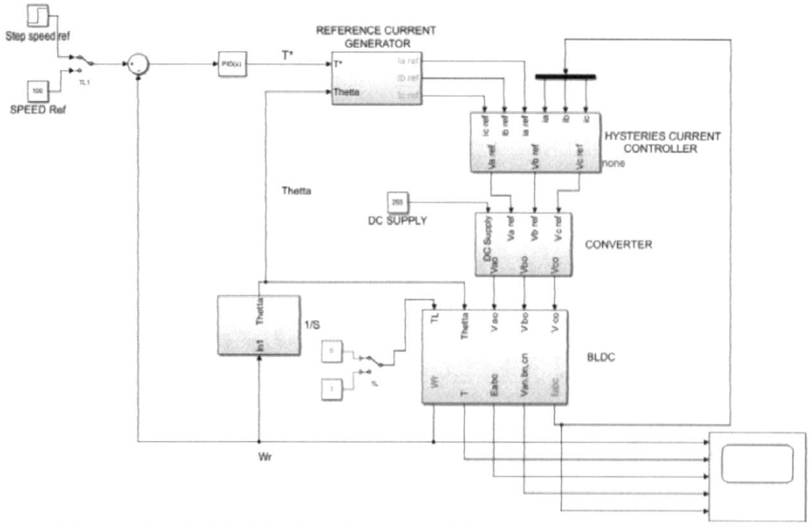

Figure 1. MATLAB Simulink model for BLDC motor Control.

2.1.1. Voltage Measurement

Using the following equation (8) the phase voltage across the terminals of the motors is determined. And as the BLDC motor has no neutral connection, equation (9) is used. Thus using these two equations the voltage across each phase is determined using equation (11) to equation (13).

$$\mathrm{V_{ao}} + \mathrm{V_{bo}} + \mathrm{V_{co}} - 3\mathrm{V_{no}} = \mathrm{R}(\mathrm{i_a} + \mathrm{i_b} + \mathrm{i_c}) + (\mathrm{L_s} + \mathrm{M})\left(\frac{d(i_a)}{dt} + \frac{d(i_b)}{dt} + \frac{d(i_c)}{dt}\right) + e_{an} + e_{bn} + e_{cn} \tag{8}$$

$$i_a + i_b + i_c = 0 \tag{9}$$

$$V_{no} = \frac{1}{3}\{(V_{ao} + V_{bo} + V_{co}) - (e_{an} + e_{bn} + e_{cn})\} \tag{10}$$

$$V_{an} = V_{ao} - V_{no} \tag{11}$$

$$V_{bn} = V_{bo} - V_{no} \tag{12}$$

$$V_{cn} = V_{co} - V_{no} \tag{13}$$

2.1.2. Current Measurement

From equation (1) the relationship between the voltage and the current equations is being obtained. And, upon integration of the equation (14), the value of current across each phase of the stator of the BLDC motor is obtained. Similarly, different block is used to find the current at different phasesi_a, i_b and i_c [33].

$$\frac{d(i_a)}{dt} = \frac{(V_{an} - R.i_a - e_{an})}{(L_s + M)} \tag{14}$$

2.1.3. Electromagnetic Torque Measurement

The electromagnetic torque, which has a unit of N-m, is measured using the help of the equation (15). Normalized Back EMF function,$f_a(\theta_r)$, $f_b(\theta_r)$ and $f_c(\theta_r)$ are trapezoidal function. Thus from equation (15) to equation (17), the generalized electromagnetic torque equation is determined. And using the help of the supporting equation (15) to equation (17), the final electromagnetic equation is determined using equation (19).

$$T_e = \{(e_{an}i_a) + (e_{bn}i_b) + (e_{cn}i_c)\}/(\omega_r) \tag{15}$$

$$e_{an} = E\, f_a(\theta_r), e_{bn} = E\, f_b(\theta_r)\, , e_{cn} = E\, f_c(\theta_r) \tag{16}$$

$$E = K_b\omega_r \tag{17}$$

$$T_e = \frac{(E.f_a(\theta_r).i_a) + (E.f_b(\theta_r).i_b) + (E.f_c(\theta_r).i_c)}{\omega_r} \tag{18}$$

$$T_e = K_b.\{f_a(\theta_r).i_a + f_b(\theta_r).i_b + f_c(\theta_r).i_c\} \tag{19}$$

2.1.4. Angular Velocity Measurement

The general torque equation for the rotating objects is expressed in equation (20), and upon simplification with the help of equation(21) the final angular velocity measurement is obtained using equation (22). [34]

$$J\frac{d\omega}{dt} + B.\omega = T_e - T_l \tag{20}$$

$$\omega_r = \omega\,.\frac{P}{2} \tag{21}$$

$$\frac{d\omega_r}{dt} = \frac{P}{2}\left(\frac{T_e - T_l - B.\omega_r}{J}\right) \tag{22}$$

2.1.5. *Creation of Back EMF (Trapezoidal Function)*

Based on the angle of rotation $\boldsymbol{\theta}_r$, the value of the normalized back EMF is taken out using the look up table in the Matlab environment, as shown in table 2.

Table 2. Logic for Look up table for rotor position.

Rotor Position Signal	**Normalized Back EMF**		
θr	$\mathbf{e_{an}}$	$\mathbf{e_{bn}}$	$\mathbf{e_{cn}}$
0° - 60°	E	−E	(6E/π)(π-θr) -E
60° - 120°	E	(6E/π)(θr-2π) +E	−E
120° -180°	(6E/π)(π-θr) -E	E	−E
180° - 240°	−E	E	(6E/π)(θr-2π) +E
240° - 300°	−E	(6E/π)(π-θr) -E	E
300° - 360°	(6E/π)(θr-2π) +E	−E	E

2.2. *Mathematical Model of Hysteresis Current Controller and Reference Current Generator*

The reference current controller takes in the input of the controller. The controller gives the required torque for the corresponding input. It uses look up table as shown in table 3, to find the reference current i_a, i_b and i_c (reference) response, based on the theta (angular position) response of the motor [35],[36],[37]. The hysteresis current controller is one of the current control strategies to control the inverter. While modeling it, a relay is used for generating the square pulses, as shown in figure 2, where the switch on point is 0.01, switch off point is -0.01. Output from the generator is on when its value 1 and the output is off when its value is -1.

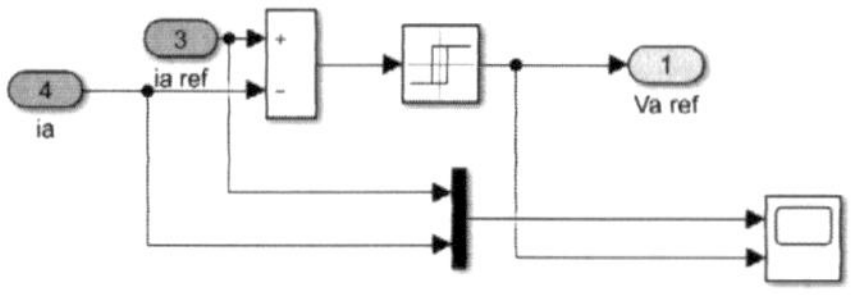

Figure 2. Reference Current model

Table 3. Look-up table for 3 phases of current.

Rotor Position Signal	**Reference Currents (amps)**		
θ r (deg)	**Ia**	**Ib**	**ic**
0° - 60°	I∗	−I∗	0
60° - 120°	I∗	0	−I∗
120° -180°	0	I∗	−I∗
180° - 240°	−I∗	I∗	0
240° - 300°	−I∗	0	I∗
300° - 360°	0	−I∗	I∗

2.3. *Mathematical Model of the 3-Phase BLDC Motor with FOC using Space Vector Pulse Width Modulation (SVPWM)*

SVPWM is a technique used in FOC control, to determine the pulse-width modulated (PWM) signals for the inverter switches in order to generate the desired 3-phase voltage supply to the motor. Here the 3 phase current (i_a, i_b and i_c)is converted into 2 phase (i_d, i_q) by using park transform, as shown in equation (23). The frequency domain is converted into time domain here. This is done with respect to

IOP Conf. Series: Materials Science and Engineering **906** (2020) 012026 doi:10.1088/1757-899X/906/1/012026

the rotating reference frame, here motor rotation angle is considered as the rotating frame. Then the PI controller is used for stabilizing the current in the d and q axis.

$$\begin{bmatrix} d \\ q \\ 0 \end{bmatrix} = \frac{2}{3}\begin{bmatrix} \sin(\theta) & \sin(\theta - \frac{2\pi}{3}) & \sin(\theta + \frac{2\pi}{3}) \\ \cos(\theta) & \cos(\theta - \frac{2\pi}{3}) & \cos(\theta + \frac{2\pi}{3}) \\ 1/2 & 1/2 & 1/2 \end{bmatrix}\begin{bmatrix} a \\ b \\ c \end{bmatrix} \tag{23}$$

The error for the PI controller of q axis is taken as the difference between the input of the speed controller and actual magnitude of value of I_d, and the error for the PI controller of d axis is taken as the difference of reference I_d as 0, and the actual I_d. Since the d axis current is the key component for torque production, this process of setting 0 as the reference helps in reducing the torque ripple from the BLDC motor. The conversion of the reference I_d and I_q currents into α and β axis currents need to be carried out for simplification of mathematical representation. As the SVPWM requires the α and β axis as the input for generating switching pattern. Hence inverse park transform is being used here. Where I_α and I_β, the angle between them is determined as shown in equation (24), and the sector is allocated according to the angle. Using this approach the mathematical model was developed as shown in figure 3.

$$\begin{bmatrix} I_\alpha \\ I_\beta \end{bmatrix} = \begin{bmatrix} \cos(\theta) & -\sin(\theta) \\ \sin(\theta) & \cos(\theta) \end{bmatrix}\begin{bmatrix} I_d \\ I_q \end{bmatrix} \tag{24}$$

The total number of possible switch configurations is 2^3=8. Where, the 6 of these configurations (001,011,010,110,100,101) correspond to different voltages, called as the basic vector. And the last two (000,111) are referred to as zero vector, as they represent zero volts on the terminals. Where 0 represent off state of switch and 1 represent on state of switch.

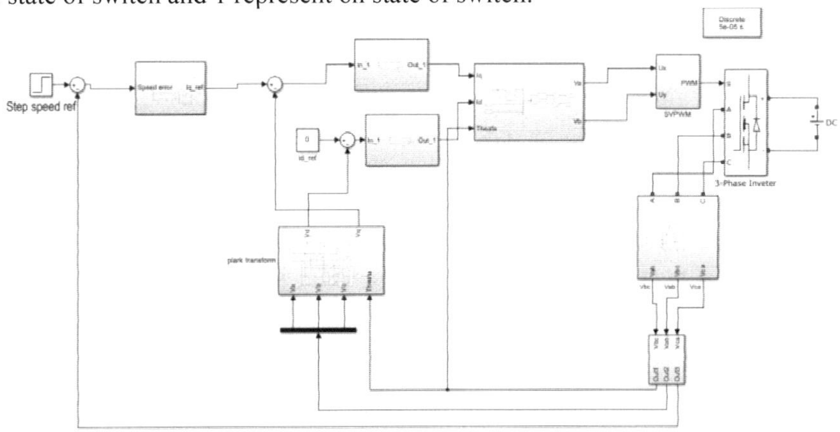

Figure 3. FOC with SVPWM Simulink Model.

Now based on the sector, I_α and I_β, the timing for each sector is determined using the Matlab function in the simulink environment. The switching frequency of the switches is kept 10 kHz. And the switching period is ΔT = 0.0001 sec. The geometric summation from the figure 4 corresponding to the phase angle, can be expressed as shown in equation (25), where T_{sv1} and T_{sv2} are the time period during which space vector SV_1 and SV_2 are selected respectively. They can be expressed in cartesian form as shown in equation (26).

IOP Conf. Series: Materials Science and Engineering **906** (2020) 012026 doi:10.1088/1757-899X/906/1/012026

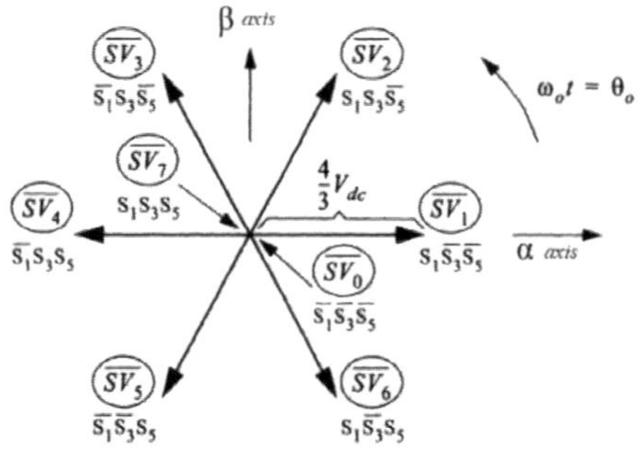

Figure 4. Space Vector Diagram.

$$V_o < \theta_o = \frac{T_{SV1}}{\Delta T/2} SV_1 + \frac{T_{SV2}}{\Delta T/2} SV_2 \tag{25}$$

$$\mathrm{V_o} \left(\cos(\theta_o) + \boldsymbol{j} * \sin(\theta_o)\right) \frac{\Delta T}{2} = \mathrm{T}_{SV1}\mathrm{V_m} + \mathrm{T}_{SV2}\mathrm{V_m}\left(\cos\left(\frac{\pi}{3}\right) + \mathbf{j} * \sin\left(\frac{\pi}{3}\right)\right) \tag{26}$$

Table 4. Equations for Space Vector Control.

$\omega_0 t = \theta_0$ (deg)	Space Vector with Active Times
0° - 60°	$T_{SV1} = \frac{V_0}{V_{dc}}\frac{\sqrt{3}}{2}\cos\left(\theta_0 + \frac{\pi}{6}\right).\frac{\Delta T}{2}$
	$T_{SV2} = \frac{V_0}{V_{dc}}\frac{\sqrt{3}}{2}\cos\left(\theta_0 - \frac{\pi}{2}\right).\frac{\Delta T}{2}$
60° - 120°	$T_{SV2} = \frac{V_0}{V_{dc}}\frac{\sqrt{3}}{2}\cos\left(\theta_0 - \frac{\pi}{6}\right).\frac{\Delta T}{2}$
	$T_{SV3} = \frac{V_0}{V_{dc}}\frac{\sqrt{3}}{2}\cos\left(\theta_0 - \frac{5\pi}{6}\right).\frac{\Delta T}{2}$
120° -180°	$T_{SV3} = \frac{V_0}{V_{dc}}\frac{\sqrt{3}}{2}\cos\left(\theta_0 + \frac{\pi}{2}\right).\frac{\Delta T}{2}$
	$T_{SV4} = \frac{V_0}{V_{dc}}\frac{\sqrt{3}}{2}\cos\left(\theta_0 - \frac{7\pi}{6}\right).\frac{\Delta T}{2}$
180° - 240°	$T_{SV4} = \frac{V_0}{V_{dc}}\frac{\sqrt{3}}{2}\cos\left(\theta_0 - \frac{5\pi}{6}\right).\frac{\Delta T}{2}$
	$T_{SV5} = \frac{V_0}{V_{dc}}\frac{\sqrt{3}}{2}\cos\left(\theta_0 - \frac{3\pi}{2}\right).\frac{\Delta T}{2}$
240° - 300°	$T_{SV5} = \frac{V_0}{V_{dc}}\frac{\sqrt{3}}{2}\cos\left(\theta_0 - \frac{7\pi}{6}\right).\frac{\Delta T}{2}$
	$T_{SV6} = \frac{V_0}{V_{dc}}\frac{\sqrt{3}}{2}\cos\left(\theta_0 - \frac{11\pi}{6}\right).\frac{\Delta T}{2}$
300° - 360°	$T_{SV6} = \frac{V_0}{V_{dc}}\frac{\sqrt{3}}{2}\cos\left(\theta_0 - \frac{3\pi}{2}\right).\frac{\Delta T}{2}$
	$T_{SV1} = \frac{V_0}{V_{dc}}\frac{\sqrt{3}}{2}\cos\left(\theta_0 - \frac{\pi}{6}\right).\frac{\Delta T}{2}$

On equating the real and imaginary parts, the solution of T_{SV1} and T_{SV2} is obtained. The table 4 shows the arbitrary target output for any phase angle, which was created from the two closest space vector components. Therefore, along with relevant expressions, these define the active time intervals for each space vector. Based on these time intervals the active pulse duration is determined and is then compared with a repeating sequence to generate the PWM waveform, which sends the supply pulses to the inverter.The inverter converts DC to AC supply, as the BLDC motor is an AC motor. Based on the switching pulses sequence the output waveform of AC supply is generated. This voltage is directly fed into the motor, for the efficient operation of the motor. As the switches in the inverter at bottom section operates exactly opposite in supply to the top section. And as the top section consists of 3 switches and bottom section consists of 3 switches for the 3 phase 2 level inverter.When the load on the motor is of unpredictable nature, as in case of Electric Vehicle (EV) the supply current from the battery keeps varying, which increases the THD. And if this parameter increases, the torque ripple increases. This is the only limitation of this mode of operation.

2.4. *Total Harmonic Distortion (THD)*

Harmonics or harmonic frequencies of a periodic voltage or current are frequency components in the signal that are at integer multiples of the frequency of the main signal. This is the basic outcome that fourier analysis of a periodic signal shows. Harmonic distortion is the distortion of the signal due to these harmonics.

A voltage or current that is purely sinusoidal has no harmonic distortion because it is a signal consisting of a single frequency. A voltage or current that is periodic but not purely sinusoidal will have higher frequency components in it contributing to the harmonic distortion of the signal. In general, the less that a periodic signal looks like a sine wave, the stronger the harmonic components are and the more harmonic distortion it will have. So, a purely sinusoidal signal has no distortion while a square wave, which is periodic but does not look sinusoidal at all, will have lots of harmonic distortion. In the real world, of course, sinusoidal voltages and currents are not perfectly sinusoidal; some amount of harmonic distortion will be present. Figure 5 provide visual comparisons, in the time domain and the frequency domain, of a sinusoidal voltage and a square wave voltage.THD could be defined as ratio of the RMS voltage of all the harmonic frequencies (from the 2nd harmonic on) over the RMS voltage of the fundamental frequency (the fundamental frequency is the main frequency of the signal, i.e., the frequency that you would identify if examining the signal with an oscilloscope) as shown in equation (27) and equation (28).

$$\mathrm{THD} = \frac{\sum_{n=2}^{\infty} V_{n\_rms}^2}{V_{fundamental\_rms}} \quad (27)$$

$$vsquare\ (t) = \left(\frac{4}{\pi}\right)\sin(2\pi ft) + \left(\frac{4}{3\pi}\right)\sin(6\pi ft) + \left(\frac{4}{5\pi}\right)\sin(10\pi ft) + \cdots + \left(\frac{4}{n\pi}\right)\sin(2n\pi ft) \quad (28)$$

THD is important in several types of systems, including power systems, where a low THD means higher power factor, lower peak currents, and higher efficiency; audio systems, where low THD means that the audio signal is a more faithful reproduction of the original recording; and communication systems, where low THD means less interference with other devices and higher transmit power for the signal of interest. Generally the EV is a miniature power system, it has various audio-video instruments and network connectivity, therefore or this system, the lower the THD, better the power factor and lower is the interference with the system.

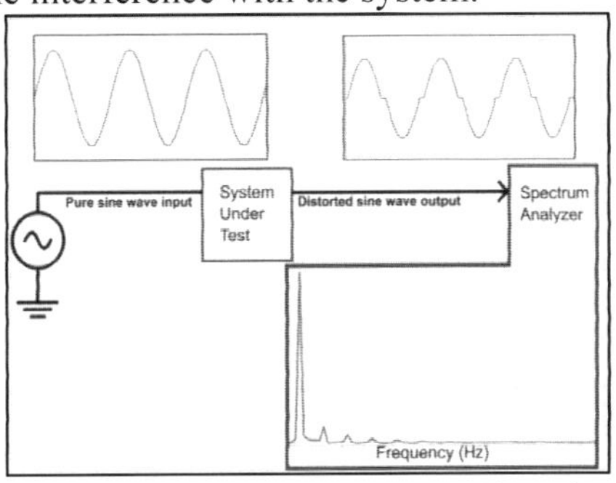

Figure 5. System that introduces crossover distortion into a signal.

2.5. Tuning of PID Controller

The designed plant is so complicated that its mathematical model cannot be easily obtained, therefore the experimental approaches to tune the PID controller is carried out. By varying the proportional gain (K_p), integral time (T_i) and derivative time (T_d) to different values, the optimal value has been estimated. This optimal value is based on the desired output from the system. For EV applications, the rise time and the settling time should be as minimum as possible, so the desired speed from the throttle to the wheel, is achieved as quick as possible. And the overshoot percentage should not be too high, as it creates sudden load to the drive wheel, which decreases the life cycle of the power train. The general layout for plant system using PID control is shown in figure 6.

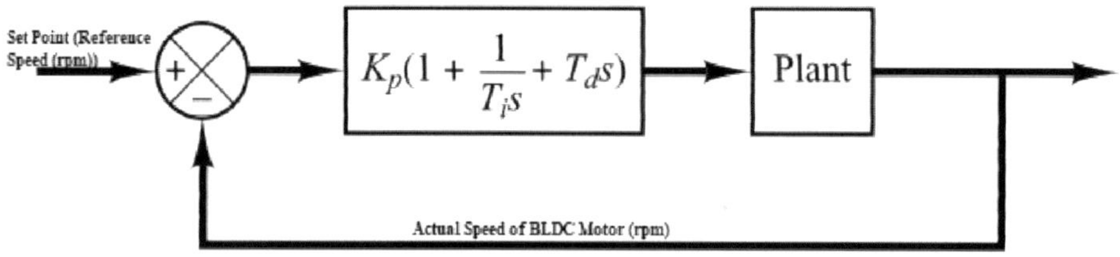

Figure 6. PID Control Loop for the model.

While tuning the following observations was noted down like, as the value of K_p, increased, the rise time decreases, the overshoot increases, and settling time is not having much influence. Similarly, with the increase the value of T_i, the rise time decreases, the overshoot increases, and settling time increases. Similarly, when T_d increases, the rise time increases, overshoot decreases and settling time decreases. By trying different iterations, the optimal value of the gains was determined, as showed in table 5.

Table 5. PID Tuning for finding optimum gain values for the PID Controller of the BLDC motor.

Sl. No.	Kp	Ti	Td	Rise Time (ms)	Over shoot(%)	Settling Time (ms)
1	0.4	6	0	26.747	26.747	186.701
2	0.5	7	0.00001	22.643	3.716	120
3	0.5	4	0.0001	41.804	1.622	205.496
4	0.5	4	0.001	51.613	1.183	120.327
5	0.5	4	0.01	92.612	1.482	272.247
6	0.4	4	0.01	85.676	3.55	317.905
7	1	4	0.04	109.04	0.5	174.108
8	0.6	4	0.01	59.691	3	275.863
9	0.8	7	0.001	25.444	1.05	174.042
10	0.9	8	0.001	23.669	2.5	221.19
11	0.8	6	0.001	26.063	1.2	167.721
12	0.8	8	0.001	22.366	1.006	110.482

3. Results and Discussion

3.1. Tuning of PID Controller

The input was given as step response of 100 to 150 rpm. Then the error from this reference to the actual speed of the motor is given as the feed to the PID controller. Then the controller sends the optimal signal to the reference current generator, to send the optimal current supply to the inverter, with the help of PWM technique the pulse width is being controlled for regulating the speed of the motor. The Sinusoidal PWM technique sends the signals to the inverter, which in turn sends the voltages, V_a, V_b and V_c to the motor. Then the actual speed of the motor is compared with the reference speed for minimization of error as shown in figure 7. Based on experiment when the Proportional gain as 0.8, integral gain as 8 and derivative gain as 0.001. the rise time is 22ms, overshoot is 1% and settling time is 110ms. These optimal gains can be directly implemented into the motor controller.

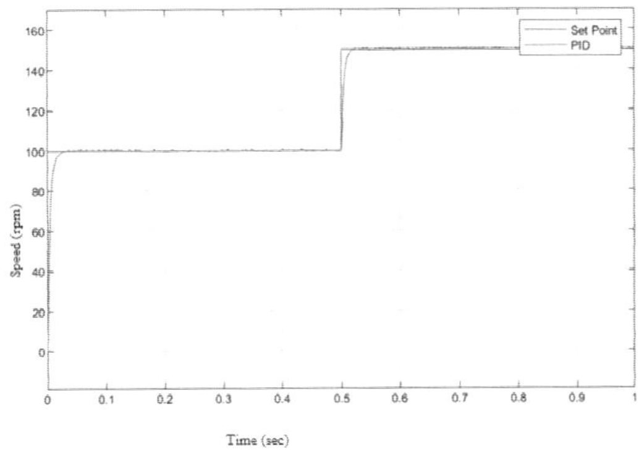

Figure 7. Speed Response chart of motor after PID tuning.

3.2. Reduction of THD

Based on the comparison between different types of control strategy used for controlling the speed of the motor, when the predefined parameters are fundamental frequency is 60 Hz, switching frequency of the inverter is 10kHz. The figure 8 shows the Total harmonic distortion of the supply current of phase A. And the value of the THD was found minimum for the SVPWM technique.

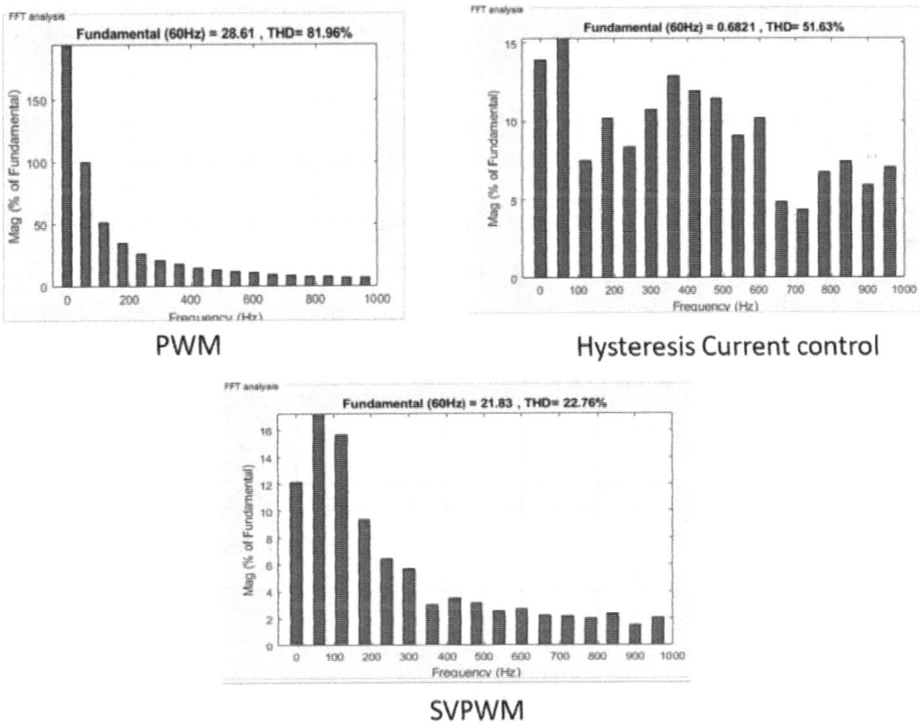

Figure 8. Clockwise from top left: PWM control based on Hall sensor, Hysteresis control and SVPWM control with FOC, SVPWM control with FOC.

4. Future Scope of Implementation of PID with Modified SVPWM Control in DSPACE Controller

Using the DSPACE MicroLabBox version 1202, the control of the speed of the BLDC motor can be carried out. Using the help of Real Time Interface (RTI) tool block which is present inside the Matlab Simulink Environment as shown in figure 9. The model in the Simulink environment was carried out, but it couldn't be tested, due to time constraint to visualize the results. [38],[39],[40],[41]

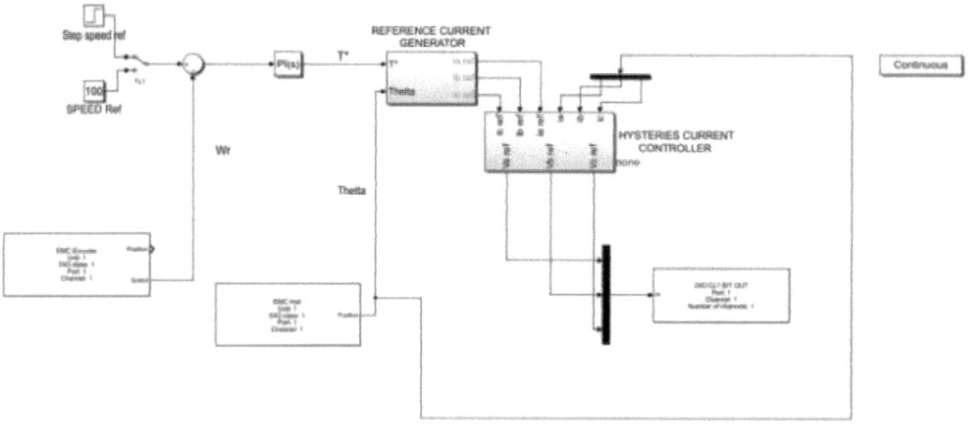

Figure 9. BLDC motor model integrated with the D-space controller to control the hardware directly with MATLAB.

5. Appendices

5.1. Motor Parameters used for Modelling

The constants that are used in the mathematical modelling of the BLDC motor using MATLAB / Simulink was developed according to the table 6 and table 7.

Table 6. Motor Parameters used for Modelling

Parameters	Value	Units	Description
Kb	1.23	(const.)	Field Flux due to permanent magnet of Rotor (Ferrite Magnet, rare earth alloys are used to make the permanent magnet pole pieces)
J	0.013	$Kg – m^2$	Equivalent Moment of Inertia of Motor and Load on Motor shaft
P	4	(const.)	Number of Poles of Rotor (can vary from 2 to 8 poles)
R	2.8	Ohm	Resistance of stator winding
Ls + M	0.00521	Hendry	Inductance (M - Mutual inductance and L_S-Self-inductance)
B	0.005	N-m-s/rad	Damping Co-efficient
DC power Supply	250	volts	Applied Voltage

5.2. Notations used in this paper

Table 7. Notations used in this paper.

Symbol	Units	Meaning
Ra,Rb, Rc	Ohm	Stator Resistance of phase A, B, C
La, Lb, Lc	Hendry	stator inductance of phase A, B, C
Van, Vbn, Vcn	Volt	Stator voltage in phase A.B,C
ia, ib, ic	Ampere	Stator Current in phase A, B,C
ean, ebn, ecn	volt	Back EMF in phase A,B, C
ωr	rad/sec	Rotor speed in electrical
ω	Rad/sec	Mechanical rotor speed
$\boldsymbol{\lambda a, \lambda b, \lambda c}$		flux linkage of phase windings a, b and c

6. References

[1] Devi, K. S., Dhanasekaran, R., & Muthulakshmi, S., 2016, May. Improvement of speed control performance in BLDC motor using fuzzy PID controller. In *2016 International Conference on Advanced Communication Control and Computing Technologies (ICACCCT)* (pp. 380-384). IEEE.

[2] Shanmugasundram, R., Zakariah, K. M., & Yadaiah, N. 2012, November. Implementation and performance analysis of digital controllers for brushless DC motor drives. IEEE/ASME transactions on mechatronics, 19(1), 213-224.

[3] Crnosija, P., Krishnan, R., & Bjazic, T. 2006, Optimization of PM brushless DC motor drive speed controller using modification of Ziegler-Nichols methods based on Bodé plots. In *2006 12th International Power Electronics and Motion Control Conference* (pp. 343-348). IEEE..

[4] Suganthi, P., Nagapavithra, S. and Umamaheswari, S., 2017, March. Modeling and simulation of closed loop speed control for BLDC motor. In *2017 Conference on Emerging Devices and Smart Systems (ICEDSS)* (pp. 229-233). IEEE.

[5] Chougale, R.G. and Lakade, C.R., 2017, September. Regenerative braking system of electric vehicle driven by brushless DC motor using fuzzy logic. In *2017 IEEE International Conference on Power, Control, Signals and Instrumentation Engineering (ICPCSI)* (pp. 2167-2171). IEEE.

[6] Varshney, A., Gupta, D. and Dwivedi, B., 2017. Speed response of brushless DC motor using fuzzy PID controller under varying load condition. *Journal of Electrical Systems and Information Technology*, *4*(2), pp.310-321.

[7] Kandiban, R. and Arulmozhiyal, R., 2012. Speed control of BLDC motor using adaptive fuzzy PID controller. *Procedia Engineering*, *38*, pp.306-313.

[8] Krishnan, P.H. and Arjun, M., 2014, March. Control of BLDC motor based on adaptive fuzzy logic PID controller. In *2014 International Conference on Green Computing Communication and Electrical Engineering (ICGCCEE)* (pp. 1-5). IEEE.

[9] Xuanfeng, S. and Xingyan, L., 2010, October. BLDC motor speed servo system based on novel P-fuzzy self-adaptive PID control. In *2010 International Conference on Information, Networking and Automation (ICINA)* (Vol. 2, pp. V2-186). IEEE.

[10] Kwon, C.J., Han, W.Y., Kim, S.J. and Lee, C.G., 2003, August. Speed controller with adaptive fuzzy tuning for BLDC motor drive under load variations. In *SICE 2003 Annual Conference (IEEE Cat. No. 03TH8734)* (Vol. 3, pp. 3118-3121). IEEE.

[11] Milani, M.M.R.A., Çavdar, T. and Aghjehkand, V.F., 2012, July. Particle swarm optimization—based determination of Ziegler-Nichols parameters for PID controller of brushless DC motors. In *2012 International Symposium on Innovations in Intelligent Systems and Applications* (pp. 1-5). IEEE.

[12] Irimia, N.D., Lazar, F.I. and Luchian, M., 2019, April. Comparison between sinusoidal and

space vector modulation techniques on the resulting electromagnetic torque ripple produced by a three-phase BLDC motor under field-oriented control. In *2019 6th International Conference on Control, Decision and Information Technologies (CoDIT)* (pp. 640-645). IEEE.

[13] Gujjar MN, Kumar P. Comparative analysis of field oriented control of BLDC motor using SPWM and SVPWM techniques. In2017 2nd IEEE International Conference on Recent Trends in Electronics, Information & Communication Technology (RTEICT) 2017 May 19 (pp. 924-929). IEEE.

[14] Holmes DG, Lipo TA. Pulse width modulation for power converters: principles and practice. John Wiley & Sons; 2003 Oct 3.

[15] Kang SJ, Sul SK. Direct torque control of brushless DC motor with nonideal trapezoidal back EMF. IEEE Transactions on power electronics. 1995 Nov;10(6):796-802.

[16] Xintong, J., Jingwei, X., Yong, L. and Yongping, L., 2009. Theoretical and simulation analysis of influences of stator tooth width on cogging torque of BLDC motors. *IEEE Transactions on Magnetics*, *45*(10), pp.4601-4604.

[17] Whiteley E, inventor; Maghemite Inc, assignee. Brushless DC dynamoelectric machine having ferrite material magnetic circuit. United States patent US 4,605,874. 1986 Aug 12.

[18] Qu R, Aydin M, Lipo TA. Performance comparison of dual-rotor radial-flux and axial-flux permanent-magnet BLDC machines. InIEEE International Electric Machines and Drives Conference, 2003. IEMDC'03. 2003 Jun 1 (Vol. 3, pp. 1948-1954). IEEE.

[19] Rashid MH, editor. Power electronics handbook. Butterworth-Heinemann; 2017 Sep 9.

[20] Sharma, Pragati K., and A. S. Sindekar. 'Performance Analysis and Comparison of BLDC Motor Drive using PI and FOC', IEEE 2016 *International Conference on Global Trends in Signal Processing, Information Computing and Communication (ICGTSPICC)*, 2016 Dec 22, (pp. 485-492).

[21] de Almeida, T. E., de Paula, G. T., de Castro, A. G., Pereira, W. C., & José, R. D. A. 2017, November. Sensorless vector control for BLDC machine. In *2017 Brazilian Power Electronics Conference (COBEP)* (pp. 1-6). IEEE.

[22] Chen, S.C. and Mbitu, E.T., 2018, April. Permanent magnet brushless motor field oriented control with dither signal injection. In *2018 IEEE International Conference on Applied System Invention (ICASI)* (pp. 1141-1144). IEEE.

[23] Kumar, B.P. and Krishnan, C.M.C., 2016, January. Comparative study of different control algorithms on brushless DC motors. In *2016 Biennial International Conference on Power and Energy Systems: Towards Sustainable Energy (PESTSE)* (pp. 1-5). IEEE.

[24] Ogata K, Yang Y. Modern control engineering. Upper Saddle River, NJ: Prentice hall; 2010 Sep.

[25] Wongkhead S, Tunyasrirut S, Permpoonsinsup W, Puangdownreong D. Simulation and Analysis of Speed Brushless Direct Current Motor Based on State Space Modeling. In2018 International Conference on Engineering, Applied Sciences, and Technology (ICEAST) 2018 Jul 4 (pp. 1-4). IEEE.

[26] Singh, C.P., Kulkarni, S.S., Rana, S.C. and Deo, K., 2013. State-space based simulink modeling of BLDC motor and its speed control using fuzzy PID controller. *International Journal of Advances in Engineering Science and Technology*, *2*(3), pp.359-369.

[27] Bondre VS, Thosar AG. Mathematical modeling of direct torque control of BLDC motor. In2017 International Conference on Innovative Research In Electrical Sciences (IICIRES) 2017 Jun 16 (pp. 1-8). IEEE.

[28] Mohankrishna C, Rajesh N, Gowd K, Ramesh A, Gupta GS. Modelling and simulation of BLDC motor using state space approach. International Journal Of Innovative Research In Electrical, Electronics, Instrumentation And Control Engineering. 2016 May;4(5).

[29] Tibor B, Fedak V, Durovský F. Modeling and simulation of the BLDC motor in MATLAB GUI. In2011 IEEE International Symposium on Industrial Electronics 2011 Jun 27 (pp. 1403-1407). IEEE.

[30] Pan, C.T. and Fang, E., 2008. A phase-locked-loop-assisted internal model adjustable-speed controller for BLDC motors. *IEEE Transactions on industrial electronics*, *55*(9), pp.3415-3425.
[31] Rodriguez F, Emadi A. A novel digital control technique for brushless DC motor drives: Conduction-angle control. InIEEE International Conference on Electric Machines and Drives, 2005. 2005 May 15 (pp. 308-314). IEEE.
[32] Jaya A, Purwanto E, Fauziah MB, Murdianto FD, Prabowo G, Rusli MR. Design of PID-fuzzy for speed control of brushless DC motor in dynamic electric vehicle to improve steady-state performance. In2017 International Electronics Symposium on Engineering Technology and Applications (IES-ETA) 2017 Sep 26 (pp. 179-184). IEEE.
[33] Ramesh, M.V., Amarnath, J., Kamakshaiah, S. and Rao, G.S., 2011. Speed control of brushless dc motor by using fuzzy logic pi controller. *ARPN Journal of Engineering and Applied Sciences*, *6*(9), pp.55-62.
[34] Ansari, U. and Alam, S., 2011, March. Modeling and control of three phase BLDC motor using PID with genetic algorithm. In *2011 UkSim 13th International Conference on Computer Modelling and Simulation* (pp. 189-194). IEEE.
[35] Rao, K.S.R. and Othman, A.H.B., 2007, November. Design optimization of a BLDC motor by Genetic Algorithm and Simulated Annealing. In *2007 International Conference on Intelligent and Advanced Systems* (pp. 854-858). IEEE.
[36] Vinodhini, R., Ganesh, C. and Patnaik, S.K., 2012, March. Genetic algorithm optimized on-line neuro-tuned robust position control of BLDC motor. In *2012 IEEE Students' Conference on Electrical, Electronics and Computer Science* (pp. 1-4). IEEE.
[37] Hu, M., Qiu, J. and Shi, C., 2011, August. A comparative analysis of fuzzy PI and PI speed control in Brushless DC motor based on dSPACE. In *2011 International Conference on Electrical Machines and Systems* (pp. 1-5). IEEE.
[38] dSPACE, Implementation Guide For Release 4.0: Real-Time Interface (RTI and RTI-MP). Documentation Guide, August 2003.
[39] dSPACE, Experiment Guide For Release 4.0: Control-Desk. Documentation Guide, August 2003.
[40] Aboulnaga, A.A., Desai, P.C., Rodriguez, F., Cooke, T.R. and Emadi, A., 2004, February. A novel, low-cost, high-performance single-phase adjustable-speed motor drive using PM brushless DC machine: IIT's design for 2003 future energy challenge. In *Nineteenth Annual IEEE Applied Power Electronics Conference and Exposition, 2004. APEC'04.* (Vol. 3, pp. 1595-1603). IEEE.
[41] Bansal, A. K., Gupta, R. A., & Kumar, R., 2011, January. Fuzzy estimator for sensorless PMBLDC motor drive under speed reversal. In *India International Conference on Power Electronics 2010 (IICPE2010)* (pp. 1-7). IEEE.

Harmonic Elimination and dc–link Voltage Balancing in Bipolar Hybrid Microgrid by ILCs: A Review

S Vasantharaj[1*], V Indra Gandhi[2]

[1]Research Scholar, Vellore Institute of Technology, Vellore
[2]Associate Professor, Vellore Institute of Technology, Vellore
[*]Corresponding author E-mail: Vasantharaj.s@vit.ac.in

Abstract: In this advanced power system scenario, the bipolar hybrid microgrid application plays a vital role, which owns both DC and AC bus instantaneously. In this grid-connected technology, renewable energy performs a major role in the generation of power demand. Due to the alternating nature of renewable sources, the grid should have stability with different energy sources like battery storage systems, vehicle to grid, etc. Hence, to retain the quality of power supply it is more significant to consider an interlinking bidirectional converter between a source and the grid. The most important indispensable component in the bidirectional link is an AC/DC bidirectional converter. This paper elaborates about the advantages and disadvantages of the prevailing AC/DC bidirectional converter topologies for the advancement in the microgrid applications. These types of three-phase ac/dc bidirectional converters are preferred to reduce the THD content in the grid side, power factor correction, and regulated dc output voltage with unidirectional and bidirectional power flow. The dc-link voltage has balanced at the dc bus with these bidirectional topologies. By use of this bidirectional converter, it has designed to perform various essential applications in the microgrid with the help of space vector modulation.

KEYWORDS:
Bipolar Hybrid microgrid, interlinking converter, THD, Space vector Modulation, Voltage Balancing, Converter, Hysteresis comparator.

Nomenclature:

C	dc-link capacitor	V_{ref}	Reference Output Voltage
CSI	Current Source Inverter	V_a	Line to neutral Voltage
EMI	Electromagnetic Interference	I_a	Line Current
BHM	Bipolar Hybrid Microgrid	V_{dcl}	DC-link Voltage
WLAN	Wireless Local Area Network	U_{ref}	Normalized Voltage Vector
SVM	Space Vector Modulation	MPPT	Maximum Power Point Tracking
ESS	Energy Storage System	BW	Bandwidth
IoT	Internet of Things	PV	Photovoltaic
DG	Distributed Generation	MG	Microgrid

Introduction:

Smart grid plays a pivotal role in this upcoming power system era by the use of computer-based remote control and automation [1–7]. This smart grid performs with two-way power flow and information flow with digital communication techniques and computer processing. During the initial state, it is used in electrical networks, from the windmills and power plants to distribute power to consumers of electricity in homes and businesses. The distributed MG [8–15], contains DG units, ESS, and distributed loads that utilize isolated mode or grid-connected mode. The VSI is a frequently used power electronic converter that connects between DGs and the grid. The Distributed generator injects active and reactive power [15–19], to control this active and reactive power interlinking inverter regulates the phase angle and the magnitude of the output voltage. By the use of appropriate control methods, this inverter compensates more power quality issues like voltage and current harmonics. In [20], it is proposed that the lower order harmonics can be eliminated by the use of proper SVM technique in the interlinking converter that is used between DGs and the grid. Similarly, in [21], it is proposed that the microgrid controls the frequency domain operation. The basic principle of this bidirectional converter is to convert direct current to alternating current and vice versa. Different applications weights distinct features and hence different types of bidirectional converters are proposed for each of its applications. For example, vehicle charger desires to be dense and fewer heaviness, while the bidirectional linkage of PV generation distress about the permanence and the reliability of its performance.

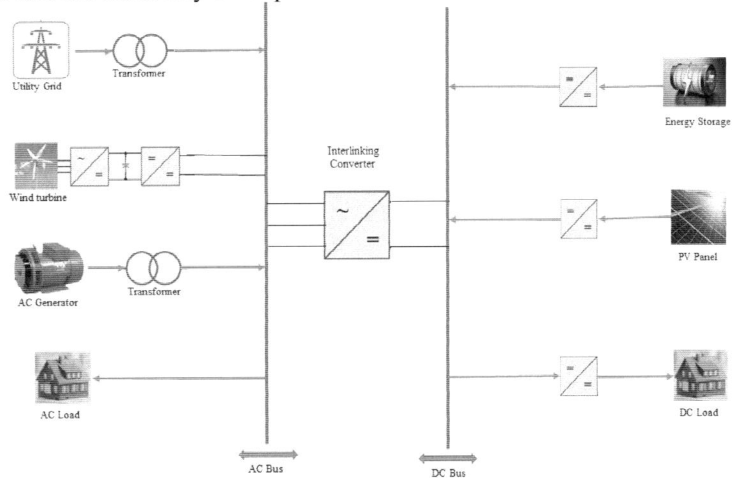

Figure 1. Diagrammatic Representation of Interlinking Converter for Bipolar Hybrid microgrid.

The bipolar hybrid microgrid with a bidirectional converter is shown in Fig. 1. Hence, different types of converter topologies are used which results in different levels of excellence and difficulty. Furthermore, the suitable bidirectional converter topology is selected for the requirement of the application. Fig. 2 shows the different methods of bidirectional converter. By use of a different type of control, methods like Particle Swarm Optimization and hysteresis current controllers are used for the elimination of lower order harmonics. In the same way, the load will be compensated by the use of droop controller in the microgrids. To eliminate the resonance under harmonic frequencies, to control active and reactive powers, improve system unbalance, balancing of V_{dcl}, and to limit the harmonic voltage and current these types of compensators are used. The voltage and current harmonics are compensated by the use of one proposed converter for the microgrid [22], which contains one distributed generator with two inverters. In these two inverters, first one is operated to compensate voltage harmonics in the load side and another inverter is used to compensate the current harmonics. Similarly, the THD of injected current, voltage through ILC, V_{dcl}, and dc voltage ripple will be evaluated.

In [23], to improve the stability of harmonic current and to reduce reactive power sharing, the output of the reference voltage of each distributed generator is injected with high-frequency harmonic voltage. By varying the BW of the control loop, it is applicable to have accurate power sharing by modifying the output impedance of individual inverter under closed loop condition. It is proposed to develop nonlinear load sharing and increase in voltage control gain with the help of improvement in distortion power (D) which is shared by the DGs. This method is used to reduce the voltage control stability.

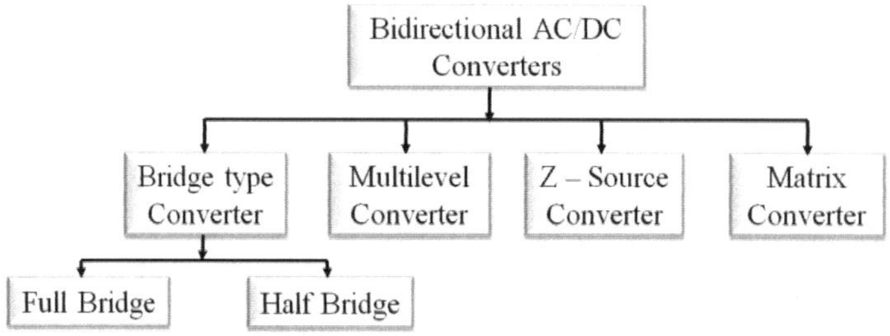

Figure 2. Different methods of the bidirectional converter.

In [24], improvement in harmonic current sharing is obtained by harmonic virtual impedance schemes. By varying the V_{ref} of the DG by the use of load current feed forward loop virtual harmonic impedance is varied. Moreover, for the difficulties associated with imbalanced power sharing in isolated microgrids a categorized control structure having more control loops has been proposed in [25].
The survey taken for balancing the dc-link poles and the same is evaluated for dc-link balancing strategies with the help of this different bidirectional converter [26–27]. The main features of this each bidirectional converter elaborated in Section 2 and explain about the elimination method for lower order harmonics with the analyses of a common mechanism for a grid-tied ILC. In section 3, carrier-based modulation and its associated V_{dcl} balancing strategies elaborated through full-scale voltage balancing competence with the different types of SVM methods. In section 4, it deals with the future scope and IoT based renewable energy sources.

2. Types of interlinking converters:

The different types of interlinking converters used in the bipolar hybrid microgrids are explained as follows,

2.1. 10 – Switch Interlinking Converter:

Fig. 3 shows the circuit diagram for 10-switch converter. An ac output voltage of 2/3 level with three-wire bipolar dc topology is proposed by 10 switch converter. The conventional voltage source converter (VSC) added with S1A to S4A switches to obtain this proposed converter. This converter comprises of 3 main legs and an auxiliary leg contains 4 switches having three different forms of switching states (P, O, N). The switching states and their illustration are shown in table 1.

Table 1. Switching states of Interlinking converter.

Switching State	ON switches		Terminal voltage
P	S1A	S1 or S3 or S5	E
O	S2A	S1 or S3 or S5	0
	S3A	S2 or S4 or S6	
N	S4A	S2 or S4 or S6	-E

The auxiliary leg of switches S2A or S3A produces a switching state O and these switches depend on former leg states. The remaining switching states are said to be understandable. Eight and twenty-seven switching vectors of two-level and three-level VSI is compared with the 10–switch converter, which has twenty-one switching vectors that contain eighteen active, and three zero vectors (PPP, OOO, NNN). The twenty-one switching vectors of 10-switch converter are split as three zero vectors, which are sited midpoint of a hexagon including six large and twelve small vectors. The dc-link poles are short-circuited due to this 10-switch converter by having this medium level vectors.
The 10-switch converter is suitable for high power applications because it contains both the advantages of 2-level and 3-level converter simultaneously. In [26], it explains about the combined topology of three-level neutral point clamped (NPC) and two level VSC. This gives the advantages of 3-level converter. 10-switch converter yields a composite of two and three-level AC voltage.

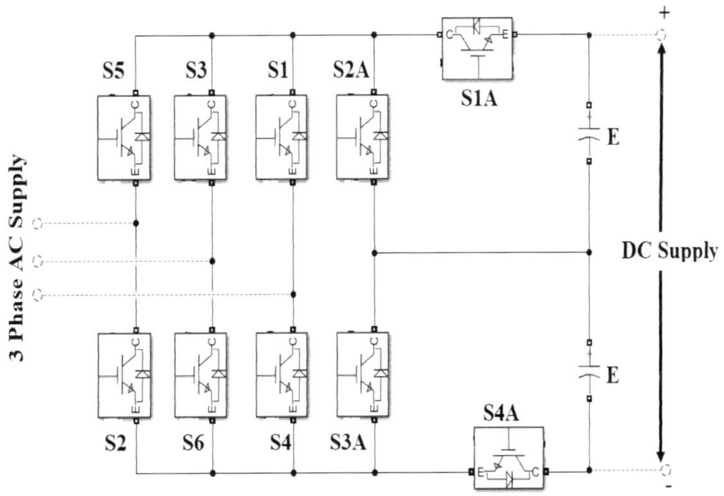

Figure 3. Circuit Diagram for 10 switch AC/DC interlinking converter.

2.2. Full Bridge bidirectional Converter:

2.2.1. System description

In [22-26], the converter dynamic response was verified by various circumstances like the difference in a voltage source, the difference in load and variations in both load, and voltage source. Fig. 4 shows a three-phase AC/DC bidirectional converter, in which the grid is coupled with the AC side of the converter and for DC side; is coupled to DC loads. The resistor with a toggle switch is used to test the variation in load. The full-bridge bidirectional converter switches perform three types of operating modes namely, rectification, inversion, and Shut-Down Modes.

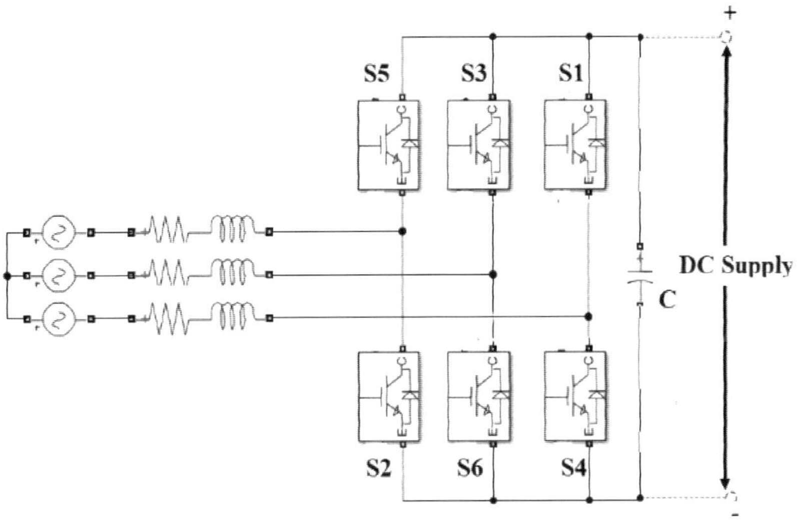

Figure 4. Circuit diagram for full-bridge AC/DC bidirectional converter.

2.2.2. Inversion mode

By this inversion mode, the bidirectional converter operates under the unity power factor (UPF) because of reactive power in the utility grid is maintained to zero. By increasing the DC source which is above the DC load then the converter can function in adjustable reactive power to change the power factor. Hence the power is transferred from dc side to ac side. The inversion mode is justified by the phase

IOP Conf. Series: Materials Science and Engineering **906** (2020) 012027 doi:10.1088/1757-899X/906/1/012027

difference of 180^0 between the line to neutral voltage V_a and the line current i_a by having proper sinusoidal waveform with low THD.

2.2.3. Shut-Down Mode

If the power delivered or absorbed by the grid is equal then this is said to be Shut-Down Mode (S-DM). At this instant, the converter is in the shutdown state. By attaining this delivered and observed power to equal, it is achieved by varying the load until it becomes to shut down mode.

2.2.4. Rectification mode

If the voltage of DC load is more than the voltage of DC source then the converter attains rectification mode (RM). By increasing the grid side voltage, rectification mode is achieved. Hence, the dc-bus voltage V_{dc} drops relative to its reference value V_{dcr}. The bidirectional converter is used to inject the power from the AC grid to the DC load side.

2.3. Bidirectional Matrix Converter

Matrix converter is one of the best exciting families in the converter. By the use of matrix converter, ac-ac power conversion is achieved without any intermediate energy storage elements. Fig. 5 shows the topology of the High Frequency Link Matrix Converter (HFLMC). Each bidirectional switch S_{xy} is self-possessed of 2 transistors S_{xy1} and S_{xy2} (x = a, b, c and y = P, N) in a mutual source structure, demands single fluctuating source. In this topology, a matrix converter operates as a CSI and supplies voltage to the high frequency transformer (HFT). Furthermore [29, 30], to shorten the description transformer is modeled has leakage inductance L on the primary side with an ideal transformer of turns ratio n. the P and N bar is connected with a single input phase which is fed by a voltage source of matrix converter shown in fig. 5.

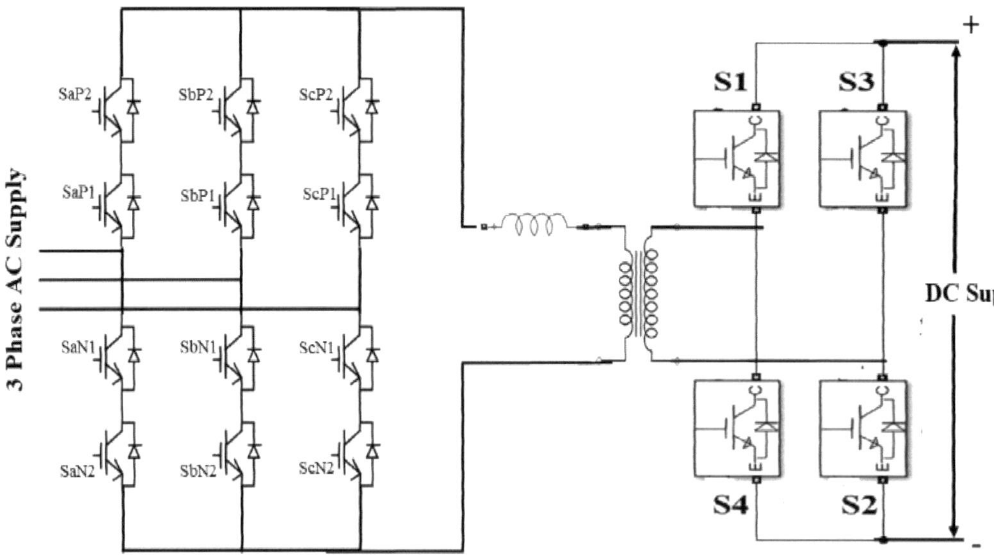

Figure 5. Circuit Diagram for AC/DC bidirectional matrix converter.

The bidirectional matrix converter performs the following advantages: 1) one-step power transformation 2) two way power transfer 3) galvanic separation 4) nonexistence of dc-link 5) voltage equivalent by the transformer turns ratio and 6) extended provision life. Same way the disadvantage of this converter is harmonic content, which is present in the current of matrix converter; an EMI filter is essential to eliminate these harmonics in the input current. Another disadvantage of this topology contains transformer having inductive nature, which allows a maximum amount of current flow in the circuit.

2.4. Dual Active Bidirectional Converter:

The major applications of Dual active bidirectional converter are renewable power generation; plug-in hybrid vehicles and supply systems because of high frequency isolation and controllability of active power flow in the distribution end. This proposed converter is used to operate all the operating modes like inversion and rectification mode of the converter that results in logical approximation of power transmission, rms current, and examination of soft-switching circumstances for power strategies.

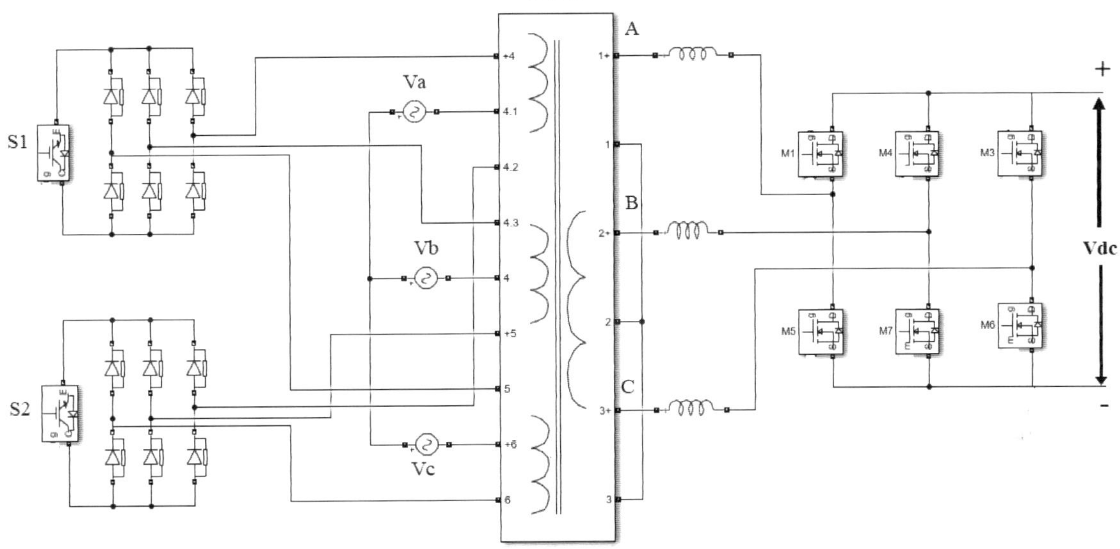

Figure 6. Circuit diagram for dual active AC/DC bidirectional converter.

Fig. 6 shows the three winding transformers for each phase with the turns ratio of 1 : n. the input phase voltages V_a, V_b and V_c are commonly connected to the two primary windings of the transformer. Three input phases V_a, V_b and V_c are connected to a common terminal of 2 primary windings of the transformer. The primary side of the three winding transformer comprises of two three-phase diode bridge in a push-pull configuration with two active switches S_1 and S_2. The inductance L that is present on the secondary side of the transformer signifies the total leakage inductance of the transformer ($L = n2L_p + L_s$). The primary and secondary winding needs external inductance that is to be connected in series in the case of the inductor L is considered more than the leakage inductance. The two level voltage source inverter (VSI) is connected on the secondary side of the transformer. Then the VSI is connected to the dc source. The phase voltage equation of the primary side is given in equation (1) with a frequency of $\omega = 2\pi f = 2\pi/T$ and amplitude V_{grid}. The primary-side switches S_1 and S_2 are switched in a complementary method with a 50% duty ratio at a frequency $f_s = 1/T_s$, such that $f_s >> f$. Accordingly, the secondary-side has high frequency square wave with line frequency which is shown in equation (2).

$$V_a = V_{grid}\sin(2\pi f t) \quad (1)$$

$$V_{AN} = \pm nV_a = \pm nV_{grid}\sin(2\pi f t) \quad (2)$$

(+ with S_1 ON, − with S_2 ON).

The advantages of this bidirectional converter are one-step transformation, no unpredictable halfway dc-link capacitor, less number of switches, modest control scheme, open-loop UPF operation, bidirectional power flow, and partial soft switching.

The comparative analysis of different types of ac/dc bidirectional converter has shown in table 2.

IOP Conf. Series: Materials Science and Engineering **906** (2020) 012027 doi:10.1088/1757-899X/906/1/012027

Table 2. Comparative analysis for AC/DC bidirectional Converters.

	Voltage source Converter	Multi-level converter	Matrix converter	10 switch converter
Types	i) Half bridge ii) Full bridge	i) Diode Clamped. ii) Flying Capacitor iii) Cascaded H – Bridge	i) Direct ii) In-direct	N/A
Switches	6	16	20	10
Capacitor	1 filter capacitor of dc side	No need	No need	2 filter capacitor of dc side
Inductor	No need	Filter inductor on ac side	Filter inductor on ac side	No need
Transformer	N/A	1	1 : n	N/A
Components Required	Less	More	More	Less
Controller Circuit	Complex	Complex	Simple	Simple
Harmonic Content	High	Less	Less	Less
Filter Circuit	High	Small	Small	Small
Voltage Stress	Low	Low		Low
DC Voltage Control	Needs additional circuit to boost rectified dc output.	Needs additional circuit to control the voltage to match desired ac output voltage with grid.	No need.	No need.
Application	Operates at low power application.	Operates in high voltage and high power applications.	Used in medium voltage application.	Used in high voltage application.
Switching Loss	High	Moderate	Low	Low
Efficiency of a converter	Less	Medium	More	More

3. Classification of Different SVM Methods:

3.1. Type I:

A dc-link voltage balancing capability is presented in this subdivision by the use of proposed SVM technique. The foremost objective of this novel SVM method is to improve dc-link voltage and decreases output of current THD and common mode voltage. Similarly, this proposed SVM does not need any passive or active elements. The SVM, consists of six sectors and these sectors are subdivided into small parts having the shapes of trapezoid and triangle. By using this SVM there will be an increase in current THD, switching configuration and neutral point voltage deviation. SVM used is operated under both balanced and unbalanced conditions which lead to a decrease in extra converter costs, volume and power loss.

3.1.1. Region determination

Initially, the corresponding SVM for three level VSI, the V_{ref} is induced to measure its angle and amplitude. Then, it is to be determined that were V_{ref} is located and to identify the sector and region for this reference voltage vector. Each sector region is similar to one another and it is validated in fig. 7. The region and the sector in which it is located will be stipulated by the angle and amplitude of the V_{ref}.

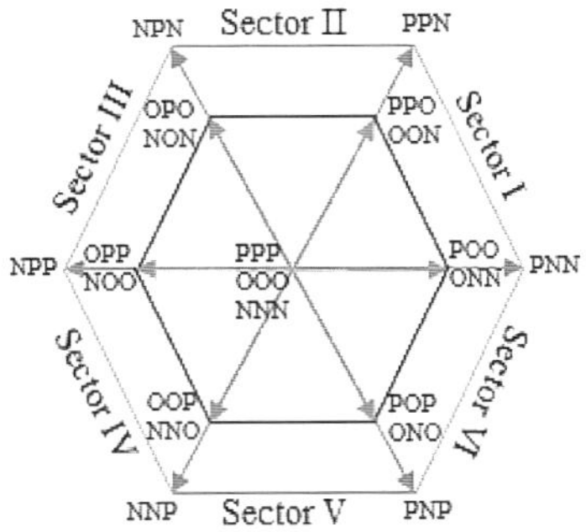

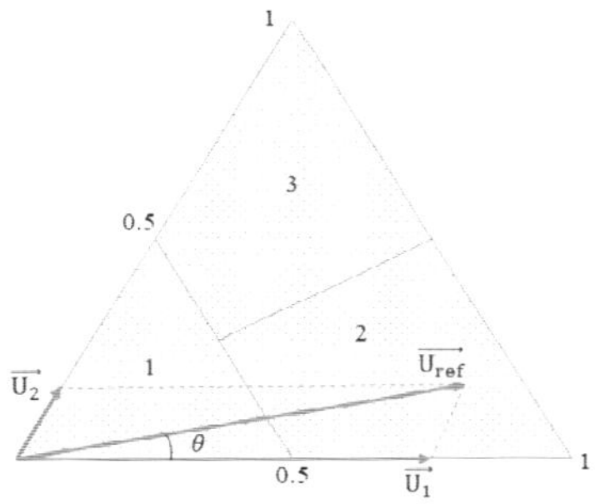

Figure 7. SVM for 10 switch converter and its sector representation for normalized voltage vector

The normalized reference voltage vector (U_{ref}) is attained to compute and locate the section of V_{ref} for the interlinking converter. The vectors U_1 and U_2 obtain the resultant vector of U_{ref} which is shown in equation (3), (4) and table 3.

$$U_1 = U_{ref}\left(Cos\theta - \frac{sin\theta}{\sqrt{3}}\right) \quad (3)$$

$$U_2 = 2U_{ref}\left(\frac{sin\theta}{\sqrt{3}}\right) \quad (4)$$

To eliminate over modulation these calculations are used. The U_{ref} angle and its amplitude are used to change the region of respective sectors with the values U_1 and U_2 maximum of 1. The U_{ref} amplitude goes in the trapezoidal area, which contains two equal areas after passing the region 1 area, which is shown in fig 7. The trapezoidal region changes from 2 to 3 when the reference voltage vector attains 30^0 which is equal to $\sqrt{3}/2$. The non-existence of medium voltage is reimbursing with the split up trapezoidal regions of 2 and 3.

Table 3. Region identification based on U_{ref}.

Case	Angle (θ)	Region
$U_1 \leq 0.5$ and $U_2 \leq 0.5$ & $U_1+U_2 \leq 0.5$	NA	1
$U_1 > 0.5$ or $U_2 > 0.5$ or	$\theta \leq 30^0$	2
$U_1 + U_2 > 0.5$	$\theta > 30^0$	3

3.1.2. Symmetrical switching sequence

This type of switching sequence is organized to minimize the THD, which is considered as the main objective of this proposed method.

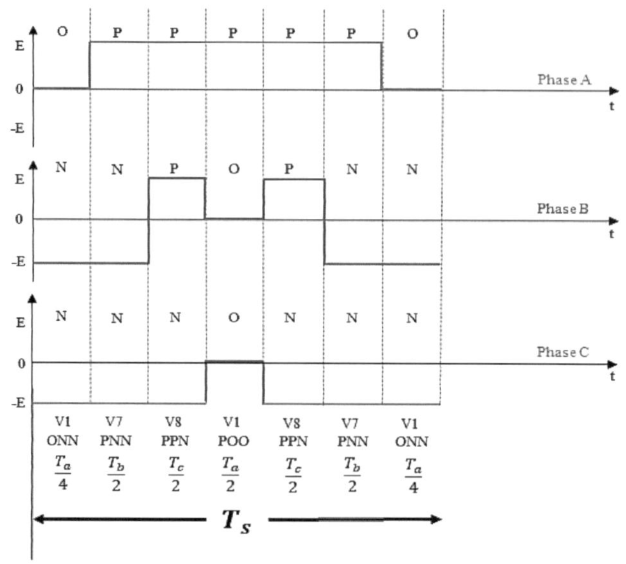

Figure 8. Switching Scquence of Sector 1.

The switching time (T_s) for sector 1 with switching sequence is shown in Fig. 8. Since fig. 7 shows identical sectors and regions, hence by having the reference of the first sector, the vector selection and switching sequence of this proposed SVM for all the six sectors eighteen regions are calculated.

3.2 Type II:

3.2.1. Sector Partition:

Table 4 shows the relation between switching stages and voltage vectors of the SVM. In this type of SVM $i_a > 0$, $i_b < 0$, $i_c < 0$, and $i_a > 0$, $i_b > 0$, $i_c < 0$ considered as an example for the rectification mode of the converter. Fig. 9 consists of six non-zero vectors, hence the SVM is divided into six sectors. The traditional partition of the SVM has six sectors that are shown in full lines and the scattered lines additionally split the sectors into 12 rendering to the direction of three-phase ac current flow.

Table 4. Voltage vectors verses Switching States

$i_a > 0, i_b < 0, i_c < 0$							$i_a > 0, i_b > 0, i_c < 0$						
S_a	S_b	S_c	S_d	S_e	Sign {vT}	V	S_a	S_b	S_c	S_d	S_e	Sign {vT}	V
0	0	0	1	1	0	0	0	0	0	1	1	0	0
0	0	1^+	1	1	0	0	0	0	1^+	1	0	-	$-\alpha^2 \frac{2}{3}\frac{n_p}{n_s} v_{dc}$
0	1^+	0	1	1	-	0	0	1^-	0	1	1	0	0
0	1^+	1^+	1	0	+	$\frac{2}{3}\frac{n_p}{n_s} v_{dc}$	0	1^-	1^+	1	0	-	$\frac{2}{3}\frac{n_p}{n_s} v_{dc}$
1^-	0	0	0	1	+	$\frac{2}{3}\frac{n_p}{n_s} v_{dc}$	1^-	0	0	1	1	0	0
1^-	0	1^+	0	1	+	$-\alpha \frac{2}{3}\frac{n_p}{n_s} v_{dc}$	1^-	0	1^+	1	0	-	$\alpha \frac{2}{3}\frac{n_p}{n_s} v_{dc}$
1^-	1^+	0	0	1	+	$-\alpha^2 \frac{2}{3}\frac{n_p}{n_s} v_{dc}$	1^-	1^-	0	0	1	+	$-\alpha^2 \frac{2}{3}\frac{n_p}{n_s} v_{dc}$
1^-	1^+	1^+	1	1	0	0	1^-	1^-	1^+	1	1	0	0

The 12 sectors are numbered according to the value of three-phase voltage v_α and v_β in stationary two phase Coordinates. Hence, the six sets of intermediate variables are given below:

$$v_{ref1} = v_\beta \tag{5}$$

$$v_{ref2} = v_\alpha \tag{6}$$

$$v_{ref3} = \frac{1}{2}(\sqrt{3}v_\alpha - v_\beta) \tag{7}$$

$$v_{ref4} = \frac{1}{2}(v_\alpha - \sqrt{3}v_\beta) \tag{8}$$

$$v_{ref5} = -\frac{1}{2}(\sqrt{3}v_\alpha - v_\beta) \tag{9}$$

$$v_{ref6} = \frac{1}{2}(-v_\alpha - \sqrt{3}v_\beta) \tag{10}$$

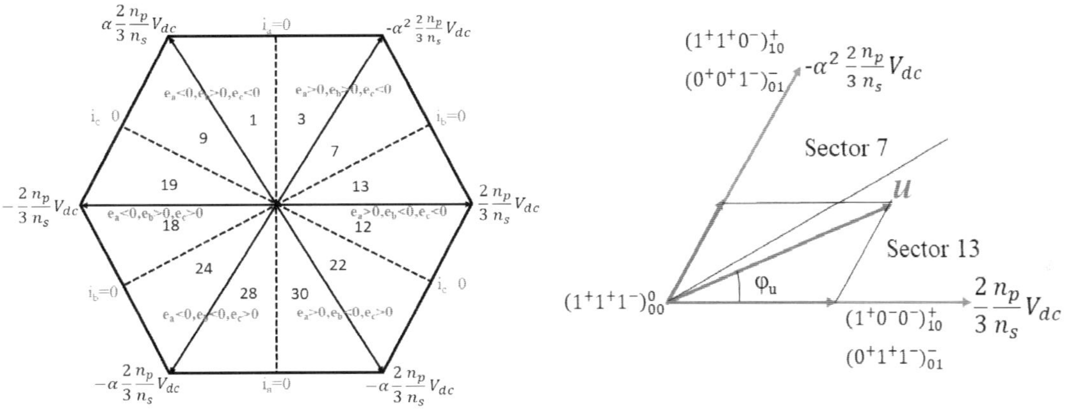

Figure 9. Sector Partition and Inverter mode of Sector 13 and 7.

If $v_{ref1} > 0$, A = 1; otherwise, A = 0. If $v_{ref2} > 0$, B = 1; otherwise, B = 0. If $v_{ref3} > 0$, C = 1; otherwise, C = 0. If $v_{ref4} > 0$,D = 1; otherwise, D = 0. If $v_{ref5} > 0$, E = 1; otherwise, E = 0. If $v_{ref6} > 0$, F = 1; otherwise, F = 0. The sector number is defined by N = A + 2B + 4C + 6D + 8E + 10F. Fig. 9 shows h sector partition and inverter mode of sector 13 and 7.

3.2.2. Vector Synthesis Method

By obeying the following laws, we can obtain the vector synthesis method: 1) retain the transformers voltage–second balancing; 2) decrease in switching loss; and 3) decrease in conduction loss.

3.2.3. Inverter Mode:

The three-phase inverter has two non-zero vectors in two different directions to produce the targeted vector with the use of SVM algorithm which is shown in sector 13 ($i_a < 0$, $i_b > 0$, $i_c > 0$). However, transformers voltage – second balancing is not guaranteed. Hence, it is considered three nonzero vectors with the sequence of switching-state are given in equation 11:

$$\begin{aligned}&(1^+0^-0^-)^+_{10} \rightarrow (1^+1^-0^-)^+_{10} \rightarrow (1^+1^-1^-)^0_{00} \rightarrow (0^+1^-1^-)^-_{01}\ \Big|t = \frac{T_s}{2}\\ &\rightarrow (0^+1^-1^-)^-_{01} \rightarrow (1^+1^-1^-)^0_{00} \rightarrow (1^+1^-0^-)^+_{10} \rightarrow (1^+0^-0^-)^+_{10}\ \Big|t = T_s\end{aligned} \tag{11}$$

In sector 7 ($i_a < 0$, $i_b < 0$, $i_c > 0$), the sequence of switching-state are given in equation 12:

$$(0^+0^+1^-)^-_{01} \rightarrow (0^+1^+1^-)^-_{01} \rightarrow (1^+1^+1^-)^0_{00} \rightarrow (1^+1^+0^-)^+_{10}\ \Big|t = \frac{T_s}{2}$$

$$\rightarrow (1^{+}1^{+}0^{-})^{+}_{10} \rightarrow (1^{+}1^{+}1^{-})^{0}_{00} \rightarrow (0^{+}1^{+}1^{-})^{-}_{01} \rightarrow (0^{+}0^{+}1^{-})^{-}_{01} \Big|t = T_s \quad (12)$$

Fig. 10 shows the inverter mode operation of driving signals in sectors 13 and 7.

3.2.4. Rectifier Mode:

In sector 13 ($i_a > 0$, $i_b < 0$, $i_c < 0$), the sequence of switching-state are given in equation 13:

$$(1^{-}00)^{+}_{01} \rightarrow (1^{-}1^{+}0)^{+}_{01} \rightarrow (1^{-}1^{+}1^{+})^{0}_{11} \rightarrow (01^{+}1^{+})^{-}_{10} \Big|t = \frac{T_s}{2}$$
$$\rightarrow (01^{+}1^{+})^{-}_{10} \rightarrow (1^{-}1^{+}1^{+})^{0}_{11} \rightarrow (1^{-}1^{+}0)^{+}_{01} \rightarrow (1^{-}00)^{+}_{01} \Big|t = T_s \quad (13)$$

In sector 7 ($i_a > 0$, $i_b > 0$, $i_c < 0$), the sequence of switching-state are given in equation 14:

$$(001^{+})^{-}_{10} \rightarrow (01^{-}1^{+})^{-}_{10} \rightarrow (1^{-}1^{-}1^{+})^{0}_{11} \rightarrow (1^{-}1^{-}0)^{+}_{01} \Big|t = \frac{T_s}{2}$$
$$\rightarrow (1^{-}1^{-}0)^{+}_{01} \rightarrow (1^{-}1^{-}1^{+})^{0}_{11} \rightarrow (01^{-}1^{+})^{-}_{10} \rightarrow (001^{+})^{-}_{10} \Big|t = T_s \quad (14)$$

The driving signals in sectors 13 and 7 under rectifier mode are shown in Fig. 10.

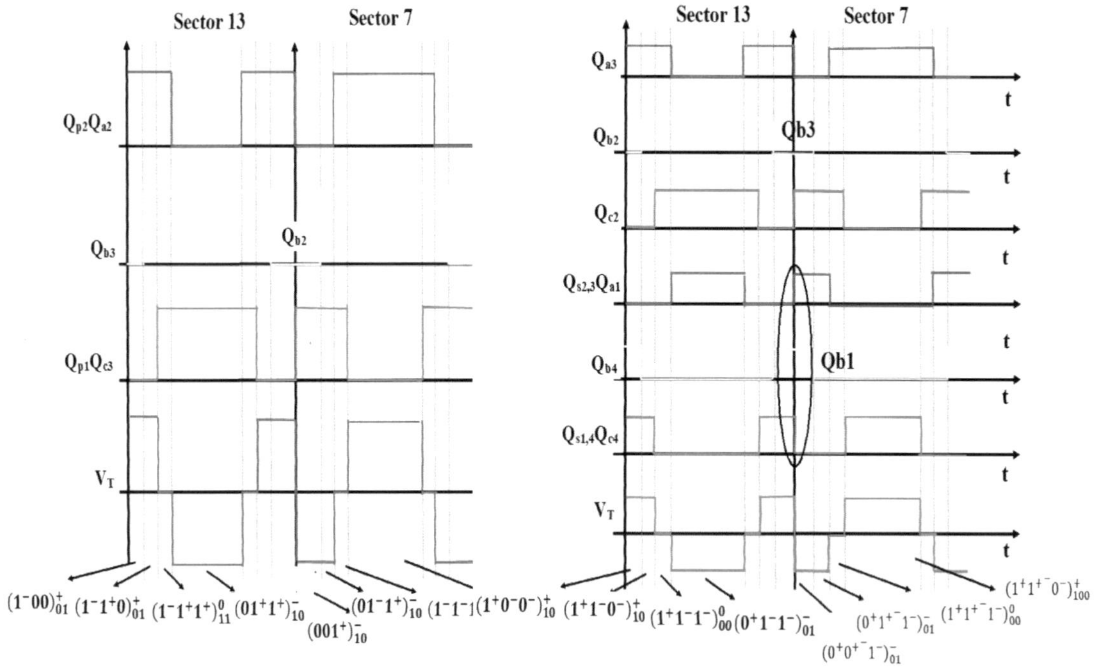

Figure 10. Driving signal of sector 13 and 7 under inverter and rectifier mode.

3.3. Type III:

In this type of SVM, the variations of active and reactive powers of for all the voltage vectors for each sector will be specified and the voltage vector for the sector I is given in the table. 5. Sector I is placed between the angles (θ) -30^0 to 0^0. Furthermore, for the twelve sectors, the angle (θ_n) will be calculated by the following equations:

$$(n-2)\frac{\pi}{6} < \theta_n < (n-1)\frac{\pi}{6} \quad (15)$$

Where, n = 1, 2, 3,.., 12.

Table 5. Active and reactive power variation for sector 1.

$\frac{dp_{inst}}{dt}$		$\frac{dq_{inst}}{dt}$	
> 0	< 0	> 0	< 0
$v_0, v_7, v_2, v_3, v_4, v_5$	v_1, v_6	v_0, v_7, v_2, v_3, v_1	v_4, v_5, v_6

Fig.11 consists of six active vectors and two zero vectors, which includes twelve sectors.

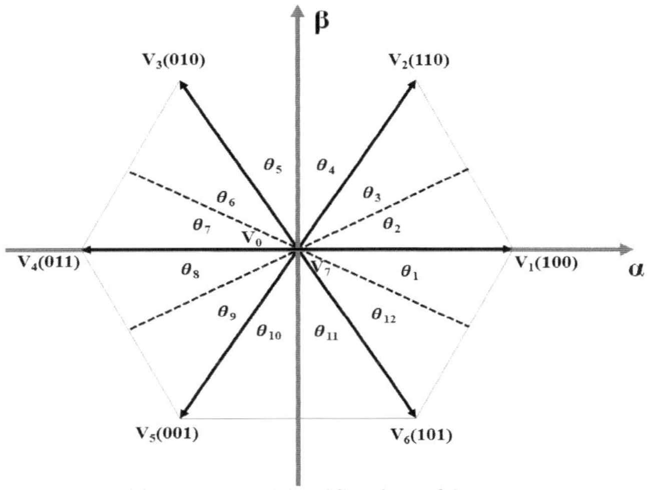

Figure 11. 12 Sector identification of SVM

3.3.1. Power hysteresis comparator

To the active and reactive power for the respective error between instantaneous and its reference value is given by the following equation.

$$\Delta_p = P_{ref} - P_{inst} \text{ and } \Delta_q = Q_{ref} - Q_{inst} \tag{16}$$

Hence, the hysteresis comparators are used to digitize the two given vectors. To obtain unity power factor Q_{ref} is set to zero. Fig. 12 shows the active and reactive power for digitized errors.

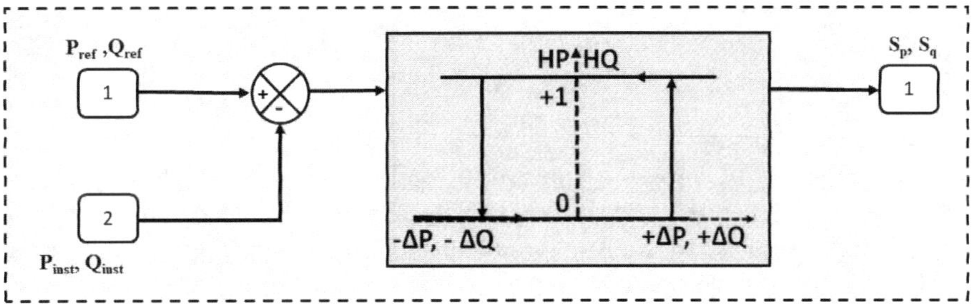

Figure 12. Hysteresis Power Comparator

$$S_p = \begin{cases} 1 \Delta_p > H_p \\ 0 \Delta_p < -H_p \end{cases} \tag{17}$$

$$S_q = \begin{cases} 1 \Delta_q > H_q \\ 0 \Delta_q < -H_q \end{cases} \tag{18}$$

Where H_p and H_q are the hysteresis bands that stipulate the control of active and reactive powers. The input of the hysteresis comparator is injected by the variation Δ_p/Δ_q, to obtain the following cases:

IOP Conf. Series: Materials Science and Engineering **906** (2020) 012027 doi:10.1088/1757-899X/906/1/012027

Case I: If the positive hysteresis bands H_p/H_q is less than the variation Δ_p/Δ_q, then the outputs of S_p/S_q, will be equal to 1. Hence, the power is amplified by receiving the driving PWM signals that is applied to the gates of the converter.

Table 6. The new switching table for DPC.

S_p	S_q	θ_1	θ_2	θ_3	θ_4	θ_5	θ_6	θ_7	θ_8	θ_9	θ_{10}	θ_{11}	θ_{12}
1	0	V_5	V_5	V_6	V_6	V_1	V_1	V_2	V_2	V_3	V_3	V_4	V_4
	1	V_0	V_0	V_0	V_0	V_0	V_0	V_0	V_0	V_0	V_0	V_0	V_0
0	0	V_6	V_1	V_1	V_2	V_2	V_3	V_3	V_4	V_4	V_5	V_5	V_6
	1	V_1	V_2	V_2	V_3	V_3	V_4	V_4	V_5	V_5	V_6	V_6	V_1

Case II: If the negative hysteresis bands $-H_p/-H_q$ is greater than the variation Δ_p/Δ_q, then the outputs of S_p/S_q, will be equal to 0. Hence, in this case the power is decreased by receiving the driving PWM signals, which is applied to the gates of the converter.

Case III: If the negative hysteresis bands $-H_p/-H_q$ and the positive hysteresis bands H_p/H_q having the variation Δ_p/Δ_q in between them then the driving PWM signal, which is applied to the gates of the converter follows the previous case.

By using this table 6, the performance of the converter is good with better THD level and by getting a pure sinusoidal form of injected current.

The comparative analysis of different types of Space vector modulation has been given in table 7.

Table 7. Comparison of Different SVM Methods.

	Type – I	**Type – II**	**Type – III**
Switching loss	Decreases	Decreases	Decreases
Conduction loss	Reduced	Reduced	Reduced
DC-link Pole	Contains both balanced and unbalanced dc-link voltages	Contains balanced dc-link voltages	Contains balanced dc-link voltages
Harmonic content	Lower order harmonics are eliminated and reduces output current THD	Reduces output current THD	Reduces output current THD
Additional Elements	Does not require active and passive elements.	Does not require active and passive elements.	Require active and passive elements.

4. Future scope:

Currently, IoT develops widespread in numerous applications, like smart homes, smart city, and smart grid. The IEEE802.11n WLAN is one of the main communication technologies pertinent to IoT applications due to elasticity and low cost, where the well-organized power management is a fundamental necessity of developing IoT applications. In future IoT based smart grids are more popular all over the world. The present scenario shows the smart grid technology and its distribution solutions for the future trend. The IoT plays a vital role in several application domains for control and monitoring of the new security issues, which is taken into an account. The concept of the smart grid is becoming a realism as it progresses from theoretical models to changing stages. Saroj Mondal, and Roy Paily [34] identify a proposed IoT based renewable energies in this smart grid era. The Internet of Things IoT nodes is applied to PV power harvesting system to attain maximum power point tracking (MPPT). Within a tracking error of 0.6%, the IoT tracks the maximum power point in PV system within 12 μs by consuming an intrinsic negative feedback loop. Furthermore, peak pursuing with an efficiency of 99% obtained with rapid changes in atmospheric conditions. By the interaction, sensing, monitoring and control of internal and external environment, the IoT is used. IoT is used to transmit the data by using sensors for detecting the things, microcontrollers to decrypt the data, and a receiver to connect with the local hub where anyone can make decisions for proper operation of the smart grid. Similarly, the IoT tracks MPP with a tracking error of 0.1% - 0.6%. The IoT application for control and monitoring of BHM is shown in below fig. 13.

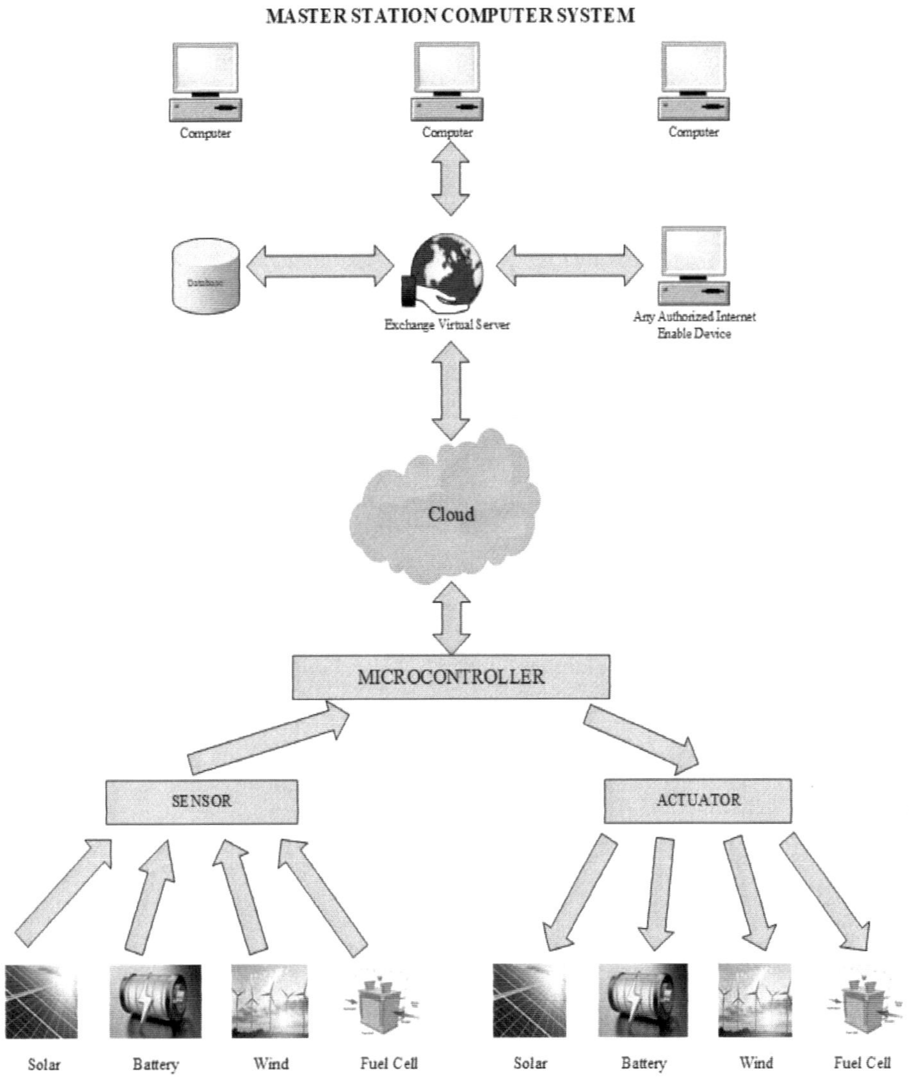

Figure 13. Structural Block diagram of an IoT based Bipolar Hybrid Mirco grid.

5. Conclusion:

A complete review on three-phase bidirectional converter for harmonic elimination has been explored with a detailed perspective on different structures to design and implement this converter for researchers and application engineers. This proposed comparison of this different bidirectional converter and SVM that is provided in this paper is used to easily identify the suitable converter for specified applications. By use of this converter in microgrid there will be a power quality improvement because of higher efficiency, lower cost with a reduced number of switches for the control of voltage and current harmonics, reduced size in a converter, lower cost and reliability in the dc-link voltages. These types of converters produce improved power quality on both ac and dc outputs. The bidirectional AC-DC converter is categorized under two stages namely single and two-stage converter. When compared to two stage converter single stage converter has fewer components, shows more efficiency and low cost, but it is not popular because of the complex optimization process and control algorithm to maintain the power quality issues. Hence, this paper elaborates on the review of AC/DC bidirectional converter for various applications. One of the new advancements in this microgrid technology is IoT, which is more reliable, controllable and easy to monitor renewable sources.

Reference:

[1] Elizabeth Noghreian, Hamid Reza Koofigar, Power Control of hybrid energy systems with renewable sources (wind-photovoltaic) using switched systems strategy, Sustainable Energy, Grids and Networks 21 (2020) 100280.

[2] C. Yin, H. Wu, F. Locment, M. Sechilariu, Energy management of DC microgrid based on photovoltaic combined with diesel generator and super capacitor, Energy Converters. Manag. 132 (2017) 14–27.

[3] E. Hossain, E. Kabalci, R. Bayindir, R. Perez, Microgrid test beds around the world: State of art, Energy Converters. Manag. 86 (2014) 132–153.

[4] J.Y. Kim, J.H. Jeon, S.K. Kim, C. Cho, J.H. Park, H.M. Kim, K.Y. Nam, Cooperative control strategy of energy storage system and micro sources for stabilizing the microgrid during islanded operation, IEEE Trans. Power Electron. 25 (2010) 3037–3048.

[5] J. Rocabert, G.M.S. Azevedo, A. Luna, J.M. Guerrero, J.I. Candela, P. Rodrguez, Intelligent connection agent for three-phase grid-connected microgrids, IEEE Trans. Power Electron. 26 (2011) 2993–3005.

[6] J. Rocabert, A. Luna, F. Blaabjerg, P. Rodríguez, Control of power converters in AC microgrids, IEEE Trans. Power Electron. 27 (2012) 4734–4749.

[7] Amir Reza Hassani Ahangar, Gevork B. Gharehpetian, Hamid Reza Baghaee, A review on intentional controlled islanding in smart power systems and generalized framework for ICI in microgrids, Electrical Power and Energy Systems 118 (2020) 105709.

[8] Mohsen Eskandari, Li Li, Mohammad. H. Moradi, Improving power sharing in islanded networked microgrids using fuzzy-based consensus control, Sustainable Energy, Grids and Networks 16 (2018) 259–269.

[9] Federico Delfino, Giulio Ferro, Riccardo Minciardi, Michela Robba, Marco Rossi, Mansueto Rossi, "Identification and optimal control of an electrical storage system for microgrids with renewables", Sustainable Energy, Grids and Networks 17 (2019) 100183.

[10] M. Parvez Akter, S. Mekhilef, N. Mei Lin Tan, and H. Akagi, "Modified model predictive control of a bidirectional ac-dc converter based on lyapunov function for energy storage systems," IEEE Transactions on Industrial Electronics, vol. 63, no. 2, pp. 704–715, Feb 2016.

[11] L. H. S. C. Barreto, D. de A. Honrio, D. de Souza Oliveira, and P. P. Praa, "An interleaved-stage acdc modular cascaded multilevel converter as a solution for mv railway applications," IEEE Transactions on Industrial Electronics, vol. 65, no. 4, pp. 3008–3016, April 2018.

[12] K. K. Gupta, A. Ranjan, P. Bhatnagar, L. K. Sahu, and S. Jain, "Multilevel Inverter Topologies With Reduced Device Count: A Review," IEEE Transactions on Power Electronics, vol. 31, no. 1, pp. 135-151, Jan. 2016.

[13] S. Hu, Z. Liang, D. Fan, and X. He, "Hybrid Ultra capacitor-Battery Energy Storage System Based on Quasi-Z-source Topology and Enhanced Frequency Dividing Coordinated Control for EV," IEEE Transactions on Power Electronics, vol. 31, no. 11, pp. 7598-7610, Nov. 2016.

[14] F. Nejabatkhah, Y.W. Li, Overview of power management strategies of hybrid AC/DC microgrid, IEEE Trans. Power Electron. 30 (2015) 7072–7089.

[15] X. Lu, J.M. Guerrero, K. Sun, J.C. Vasquez, R. Teodorescu, L. Huang, Hierarchical control of parallel AC-DC converter interfaces for hybrid microgrids, IEEE Trans. Smart Grid 5 (2014) 683–692.

[16] Ling Gu and Ke Jin," A Three-Phase Bidirectional AC/DC Converter With Y–Δ Connected Transformers", IEEE TRANSACTIONS ON POWER ELECTRONICS, VOL. 31, NO. 12, DECEMBER 2016.

[17] Rohit Baranwal, Kartik V. Iyer, Kaushik Basu, Gysler F. Castelino, and Ned Mohan," A Reduced Switch Count Single-Stage Three-Phase Bidirectional Rectifier With High-Frequency Isolation", IEEE TRANSACTIONS ON POWER ELECTRONICS, VOL. 33, NO. 11, NOVEMBER 2018.

[18] Hammad Armghan, Ming Yang, Ammar Armghan, Naghmash Ali, M.Q. Wang, Iftikhar Ahmad, "Design of integral terminal sliding mode controller for the hybrid AC/DC microgrids involving renewables and energy storage systems", Electrical Power and Energy Systems 119 (2020) 105857.

[19] Farzam Nejabatkhah and Yun Wei Li, "Overview of Power Management Strategies of Hybrid AC/DC Microgrid", IEEE TRANSACTIONS ON POWER ELECTRONICS, VOL. 30, NO. 12, DECEMBER 2015.

[20] Hurng-Liahng Jou, Kuen-Der Wu, Jinn-Chang Wu, You-Zu Lin, Li-Wen Su, "Asymmetric isolated unidirectional multi-level DC-DC power converter", Engineering Science and Technology, an International Journal 22 (2019) 894–898.

[21] S. Hu, Z. Liang, D. Fan, and X. He, "Hybrid ultra capacitor battery energy storage system based on quasi-z-source topology and enhanced frequency dividing coordinated control for ev," IEEE Transactions on Power Electronics, vol. 31, no. 11, pp. 7598–7610, Nov 2016.

[22] Mahmoud A. Sayed , Kazuma Suzuki, Takaharu Takeshita and Wataru Kitagawa," PWM Switching Technique for Three-Phase Bidirectional Grid-Tie DC–AC–AC Converter With High-Frequency Isolation", IEEE TRANSACTIONS ON POWER ELECTRONICS, VOL. 33, NO. 1, JANUARY 2018.

[23] H. Wu, S. Wong, C. K. Tse, and Q. Chen, "Control and modulation of bidirectional single-phase acdc three-phase-leg SPWM converters with active power decoupling and minimal storage capacitance," IEEE Transactions on Power Electronics, vol. 31, no. 6, pp. 4226–4240, June 2016.

[24] P.C. Loh, D. Li, Y.K. Chai, F. Blaabjerg, Autonomous control of interlinking converter with energy storage in hybrid AC-DC microgrid, IEEE Trans. Ind. Appl. 49 (2013) 1374–1382.

[25] Nan Jin, Shiyang Hu , Chun Gan and Zhibin Ling," Finite States Model Predictive Control for Fault-Tolerant Operation of a Three-Phase Bidirectional AC/DC Converter Under Unbalanced Grid Voltages", IEEE TRANSACTIONS ON INDUSTRIAL ELECTRONICS, VOL. 65, NO. 1, JANUARY 2018.

[26] Parviz Najafi, Abbas Hooshmand Viki, Mahdi Shahparasti, " Novel space vector-based control scheme with dc-link voltage balancing capability for 10 switch converter in bipolar hybrid microgrid", Sustainable Energy, Grids and Networks 20 (2019) 100256.

[27] Faouzi Tlili, Ameni Kadri, Faouzi Bacha," Advanced control strategy for bidirectional three phase AC/DC converter", Electric Power Systems Research 179 (2020) 106078.

[28] A. L. Julian and G. Oriti, "A novel clamp circuit for a regenerative rectifier using ac/ac matrix converter theory," IEEE Transactions on Industry Applications, vol. 41, no. 1, pp. 68–74, Jan 2005.

[29] Diogo Varajao, Rui Esteves Araujo, Luis Miguel Miranda and Joao A. Pecas Lopes, "Modulation Strategy for a Single-Stage Bidirectional and Isolated AC–DC Matrix Converter for Energy Storage Systems", IEEE TRANSACTIONS ON INDUSTRIAL ELECTRONICS, VOL. 65, NO. 4, APRIL 2018.

[30] Ling Gu and Ke Jin," A Three-Phase Isolated Bidirectional AC/DC Converter and its Modified SVPWM Algorithm", IEEE TRANSACTIONS ON POWER ELECTRONICS, VOL. 30, NO. 10, OCTOBER 2015.

[31] M. Y. Lee, P. Wheeler, and C. Klumpner, "Space-vector modulated multilevel matrix converter," IEEE Transactions on Industrial Electronics, vol. 57, no. 10, pp. 3385–3394, Oct 2010.

[32] Jonathan E. Bosso, Marcelo Llomplat, Germán Gustavo Oggier, Guillermo Oscar García, "Isolated bidirectional DC-to-three-phase AC converter for integration of renewable energy sources to electric grid", IET Power Electronics, on 28th June 2019.

IOP Conf. Series: Materials Science and Engineering **906** (2020) 012028 doi:10.1088/1757-899X/906/1/012028

Development of Automatic Tube Oil Skimmer

Vignesh. T[1], Bhuvaneswari M[2], S.Manojkumar[3], S.Kishore[4], N.Bommuraj[5]

Department of Mechatronics Engineering
Sri Krishna College of Engineering and technology, Coimbatore – 641008
vigneshT@skcet.ac.in , bhuvaneswarim@skcet.ac.in, 16eumt046@skcet.ac.in, 16eumt054@skcet.ac.in, 17eumt502@skcet.ac.in

Abstract. The paper work represent the removing oil from the water which is wasted from the machines. The machines get more heated while running, oil and water is used to cool the machine. This investigations is removing oil from water surface is done by skimmer. The oil skimmer is having tube under different operation condition (thickness of oil film, temperature of oil, rotational speed of tube, water PH, oil tube and tube material tube). The oil skimmer has floating tube with fixed length, flexible and discrete segmented model. The design of the wheel according to the tube size and tube length is fixed according to the tank size. Wheel is connected with the belt which is immersed in the oil and water mixer tank. While running the pulley wheel belt is coming upper, the wiper is attached in the frame. Wiper removes the oil coming from the belt. This method is most efficient method to solve the human fatigue and cost effective method.

1. INTRODUCTION

Now a day's industry running in the whole day and year. So the machine will get heated while running. To remove the heat in the machine the coolant is used. Most the industry used oil and water mixer as the coolant. The expansion of new technology and importance of oil skimmer is evolved [1]. Skimmer medium as 'belt and metal disc' it is run above the water surface it is brought [2]. Belt is used under different working condition like thickness of oil film, temperature of oil, speed of the belt, type of oil used etc, oil recovery rate is based on this operating [3]. The recovery vessels is used to oil recovery vessels are available on the open sea but unworkable in narrow sea and it is developed a handy oil recovery system which is portable and used for supplementing the weak point of the oil recovery vessel [4].There is difficult to recover the spilled oil on the water in dangerous areas, so this method is clean the surface using unmanned vehicle and also dynamic inclined plane (DIPS) place in front of USV, it is run very high speed and accuracy to the destination [5]. This system explain the recovery oil from the coastline which is very long by using truck. The barge crane makes to recuperate oil at the quay and loads the system on its floor. The two equipment is used one is skimmer operated by crane and another is buoy system for stretching oil boom automatically [6] During transportation crude oil it requires the safety measure is port terminals and necessary to develop oil separate system belt skimmer is developed and it consist of floating disc and belt. Which is operate using Bluetooth [7]. The monitoring system used the IOT platform to get the data and is monitored continuously it is controlled by the mobile phones. The many open source hard ware kit is used to develop the prototype model. The oil skimmer is dragged by two connecting vehicles it is mostly used in oil skimming, it drawn from the oil from water surface. [8]. It requires the vacuum blower with high power capacity acquire a gas liquid in the medium. This method is used in the well and also wavy sea it has a wide range of viscosity.

2. METHOAD AND MATERIAL

Before get in to the costly process like chemical and other treatment this oil skimmer is used to do pre-process of removing the oil from the water. This skimming mechanism need the basic three properties like specific gravity, affinity and surface tension. There are many constrains taken while selecting the skimmer. The major property is the viscosity and the adhesiveness of the oil is skimmed. In this the tank is made of aluminium plates and plastic tube is used. The metal pulley wheels is used to connect the belt.

3. CAD MODELING

To visualise the idea many cad tool are used. In this skimming dive the creo software used to create the model. The various design considerations are followed while designing the model. The figure 1 shows the 3d model of the oil skimmer

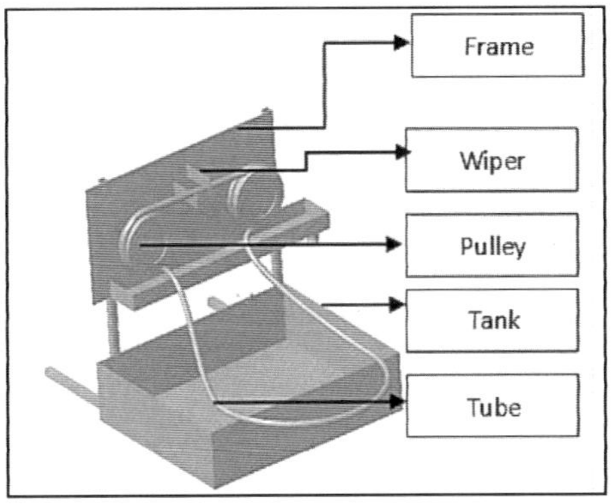

Figure 1. 3D model of tube oil skimmer

The manufacturing the model, drafting is majorly used. The drafted model of the oil skimmer is shown in the Figure 2, and various views also showed.

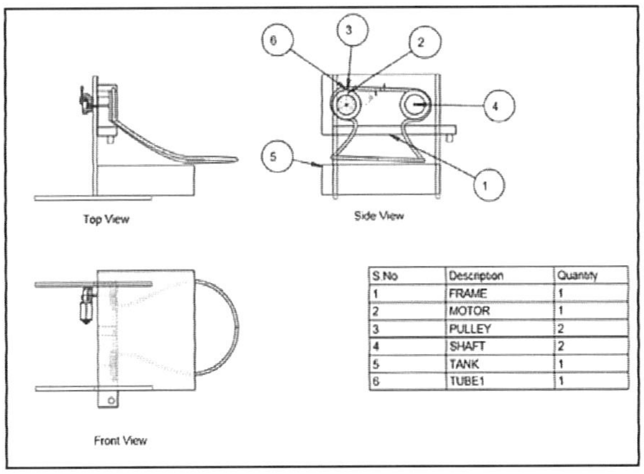

Figure 2. Assembly view of tube oil skimmer

4. CALCULATIONS

4.1 Friction on wiper:

IOP Conf. Series: Materials Science and Engineering **906** (2020) 012028 doi:10.1088/1757-899X/906/1/012028

$f=\mu\times N$

Ƒ-Friction force

μ-0.8 Mild steel & Rubber Coefficient

N-Normal force

P=F.V

F-Force

P-power

V-Velocity

F=8/0.3327N

$f=\mu\times N=0.8\times20=22N=22$

4.2 Pulley Calculation:

Recommended series of pulley diameter =150mm

Mild Steel =7.85g/⟦cm⟧^3

Volume of pulley = π r^2h

r –radius of pulley

h - Thickness of pulley (DDB-7.55)

$V = \pi \times (0.075\times0.075) \times0.020$

V=0.00035m^3

e =m/v

m =e × v

m=785 ×"0.00035" =0.27kg

4.3 Velocity:

V=D/T D- Distance, T-Time D=3m (Tube distance)

=3/10

V=0.3m/s

ω =v/r ω-Angular Velocity

ω =0.3/0.075 =4 rad

P=T × ω P-Power

P1.9 × 4 =8w

ω =2π/60 N = ("4" ×60)/ (2×π) = 38rpm

4.4 Volume of tank:

V = length (l) ×width (w) ×height (h)

V= 0.60 ×0.50 ×0.15 Golden Ratio = (L)/ (D) =1.6

V = 0.045m^3

Volume rate:

$V = t \times D \times \pi \times d \times N$

t- Thickness

D-diameter of tubed- diameter wiper

N-rpm

$V = 0.001 \times 0.012 \times \pi \times 0.012 \times 38$

V= 0.28ml/min

5. HARDWARE IMPLEMENTATION

The micro controller is the chip which used to program and control the hardware setup. It has own ram and Rom memory's for storing the program and execution. It has a various input and output pin for getting and sending the information to the other devices. Serious ports is used to send the data to the output devices in serially. It has parallel ports also. Timer and counters are used to maintain the clock timing to the device. This connect to the real time world timing. Analog and digital converter are used to convert the signal to one form to another form. This all features are intergrade in the single chips. When the data sent via serial or parallel ports the baud rate is the major parameter to consider. Both input and output device need a same baud rate then only the data is send without delay and losses. Microcontroller AT89C51 is used in this project, it has a high performance and low power. It has 8 bit of C MOS and 4k Programmable flash memory. It having a read only memory for the erasable data's. It is manufactured by Atmel high density non-volatile memory technology. The flash memory used to reprogram the controller. To combine 8 bit CPU with flash memory it's become more power full microcomputer.

Figure 3. PIC 16F877A Microcontroller

5.1 DC Motor

The DC motor used is 12V motor which is used to run the motor in this project. Dc motor is used to convert the one form of energy to another form like electrical to mechanical. It has 5 principal's field system, armature core, armature winding, Brushes and commutator. Figure 4 shows the DC motor used in the oil skimmer.

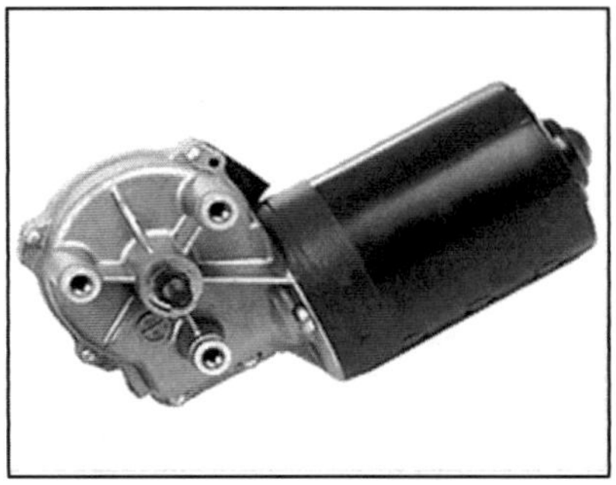

Figure 4. DC Motor

5.2 Relay

A relay is a type of stich. It is electromagnetic switch. It operated by less electric current but it turns on or off the larger current. This relay used to convert high voltage into 5V for micro controller. Figure 5 show the relay used in the oil skimmer.

Figure 5. Relay

5.3 LCD Display

Liquid Crystal Display (LCD) it's a type of display it used liquid crystal for primary operations. LED also used in various smart phones and television system other computer etc. it has various use cases for business and consumers. The LCD 16*2 used in this works. It's given in the figure 6.

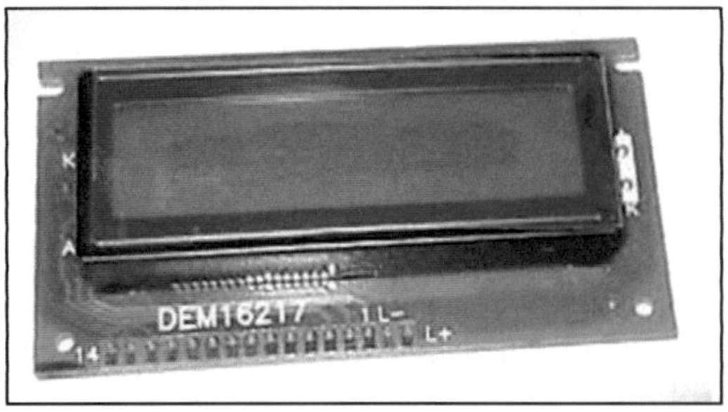

Figure 6.LCD Display

5.4 Specification of electronics components

Table 1. Shows the Specifications of Electric Component

Parameter	Value
Frame	Mild steel
Dc Motor	Torque: 10 kg-cm
PIC16F877	Micro controller
Transformer	12v
Relay	5v
Float sensor	5v

5.5 Microcontroller PIC 16F877A

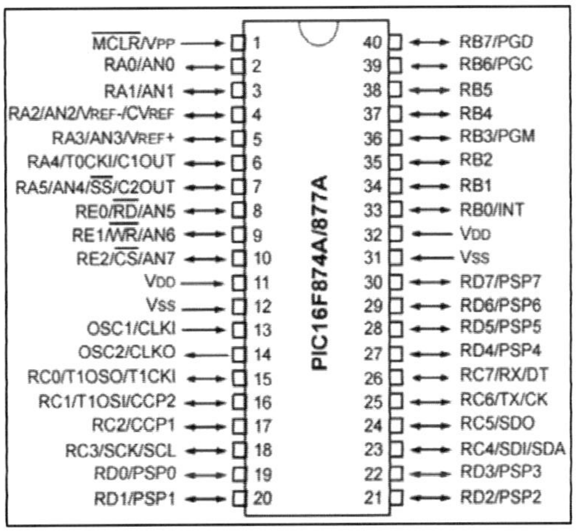

Figure 7. Microcontroller PIC 16F877A

The microcontroller having the EEPROM, it has flash memory and non-volatile ROM. To interface with the input and output device it having 40 pins in this 33 pins used for input and output. Other pins used conversions purpose like analog to digital. To generate the pulse the chip has PWM pin. This Chip having timer, receiver and transmitter circuits. The figure 3 shows the architecture of the pic microcontroller

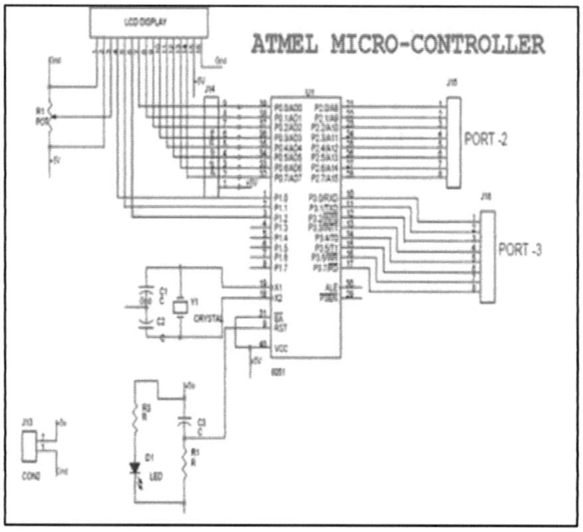

Figure 8. PIC controller connected with sensor and display

From fig.7 the +5V power supply given to controller, it is run by +5V power supply only. The micro controller consist of 4 ports and 40 pin configuration. The ports are port0, port1, port2, port3. Each ports is having 8 pins. From the port zero data line is taken for the LCD connection. The control line of LCD taken from the port 1, 0, 1 and 2 pins. To adjust the brightness of the LCD the small pot variable resister is used. To produce a clock pulse crystal oscillator is used. There are many oscillators used in the industry but they will produce the varying voltage and fluctuation in temperature. If the voltage and temperature fluctuation it is affect the accuracy. The ports two and three is used as input and output purpose

6. RESULT AND DISCUSSION

The oil skimmer model is fabricated and demonstrated in the industry. This proposed model is less and mist cost effective. It has very less weight compare to other models. Both wheels are designed in equal diameter based on the requirement and design calculations. All components in this model is designed based on the requirement and design calculations. In this wiper mechanism used it is most effective method to filter the oil. It has more accuracy. The figure 10 shows the fabricated image of oil skimmer

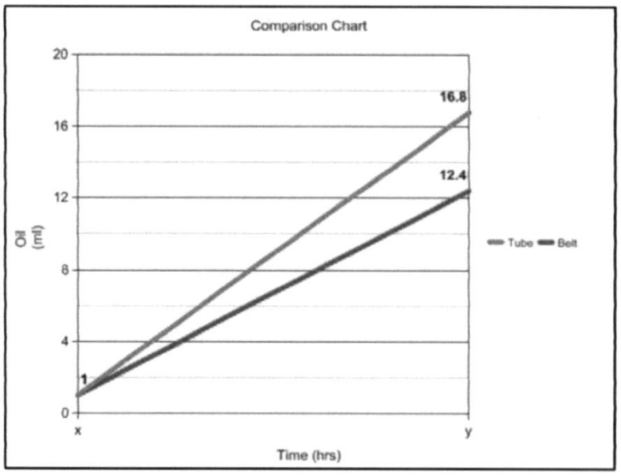

Figure 9. Comparison Graph

Figure 10. Oil Skimmer Fabricated Image

7. CONCLUSION

Oil skimmer system prototype is tested and evaluated in the industry circumference. The recovery test has done in the industry. The system gives the good performance result. The system works well in harsh surface. This method is more cost efficient and less material requirement. This system is designed and tested and it shows that it can regain most of the oil from water

References

[1] S. Siva and P. Praveen, "Design and Fabrication of Belt Type Oil Skimmer," vol. 5, no. 07, pp. 1–5, 2017.

[2] M. A. Dawood and R. J. Algawi, "Study of operating conditions for oil skimmer apparatus from water," 2017 Int. Conf. Environ. Impacts Oil Gas Ind. Kurdistan Reg. Iraq as a Case Study, EIOGI 2017, vol. 2018-January, pp. 65–70, 2017, doi: 10.1109/EIOGI.2017.8267627.

[3] N. W. Skimmer, "Recovery Vessel Dr.Kaiyo," pp. 5–9, 2008.

[4] J. Wang, F. Ren, Z. Li, Z. Liu, X. Zheng, and Y. Yang, "Unmanned surface vessel for monitoring and recovering of spilled oil on water," Ocean. 2016 - Shanghai, no. 2, pp. 3–6, 2016, doi: 10.1109/OCEANSAP.2016.7485405.

[5] M. Yoshie, I. Fujita, and K. Takezaki, "Field simulation of oil skimming system package for crane barges in emergency," Ocean. MTS/IEEE Kobe-Techno-Ocean'08 - Voyag. Towar. Futur. OTO'08, 2008, doi: 10.1109/OCEANSKOBE.2008.4530997.

[6] S. Sunil et al., "Floating Oil Skimmer with Garbage Collector," vol. 5, no. 04, pp. 33–36, 2017.

[7] W. Choi, D. Yoon, T. Kim, J. Lee, and A. Background, "Won-Seok Choi 1 , Dong-Geul Yoon 1 , Tae-Eon Kim 2 and Jang-Myung Lee 3 *," no. Mfi, pp. 458–463, 2017.

[8] I. Fujita, "An Air," vol. 00, 2006.

IOP Conf. Series: Materials Science and Engineering **906** (2020) 012029 doi:10.1088/1757-899X/906/1/012029

Dynamic Analysis of Sensorless Controlled Industrial Pump Drive

K. Anil Kumar[1], M. Harshit[2], R. Raja Singh[3]

[1] Department of Electrical Engineering, MNIT Jaipur, Rajasthan, India
[2] Department of Electrical Engineering, Indian Institute of Technology Roorkee, Uttrakhand, India
[3] Avanced Drives Laboratory, Department of Energy and Power Electronics, Vellore Institute of Technology, India

Correspondance: hmohann@gmail.com

Abstract. Industrial pumps operating with variable speed provides numerous benefits including valve elimination, wide range of flow rate, single stage higher head and energy saving. The realization of variable speed pump operation is carried out using an induction motor, power electronic switches, current sensors and controller without speed feedback is known as sensorless variable speed control. This paper implements a field-oriented control based sensorless scheme on 150 kW induction motor drive system, which is integrated to centrifugal pump load in Matlab/ Simulink environment. The centrifugal pump is modelled mathematically in the same environment for obtaining the identical torque speed characteristics.

Keyword: Induction Motor, Industrial Pump Drive, Centrifugal Pump, Sensorless Control, Field Oriented Control, Energy Efficiency.

1. Introduction

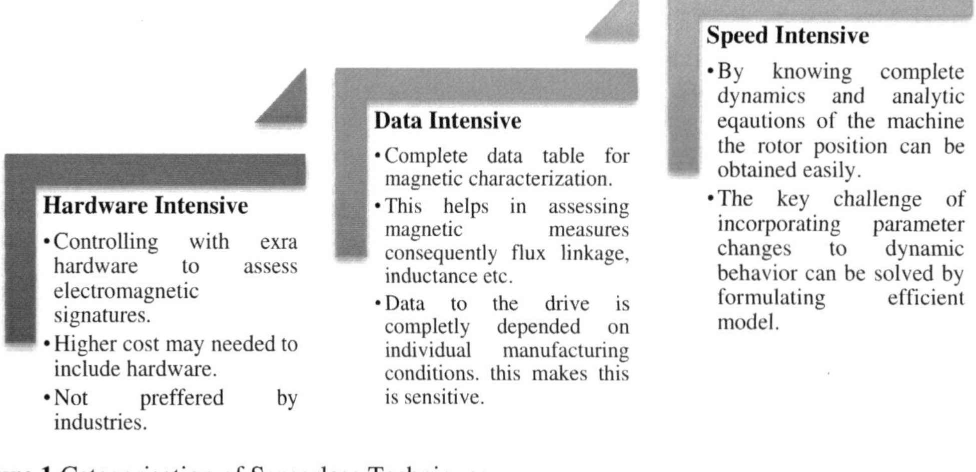

Figure 1 Categorisation of Sensorless Techniques

Continuous industrial process needs extensive research in the development of drives systems. In view of sustaining high harsh environments, decreasing cost and sometimes to increase system reliability, sensorless control of ac drives are preferred. The sensorless drives have no mechanical sensors for identifing the rotor position and speed without deteriorating its dynamic performance. Controlling of drives without sensors needs either extensive hardware or high data on drive properties with high-speed capability. These drives are broadly classified in to three types as shown in figure 1 and the adoption of various strategies on different machines and applications are reviewed in [1]. In most of the industrial applications, the induction motors are widely preferred due to its lower cost, higher performance, lesser parts with simplified design and reliability.similarly, pumps drives are widly notable in many industries[2], [3], [4]. Overall this paper organization are as follows, section 2 discuss the pumping applications in industries and the benefits of variable speed operation. Section 3 describes the operation of sensorless controlled pump drives implementation. Section 4 provides the detailed mathematical modelling of induction motor based pump vector control drive system woth speed and flux estimation and centrifugal pump. In section 5, the simulation of sensorless controlled pump drive is performed in Matlab Simulink environment and followed by the results are disused in detail. Finally, the paper concluded in section 6.

2. Industrial Pumping

Industrial pumps have significant role in industriual process such as handling the liquids under different pressure and transfering of liquids water, chemicals, petroleum, oil and semi solid products like sludge, food, etc., the maximum percentage of pumps used in industries are centrifugal and positive displacement pumps. Industrial pump applications have benefited from the use of different forms of variable speed. Over the last few decades, the electronic drives used with speed motors have improved their cost to performance ratio and are now considered the variable speed technology of choice for many pump applications. Due to the advantages of this speciality drive systems, use in process plant environments has increased. Benefits include pump performance flexibility, ability of a single pump to cover a broader range of flows and heads, higher allowable head per stage, energy savings, and possible throttling or bypass valve elimination. Beginning in the 1960s, electronic drives were developed for varying the speed of AC induction motors, the most commonly used industrial motors. These devices, commonly called variable frequency drives or VFDs, use solid-state semiconductors to transform fixed frequency AC power available from the electric utility into varying frequency power. This transformation allows variable speed motor operation and operation of motors at higher than the standard speed. The key to the advancement of VFDs has been the development of solid-state semiconductors of higher amperage and voltage levels. This technology is providing continued advancement in the most popular VFD style today, the pulse width modulated, or PWM, drive. The continued advancements of VFD technology have provided wide application possibilities. VFD usage today is widespread in several industrial applications. The fundamental variable speed pump-system interaction provides many significant benefits; they are;

2.1. Performance Flexibility

Centrifugal and positive displacement pumps can cover a broad range of hydraulic conditions, producing exactly the desired flow and head shown in figure 2. If a user is unsure of required flow and head at pump selection, time, variable speed can be used in the field to produce exactly the required conditions similarly [2], [3]. If a user needs for several pumps with similar but not identical duty points, identical pumps shall be used for part interchangeability and fine-tune their performance with speed. This flexibility also allows for changes in specific gravity and viscosity. From a pump manufacture's point of view, the use of this range ability can allow a reduction in the number of models stocked to cover a given hydraulic envelope.

2.2. Higher Head Per-Stage

Pumps with higher speed produce higher flow or in case of centrifugal pumping, higher head.Thus the employment of higher speed pumps can replace multiple stage pumps[4]. When pumps are operated at higher speeds, radial bearing and other loads can increase dramatically, often with the square of the speed. Net positive suction head of centrifugal pump also the increases with the square of the speed. Cavitation damage can be more severe. The variation in speed changes flow, hence valve can be eliminated. Hence, this saves capital cost, maintenance, the pressure drop across the valve (often 10 percent of the total required pressure rise) and valve stem leakage. For instance, the centrifugal pump can be operated to a noticeable lower flow if a valve or other form of restriction is placed within 5 feet of the pump discharge flange. However, the minimum allowable stable flow is reduced linearly with speed and thus, the lower flow may be available from a VFD operated pump with no valve than from a fixed speed pump with valve.

2.3. Energy Savings

The lack of throttling valve removes pressure drop hence loss of energy be compensated with larger flow or higher height [5]. The centrifugal pump curve as shown in figure 2b consists of a fixed speed pump using discharging throttling that would produce head H, at Q2 flow rate to deliver the desired H2 head at Q3 flow rate to the system. Alternatively, by using variable speed, only Q2 flow rate at H2 head need be produced if the proper speed is selected. Hydraulic horsepower saving from using variable speed for variable speed positive displacement pumps is in cases with large head changes and for positive displacement pumps in cases with large flow changes. To use these chart, locate the desired off-design point and read from the chart the percentage amount of energy saved when reaching an off-design point by a variable speed, rather than by discharge throttling or flow bypass. Any gearbox or other ancillary loads may vary with speed differently. Offsetting these saving are increased losses in the VFD and motor at lower speeds and loads.

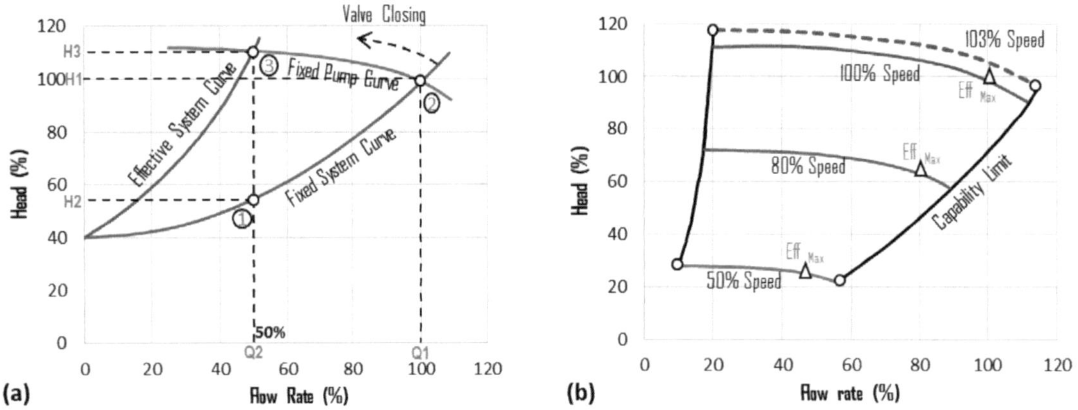

Figure 2. Centrifugal pump operation (a) variable speed and (b) fixed speed

2.4. Determine Net Positive Suction Head (NPSH) and Efficiency Vs Speed

Centrifugal pump NPSH is proportional to the square of the speed hence efficiency of the pump as shown in figure 3. Minimum and maximum efficiencies remain unchanged, but the flows at which they occur are shifted by the amount of the speed change. Pump selection depends on NPSH, flow and speed hence care must be taken while its selection Since NPSH, is increased at higher speed, inducers may be required to reduce NPSH, to available levels [6].

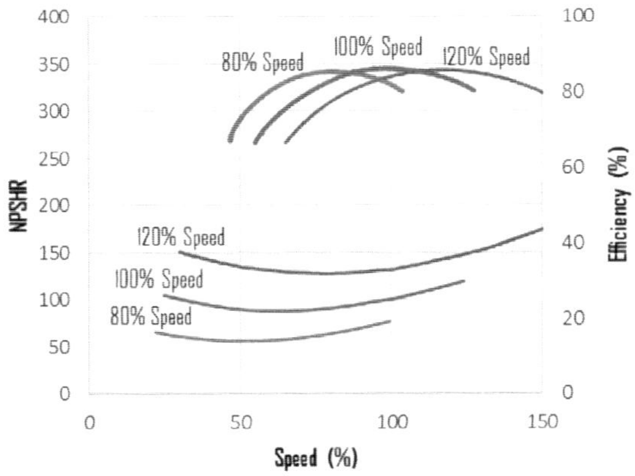

Figure 3. NPSHR – Efficiency Vs Speed

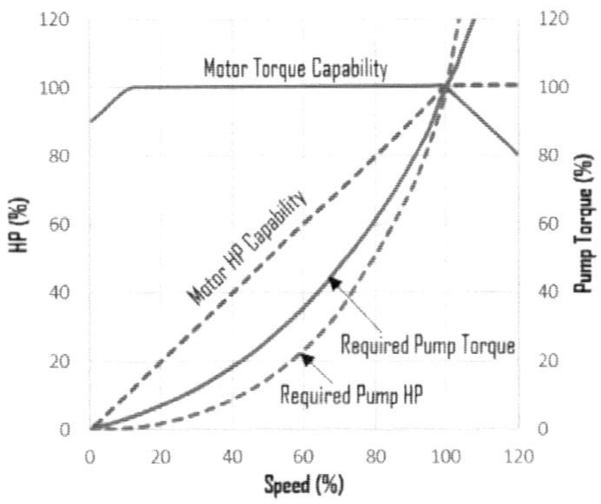

Figure 4. Pump capability curve

2.5. *Match pump torque with motor torque*

Available motor torque must exceed required pump torque at start-up and every speed. The motor torque capability remains constant as speed is lowered, while the centrifugal pump torque required always falls rapidly, its exact value being dependent upon the shape of the system curve. This relationship for the case of zero static lift is shown in figure 4. Starting torque for a centrifugal pump is generally small and easily provided for by a variable speed motor. On occasion though, some forms of centrifugal pumps with high suction pressures may have a high instantaneous breakaway torque required.

2.6. *Avoidance of Lateral Critical Speed*

The pump standard API 610 states that, depending on the unbalanced response amplification factor, a pump may not be operated between 85 percent and 105 percent of its predicted critical speed. Adherence to these rules can block out a large portion of the allowable performance envelope, as shown in figure 5. This potential problem indicates the need for careful application of variable speed technology with additional knowledge of the customer's system and desired duty points to ensure a careful matching of pump performance to the desired range of flow and head. Fortunately, most pumps are of a stiff staff design and will operate below their first lateral critical speed. However, if faced with this block out, a pump vendor may be able to change the mechanical design to raise or lower the critical to provide full range speed adjustment.

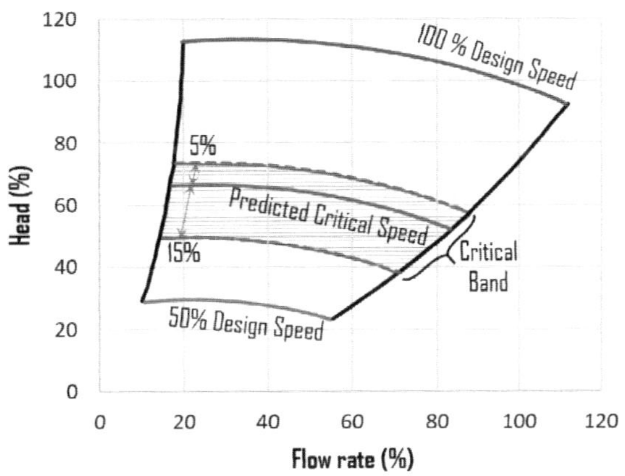

Figure 5. Critical speed band curve

2.7. Account for Torsional Critical Speeds

Torsional critical speeds are resonant frequencies at which motor and driven equipment shafts can begin to oscillate with angular displacement as a result of torsional excitation. VFDs can cause torsional excitation known as torque ripple. Typically, torque ripple is not a concern for units below 200 hp. In fixed speed systems, torsional excitation may occur at some multiple of fixed motor speed such as 2 x running speed due to coupling misalignment. If this excitation is at a torsional critical frequency, the equipment can suffer serious damag e depending on the amplitude of the excitation and the resulting stress on the system. Frequently, however, the pump design or coupling angular stiffness can be changed to avoid this fixed speed induced torsional problem. With variable speed, however, the situation is complicated considerably since the base speed changes and the VFD causes torque ripples due to electrical harmonics at multiples of base frequency or speed.

3. Sensorless Controlled Pump Drive

The annotations for the use of induction motors (IM) are lower cost, simplified design with fewer parts, good performance under high power application, higher reliability and less chance of raw material price fluctuations and shortages. For sensorless applications by using IM, several algorithms have been developed, where the most significant part is at zero frequency and hence near to zero frequency a lot of research presented in various papers [1,2]. In IM, the rotor position can be estimated with tracking changes in reluctance due to rotor slotting's. However, in general, IMs are designed to close slotting's and skewing's hence minimizing variations by having them [9]. Hence sufficient tracking of the IM necessities special design machine that shows significant variations to detect slotting's. Closing and skewing these slots may deteriorate direct-on-line staring behaviour. But with vector controlled power converters these deteriorations be avoided. Nevertheless, IMs are manufactured at low-profit margins, and there is a resistance to solutions that require specially designed machines [11]. Now the rotor position can be tracked by detecting changes in reluctance. This is different from zero excitation frequency injection method (ZEFIM) and gives better torque and flux control than earlier without any restrictions any even zero frequency[6]. In simple principle, specially designed machines are necessary unlike all common machines. Although an auxiliary observer is necessary to derive exact speed estimate, there are many full-torque zero-speed applications in which either a speed signal is not required or only required at low accuracy. These include pumping, traction and marine drives, dynamometers, self-locking and brake release drives, and many high-power industrial drives. Previous research into flux-position-tracking ZEFIM drives has focused mainly on the signal injection and signal processing techniques [10], [7], [8]. There has been relatively little research investigating the actual closed-loop sensorless performance.

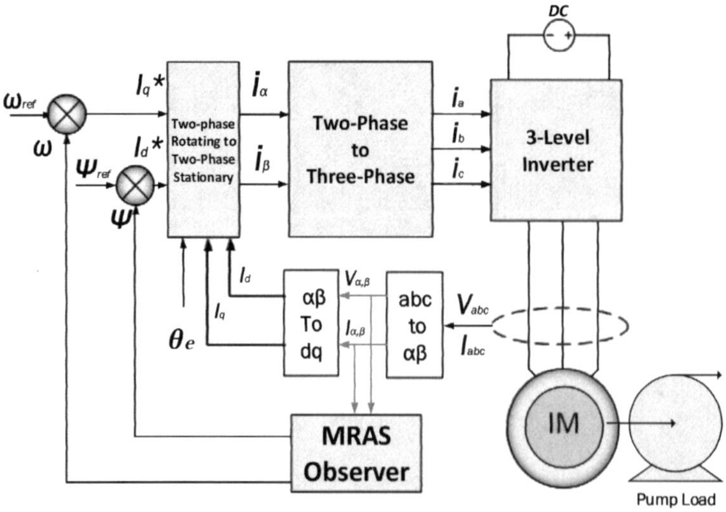

Figure 6: Schematic of Sensorless Controlled Pump Drive

4. Mathematical Modelling of Sensorless Controlled Pump Drive

4.1. Vector Control of Induction Machine

The proposed vector control scheme, the induction machine is modelled in a synchronously rotating reference frame known as rotor flux oriented frame of reference and hence denoted by "RF". The voltage equations of stator and rotor for the induction machine in this frame are;

$$\vec{v}_s^{RF} = r_s\vec{i}_s^{RF} + \frac{d\vec{\psi}_s^{RF}}{dt} + j\omega_{RF}\vec{\psi}_s^{RF} \tag{1}$$

$$\vec{v}_r^{RF} = r_r\vec{i}_r^{RF} + \frac{d\vec{\psi}_r^{RF}}{dt} + j(\omega_{RF}-\omega_M)\vec{\psi}_r^{RF} \tag{2}$$

Where, v, i and ψ denotes the voltage, current and flux respectively. The subscript s & r are for the stator and rotor and ω_M is the rotor mechanical speed. ω_{RF} represents the speed at which reference frame is rotating. The corresponding flux and torque equations can be written as;

$$\vec{\psi}_s^{RF} = L_s\vec{i}_s^{RF} + L_m\vec{i}_r^{RF} \tag{3}$$

$$\vec{\psi}_r^{RF} = L_r\vec{i}_r^{RF} + L_m\vec{i}_s^{RF} \tag{4}$$

$$J\frac{d\omega_M}{dt} = T_e - T_L = \frac{3}{2}.p.Re\,[j.\vec{\psi}_s^{RF}.\vec{i}_s^{RF}] - T_L \tag{5}$$

Where, J is the moment of inertia, T_e is the electromagnetic torque developed and T_L is the load torque. Since the decoupled control is required in vector control, hence the flux equations can be transformed in dq reference frame as;

$$\frac{d}{dt}\begin{bmatrix}\psi_{sd}^{RF}\\ \psi_{sq}^{RF}\\ \psi_{rd}^{RF}\\ \psi_{rq}^{RF}\end{bmatrix} = \begin{bmatrix} -r_s/\sigma L_s & \omega_{RF} & r_sL_m/\sigma L_sL_r & 0\\ -\omega_{RF} & -r_s/\sigma L_s & 0 & r_sL_m/\sigma L_sL_r\\ r_rL_m/\sigma L_sL_r & 0 & -r_r/\sigma L_r & \omega_{RF}-\omega_M\\ 0 & r_rL_m/\sigma L_sL_r & \omega_M-\omega_{RF} & -r_r/\sigma L_r\end{bmatrix}\begin{bmatrix}\psi_{sd}^{RF}\\ \psi_{sq}^{RF}\\ \psi_{rd}^{RF}\\ \psi_{rq}^{RF}\end{bmatrix} + \begin{bmatrix}1 & 0\\ 0 & 1\\ 0 & 0\\ 0 & 0\end{bmatrix}\begin{bmatrix}v_{sd}^{RF}\\ v_{sq}^{RF}\end{bmatrix} \tag{6}$$

And the corresponding current equations will be;

$$i_{sd}^{RF} = \frac{1}{\sigma L_s}\psi_{sd}^{RF} - \frac{L_m}{\sigma L_sL_r}\psi_{rd}^{RF} \tag{7}$$

$$i_{sq}^{RF} = \frac{1}{\sigma L_s}\psi_{sq}^{RF} - \frac{L_m}{\sigma L_sL_r}\psi_{rq}^{RF} \tag{8}$$

$$i_{rd}^{RF} = \frac{1}{\sigma L_r}\psi_{rd}^{RF} - \frac{L_m}{\sigma L_sL_r}\psi_{sd}^{RF} \tag{9}$$

$$i_{rq}^{RF} = \frac{1}{\sigma L_r}\psi_{rq}^{RF} - \frac{L_m}{\sigma L_sL_r}\psi_{sq}^{RF} \tag{10}$$

Now by using the above equations the torque equation in dq reference frame can be assessed as;

$$J\frac{d\omega_M}{dt} = T_e - T_L = \frac{3}{2}\frac{L_M}{L_s}[\psi_{sd}^{RF}i_{rq}^{RF} - \psi_{sq}^{RF}i_{rd}^{RF}] - T_L \tag{11}$$

4.2. Speed and Flux Estimation

The Model Reference Adaptive System (MRAS) is used for the estimation of speed and flux. The reference system consists of the estimation of flux derivatives by using voltages, currents and machine parameters which is independent of speed as;

$$p\psi_{dr} = \frac{L_r}{L_m}(v_{ds} - r_si_{ds} - \sigma L_spi_{ds}) \tag{12}$$

$$p\psi_{qr} = \frac{L_r}{L_m}(v_{qs} - r_si_{qs} - \sigma L_spi_{qs}) \tag{13}$$

Whereas the adaptive system consists of the estimation of flux derivative by using currents and machine parameters which is dependent on speed as;

$$p\hat{\psi}_{dr} = -\hat{\omega}_r\hat{\psi}_{qr} - \frac{1}{T_r}\hat{\psi}_{dr} + \frac{L_m}{T_r}i_{ds} \tag{14}$$

$$p\hat{\psi}_{qr} = -\hat{\omega}_r\hat{\psi}_{dr} - \frac{1}{T_r}\hat{\psi}_{qr} + \frac{L_m}{T_r}i_{qs} \tag{15}$$

The error between these two is further used along with PI controller to estimate the speed. The corresponding equations are;

$$K = \psi_{qr}\hat{\psi}_{dr} - \psi_{dr}\hat{\psi}_{qr} \tag{16}$$

4.3. Centrifugal Pump

The loading characteristics for the centrifugal pumps are dissimilar from other loads. The centrifugal pump load torque T_L can be modelled by using the equations below as;

$$T_L = {}^{g}/_{\omega} \tag{17}$$

$$P_H = 1000 * QgH \tag{18}$$

$$Q = {}^{T\omega}/_{gH} * 1000 \tag{19}$$

Where, the H is the head in meters, Q is the discharge rate in litre/sec, g is gravitational constant, P_H is the hydraulic power and ω is the speed in rad/sec.

5. Simulation of Sensorless Controlled Pump Drive

The figure 7a shows the illustration for the sensorless speed control of induction motor drive for the pumping application. The centrifugal pump model is used to generate the reference speed and torque commands as these are in proportion to each other. The figure 7b shows the Simulink representation for the vector control scheme, where the reference speed & torque commands of the centrifugal pumps used to generate the torque producing quadrature axis component of current and reference flux is used to generate the flux producing direct axis component of current as shown. These components of currents

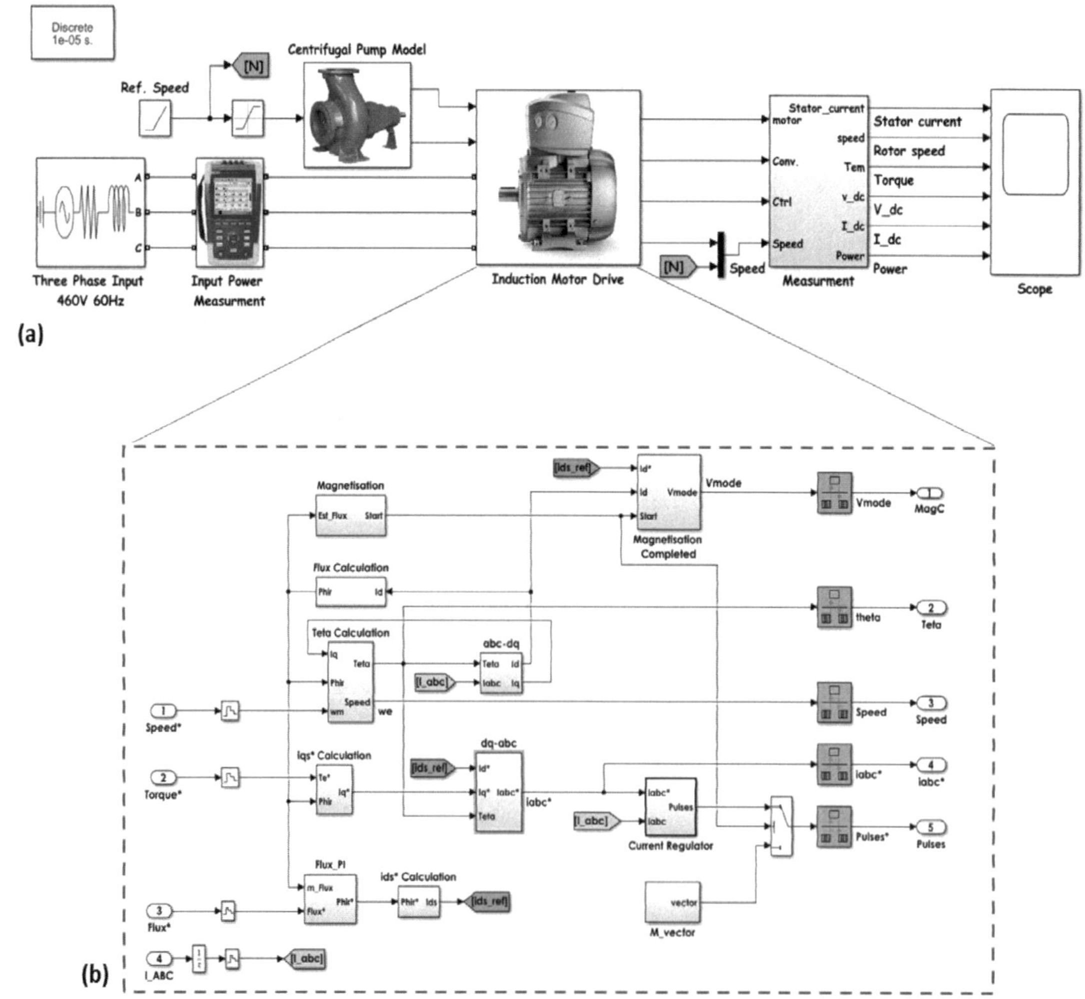

Figure 7: Simulation of Sensorless Controlled Drive with Centrifugal Pump Model
a) Overall simulation model b) Sensorless vector control of induction motor drive

IOP Conf. Series: Materials Science and Engineering **906** (2020) 012029 doi:10.1088/1757-899X/906/1/012029

are further transformed to the three phase systems, where by comparing with the actual three phase currents, the switching pulses are obtained. As the vector control is observed in rotor field oriented reference frame and corresponding to which the angle or rotation is calculated in theta calculation block. This makes the decoupled control for the torque and flux. In the next section, the simulation results for this scheme are observed and discussed.

5.1. Result and Discussion

To analyse the operating performance of the sensorless vector controlled induction motor drive based pumping system, the electrical and mechanical parameters are recorded under different operating conditions, shown in figure 8 for a 200Hp induction machine. The parameters including stator current, speed, torque, dc-link voltage, dc-link current and input power are taken into account for investigation. The centrifugal pump drive is considered for analysis where the torque is directly proportional to the square of the speed, and hence the speed is varied to observe the performance of the drive. Once the drive is energized, the reference speed (1800 RPM) is fed to the control scheme with the acceleration rate of 500 RPM per second, at t=0.5 sec. Thus the speed ramps up from zero speed to 1800 RPM in 3.6 sec, shown in figure 8b. Similarly, the speed reduction comment is provided to the controller at 4.5s, and thereby, the speed is deaccelerated from 1800 RPM to 1300 RPM in 0.2 sec.

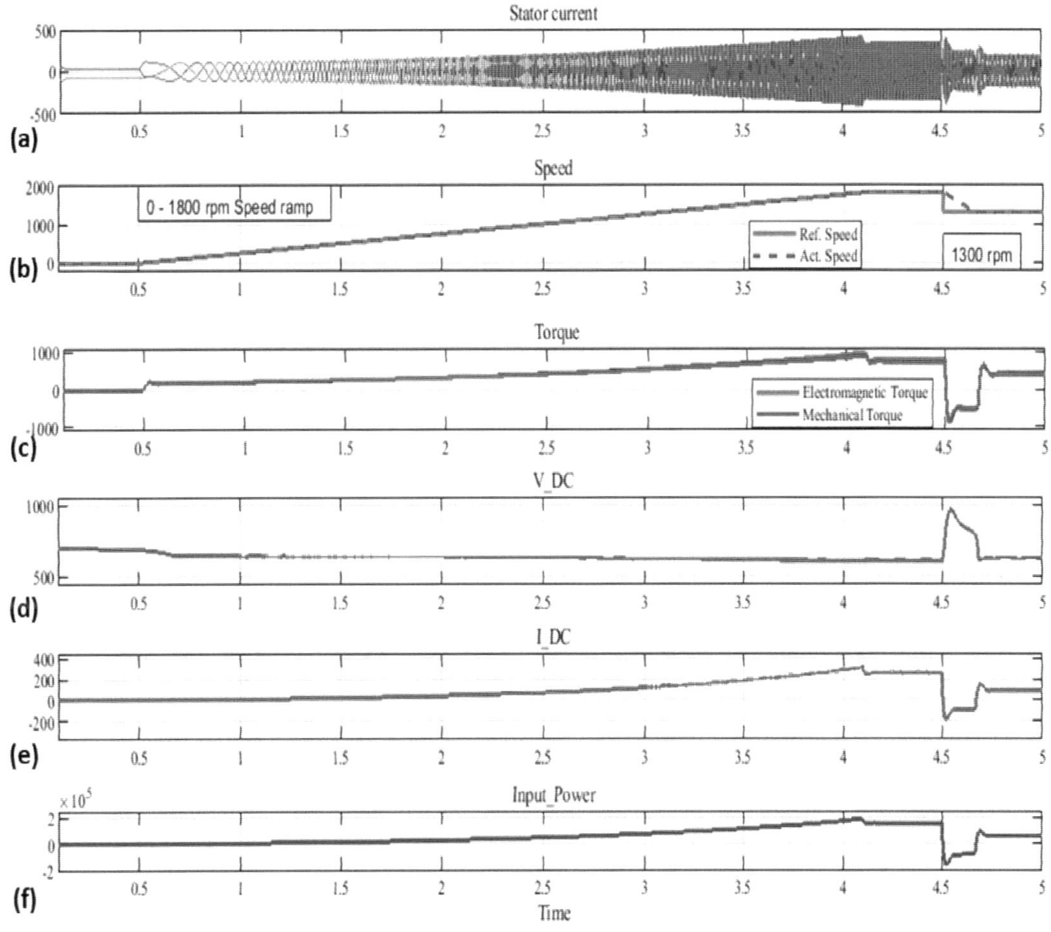

Figure 8: Simulated results of sensorless pump drive
(a) stator current, (b) reference speed and actual speed, (c) electrical torque and mechanical load torque, (d) converter dc-link voltage, (e) converter dc-link current and (f) input dc-link power

During this investigation, the stator current increases gradually as the reference speed is increased to its rated speed, up to time t=4.1 sec. Once the reference speed is achieved, the stator current settles to its corresponding loading current. Later at time t=4.5 sec, when reference speed is decreased, the stator current settles to a new lower value after transients as shown in figure 8a. As in the centrifugal pump, the torque varies in proportion with the speed, the developed electromagnetic torque increases for the first phase of speed transition to maintain the mechanical torque demand, which later at time t=4.5 sec decreases to a new value corresponding to the speed, shown in figure 8c. At time t=4.5 sec, the negative developed torque is observed as the reference speed reduces and hence this power is fed back to the system, which can be utilized for other system requirements. The key measures are dc-link voltage figure 8d and currents figure 8e, where the variation is observed following the loading and hence can be justified. The negative change in the dc-link current at time t=4.5 sec is because power is returned to the system as discussed. figure 8f shows the plot of incremental input power during the initial phase of attending the reference speed, which settles down to a constant value once reached. Later the input power requirement decreases as the reference speed is reduced and the transient at t=4.5 sec, shows the corresponding power back to the source.

6. Conclusion

Industrial pumps operating with variable speed drives offers several benefits including valve elimination, wide range of flow rate, single stage higher head, energy saving etc. The realization of variable speed operation is carried out using an induction motor, power electronic switches, current sensors and controller without speed feedback is known as sensorless variable speed control. A brief review for the VFD drives used in pumps and their speed control algorithms are discussed in this paper. Along with this in this paper an indirect field-oriented sensorless control scheme on 150 kW induction motor drive system is implemented, through centrifugal pump load. The Matlab/ Simulink is used to simulate the scheme and the various results for this type of pump loading are presented. Overall from the annotations it can be concluded that the use of variable speed drives for pumping applications offers advantages in terms of energy saving etc. however further research is supplementary such as finer low & wide ranging speed control, effective consumption of energy during regeneration etc.

References

[1] Harshit Mohan, Mukesh Kumar Pathak & Sanjeet Kumar Dwivedi (2019) Sensorless Control of Electric Drives – A Technological Review, IETE Technical Review, DOI: 10.1080/02564602.2019.1662738.

[2] M. I. Abdelwanis and F. Selim, "A sensorless six-phase induction motor driving a centrifugal pump system," 2017 Nineteenth International Middle East Power Systems Conference (MEPCON), Cairo, 2017, pp. 242-247, doi: 10.1109/MEPCON.2017.8301190.

[3] A. T. Mohsin and I. M. Abdulbaqi, "Analysis of an irrigation pump driver fed by solar PV panel," 2018 1st International Scientific Conference of Engineering Sciences - 3rd Scientific Conference of Engineering Science (ISCES), Diyala, 2018, pp. 92-97, doi: 10.1109/ISCES.2018.8340534.

[4] Z. Massaq, A. Abounada, G. Chbirik, M. Ramzi and A. Brahmi, "Double Stage Solar PV Array Fed Sensorless Vector Controlled Induction Motor for Irrigational Purpose," 2019 7th International Renewable and Sustainable Energy Conference (IRSEC), Agadir, Morocco, 2019, pp. 1-6, doi: 10.1109/IRSEC48032.2019.9078149.

[5] M. T. Joy and J. Böeker, "Sensorless Control of Induction Motor Drives Using Additional Windings on the Stator," 2018 IEEE 9th International Symposium on Sensorless Control for Electrical Drives (SLED), Helsinki, 2018, pp. 162-167, doi: 10.1109/SLED.2018.8486038.

[6] M. Comanescu and P. S. Altoona, "Development of a flux, speed and rotor time constant estimation scheme for the sensorless induction motor drive," 2017 IEEE International Symposium on Sensorless Control for Electrical Drives (SLED), Catania, 2017, pp. 213-218, doi: 10.1109/SLED.2017.8078452.

[7] H. Mohan, S. K. Dwivedi and M. Kumar Pathak, "Control of Induction Motor Drives –

Technological Advancements," 2018 8th IEEE India International Conference on Power Electronics (IICPE), JAIPUR, India, 2018, pp. 1-8.

[8] F. Alonge, F. D'Ippolito and A. Sferlazza, "Sensorless Control of Induction-Motor Drive Based on Robust Kalman Filter and Adaptive Speed Estimation," in IEEE Transactions on Industrial Electronics, vol. 61, no. 3, pp. 1444-1453, March 2014.

[9] F. Alonge, T. Cangemi, F. D'Ippolito, A. Fagiolini and A. Sferlazza, "Convergence Analysis of Extended Kalman Filter for Sensorless Control of Induction Motor," in IEEE Transactions on Industrial Electronics, vol. 62, no. 4, pp. 2341-2352, April 2015.

[10] R. Kumar, S. Das and A. K. Chattopadhyay, "Comparative assessment of two different model reference adaptive system schemes for speed-sensorless control of induction motor drives," in IET Electric Power Applications, vol. 10, no. 2, pp. 141-154, 2 2016.

[11] S. M. Allam, "Performance Analysis and Enhancement of a PV Fed Sensorless BLDC-Motor for an Efficient Water-Pumping System," 2019 21st International Middle East Power Systems Conference (MEPCON), Cairo, Egypt, 2019, pp. 776-783, doi: 10.1109/MEPCON47431.2019.9008174.

[12] A. Pal, G. D. Srivastava and R. D. Kulkarni, "Simulation of Sensorless Speed Control of Induction Motor Using Direct Torque Control Technique Employing Five Level Torque Comparator and Twelve Sector Method," 2019 International Conference on Nascent Technologies in Engineering (ICNTE), Navi Mumbai, India, 2019, pp. 1-6, doi: 10.1109/ICNTE44896.2019.8945936.

IOP Conf. Series: Materials Science and Engineering **906** (2020) 012030 doi:10.1088/1757-899X/906/1/012030

Design and implementation of Sun Tracking Solar Panel and Smart Wiping Mechanism using Tinkercad

[1]Keyur Patel, [1]Sanket Borole*, [2]Ketan Ramaneti, [1]Atharv Hejib, [1]R. Raja Singh

[1]School of Electrical Engineering, VIT Vellore, India
[2]School of Electronics Engineering, VIT Vellore, India
*Corresponding author's email: sanketsubhash.borole2017@vitstudent.ac.in

Abstract. In today's times, electricity has become very essential for mankind. Increasing demand for electricity leads to the depletion of the coal and fossil fuel available to humans. As fossil fuels are bound to get exhausted and also not being a clean source of energy is another added disadvantage. In this era of enormous energy demand, wide research and development of renewable sources of energy is mandatory. The use of solar energy can be advantageous to great extent. To make most of the power from photovoltaic cells, development should focus on getting greater efficiencies using Solar panel arrays. Dust on the surface of the panel and varying intensity of the sunlight contributes to a significant amount for reduction in efficiency. In this paper, an integrated solution for both the dust accumulation problem and the sun's continuous movement has been proposed, which leads to an increase in the efficiency of the panel. The consequences of dust and intensity of sunshine on the efficiencies of the Photo-voltaic panels are highlighted. The paper gives a brief description of the planning and construction of microcontroller-based cleaning and tracking system. The solution proposed is visualized in Tinkercad, which is an online simulator to design CAD models and electronic circuits. The designed model is practically implemented as a hardware prototype for validation.

1. Introduction

Energy is inheriting a crucial role in the world of development and urbanization. The primary source through which power can be harnessed has always been the nature itself. Non-renewable sources of electricity such as coal, oil, and natural gas, are commonly used to operate the base load which is majority of the power usage. However, these resources are non-replenishable. These resources account for majority of the climate change problems faced by the humanity. There is a way out through this. Using renewable energy resources such as solar, wind, hydropower, etc can drastically reduce the affects that the power needs have on the environment. Solar energy has risen as a potential and efficient way to generate electricity in the future. There's nothing wrong in saying that renewable energy will occupy more than 80% of the total energy and nearly 60% will be comprised of solar energy [1]. Amidst the rush of finding sustainable solutions, Solar energy generation is flourishing with great potential. With the development and research in mind, India has promised to achieve 100 GW of solar energy by the year 2022.

Photovoltaic systems are on the rise and also the investments and risks related to these systems are increasing [2]. Solar energy stands for generating energy using the energy of the sun such as heat and light. A photovoltaic cell converts the sunlight into electrical energy through the process of photovoltaic effect. Solar power plants use solar panels which have hundreds of photovoltaic cells in it. The solar panel's array efficiency has to be maximized to make solar energy a largely and widely used source of energy [3]. The generation of electricity using photovoltaic cells has few drawbacks. To get the most out of the solar panels, the panels must be facing perpendicular to the sunlight that is incident on them and also the surface of the panel must have any dust on it. The automatic sun tracking system is a solution for this issue [4]. If the incident light is not perpendicular to the surface, the panel will not give its maximum power output. Real time sun tracking has to be done to get maximum possible efficiency [5]. There is drastic decrease in the power output of the panel, if there is sand accumulated on the surface of the panel [6].

A solution has been proposed in this paper to effectively harness the solar radiation for the majority part of the day and convert it into electricity. An additional feature has been incorporated to clean the surface of the solar panel and solve the dust accumulation problem. Solar panels are placed on house roofs to give more exposure to sunlight. Dust gets accumulated very quickly on the surface. As a result, the intensity of the solar radiation is reduced, and hence the panel's efficiency drops. In [7], the authors predicts system soiling losses of up to 27%. Solving these problems will give enhanced efficiency for the panel and hence, lower payback periods.

A variety of solutions are proposed by researchers in the field of solar photovoltaic generation. Passive tracker solutions have been developed which are dependent on thermal expansion of materials or on the principle of a pressure difference between two points. They are used to having no gears, motors, or sensors in their tracking process. Solar tracker solutions are also classified on the base of the rotational axes/degree-of-freedom. Broadly, they are classified into a single axis and dual axis, out of which dual axis is found to be more efficient [8]. Single-axis trackers have a single axis of rotation and they give 15% to 30% more efficiency compared to the static solar cells. The next classification in the field of solar trackers comes in the form of dual-axis solar tracking. As they have two axes of rotation, they can track the sun throughout the day [9]. They aren't affected by the varying seasons or the movement of the sun. As they have a greater degree of freedom, these types of systems are able to have 40% more efficiency compared to their static counterparts. In [10], the authors have expected to have an increase of 65% in their efficiency using the helianthus model. Though, tracking helps in increasing the efficiency of the system, if dust cleaning and sun tracking is incorporated in one system, a much better efficiency can be extracted [11]. The author presented an automatic solar tracking system in which the author designed a solar panel tracking system based on microcontroller and observed that single axis tracker increases efficiency by 30% compared to the static panel system. In [12] a solar tracker has been made which uses Programmable Logic Controller (PLC) as the controller to track the sun.

To get the peak efficiency of the solar panels, researchers have also explored a few methods to get rid of the dust accumulation on the surface of the panel. Self-cleaning nano-films have been developed which have anti-soiling properties [13]. Nano-film solutions are very useful in regions with water scarcity as they need very minimal water or often no water to clean the surfaces. Another breakthrough in the technology of cleaning PV Panels has been observed using Electrostatic methods [14]. A piece of detachable cleaning equipment is employed for the removal of dust that accumulates on PV panels. It is performed using electrostatic standing waves which were generated using negligible power [15].

In this paper the proposed solution has been implemented on Tinkercad. Tinkercad has a very user-friendly interface for designing circuits and CAD models. It also has a wide range of components such as Arduino UNO, Integrated circuits, and various other electronic components. Along with the software implementation of the model, a hardware model has also been made. The hardware model is aimed to the increase in efficiency of a solar panel by employing Sun-tracking and smart dust wiping Mechanism. Model establishes a proof of concept for the model which can enhance

IOP Conf. Series: Materials Science and Engineering **906** (2020) 012030 doi:10.1088/1757-899X/906/1/012030

the efficiency. Overall organization of the paper is given as follows. Section 2 discusses about the design and implementation strategies of sun tracking and smart cleaning mechanism. In section 3, the prototype development of the proposed sun tracking and smart cleaning mechanism is discussed. Followed by it, the results and discussion with different cases are given in section 4 and finally concluded with future scope in section 7.

2. Design of Sun tracking and smart wiping mechanism

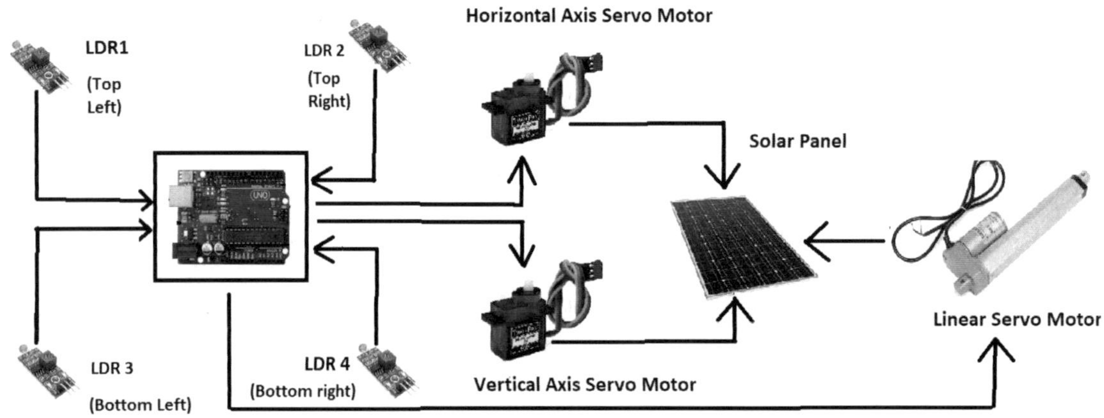

Figure 1. Pictorial representation of sun tracking system with smart wiping mechanism

Pictorial representation of sun tracking system with smart wiping mechanism is shown in Figure 1. As it is evident from Figure1 a total of four LDR's are present which sense the light falling on them from four different directions. The intensity values are given to the controller. The controller analyses those values and gives the output to the corresponding servo motors to move in the respective direction for the required angle. These actuators then turn the solar panel to maximize the sunlight falling on it. For smart cleaning mechanism a linear servo motor is connected which would be programmed such that after a duration of four hours the wiper connected would wipe the panel.

The sun tracking system with smart wiping system is designed in Tinkercad online environment which is shown in Figure 2. Each of the components used has been specified in the table above. In the circuit above, 4 LDRs have been connected to take the intensity of the sunlight from all the directions. The LDRs pin are connected to the analog pins of Arduino UNO control board. A better resolution of the values of the intensities can be received by using the analog pins. Two servo motors namely, horizontal servo motor and vertical servo motor is connected to the Pulse Width Modulation pins of the microcontroller. Along with the tracking actuators, the cleaning actuator is connected to the DC output of the microcontroller to get constant motion.

2.1 Sun tracking and wiping mechanism

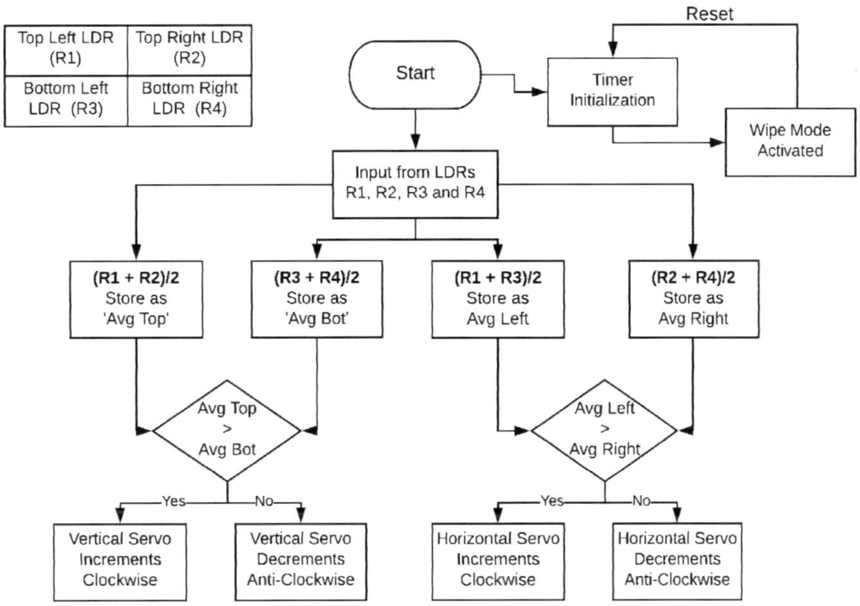

Figure 2. Flowchart for sun tracking mechanism

The algorithm is made to sustain autonomously. The sun tracking algorithm works by comparing the average value of sunlight falling on the 4 LDRs. The flowchart shown above is a clear depiction of the algorithm coded on the controller. The terminology of the LDRs based on the location. Four distinct LDRs are named as follows: top left, top right, bottom left, and bottom right.

Initially the algorithm calculates the average of the intensity of light falling on top, bottom, right, and left, respectively which gives a sense of orientation of Sun. The algorithm starts with comparing the average intensity of top and bottom LDR, i.e., comparing the average of top left and top right LDRs with the average of the bottom left and bottom right LDRs. If the average intensity of the top LDRs is less than the average intensity of the bottom LDRs, the vertical servo rotates anti-clockwise. The vertical servo rotates clockwise when the average intensity of light on the top is more than the average intensity of sunlight at the bottom.

Similarly, If the average intensity of the left LDRs is less than the average intensity of the right LDRs, the horizontal servo rotates anti-clockwise. The horizontal servo rotates clockwise when the average intensity of light on the left is more than the average intensity of sunlight at the right.

Here the linear motor is attached on the edge of the panel. The wiper is arranged in such a way that it can clean the panel's surface in a single swipe. The controller has been programmed such that after an interval, of 4 hours in this case, the controller would power the linear servo motor to perform a certain number of wipes.

IOP Conf. Series: Materials Science and Engineering **906** (2020) 012030 doi:10.1088/1757-899X/906/1/012030

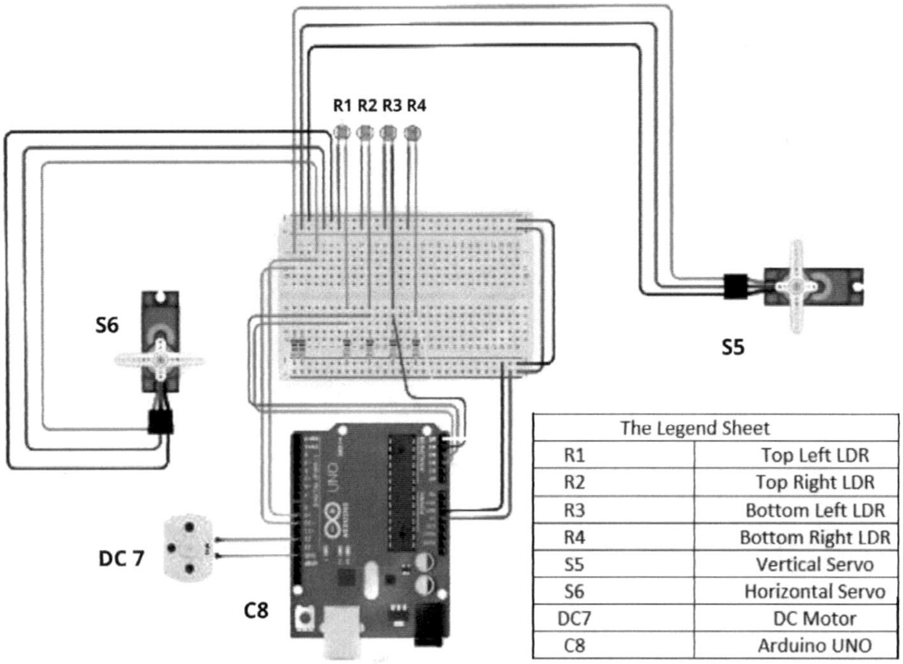

The Legend Sheet	
R1	Top Left LDR
R2	Top Right LDR
R3	Bottom Left LDR
R4	Bottom Right LDR
S5	Vertical Servo
S6	Horizontal Servo
DC7	DC Motor
C8	Arduino UNO

Figure 3. Design of sun tracking system with smart wiping mechanism

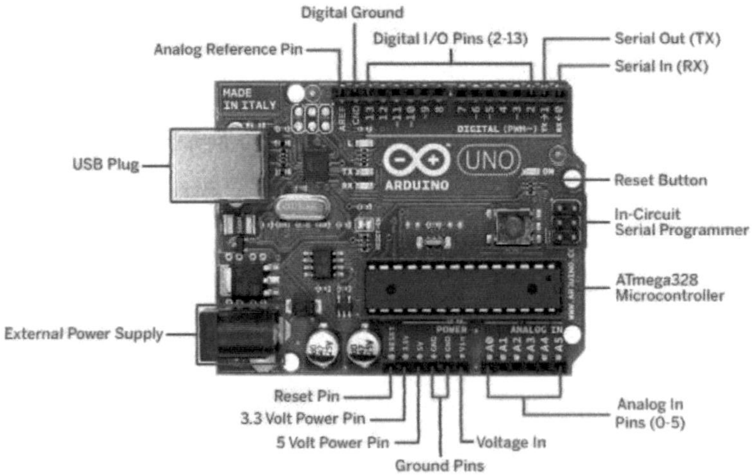

Figure 4. Pin diagram of Ardiuno UNO

3. Implementation of sun tracking and wiping mechanism

The prototype is built using two dc servo motors for tracking and a linear servo motor used for the cleaning mechanism. The servo motors are separately powered by the controller itself due to the low power output of the solar panel. But, for practical purposes, the servo motors cannot be directly powered by the controller because the motors (~5 to 6 Watts) can damage the controller. The wiper is placed on the edge of the panel. The wiper wrapped with a piece of fabric acts as a mop to draw the dust out of the panel's surface. These dc motors controlled are using the microcontroller through sophisticated control algorithms. The angular servo motor used for the sun-tracking mechanism is employed to act as a robotic arm. One servo motor is used for the Yaw movement, and the other servo is used for the Pitch movement of the panel.

The prototype of the sun tracking solar panel with smart wiping is implemented with various hardware components. *The* description *and* detailed function of each component *are* as follows.

3.1. Arduino UNO Controller
Arduino UNO microcontroller has been used to control the overall process. It has fourteen input/output pins. Among them six pins are dedicated to PWM or Pulse Width Modulation. Because of the ease of use, Arduino UNO has been considered the ideal controller for the prototype. The function of the controller is to track the Sun's movements and send the signal to the actuators to rotate accordingly. Along with sun tracking, the controller also performs the self-cleaning task through the movement of an external actuator.

3.2. Solar Panel
The rating of the prototype solar panel is 6V, 100 mA. The panel is attached to the rotary actuator. A solar panel has numerous photovoltaic (solar) cells arranged in an order. When the sunlight is incident on the surface of the photovoltaic cell, the energy is transformed by this cell into electrical potential across the terminals, i.e., electrical power. The absorbed energy makes the electrons jump from one energy level, i.e., orbit to another energy level inside the atom, thus releasing energy. For residential purposes, panels with ratings of 24V/330 W are normally used.

3.3. LDR
Light Dependent Resistors are the critical element in the prototype. LDR is a type of resistor whose resistivity is inversely proportional to intensity of sunlight incident on the surface. That means, in the dark when the intensity of light is negligible the LDRs show high amount of resistance, whereas when the intensity of light is higher in the day the LDRs present low resistance. The LDRs are placed on the edge of the panel and move along with the panels to track the sun. While using LDRs, the value is taken as an analog input to get an accurate value when the reading values are combined.

3.4. Actuator – Servo Motor
Servo motor-based actuator is used in this prototype. The role of servo motor in the prototype is to rotate the panel as per the control signals. Rotating the servo motor is done using a PWM signal provided to the motor's control pin. The controller generates the PWM signal. The control system used to operate the servo motor implies an error-sensing feedback control, which helps the servo motor to precisely land on the desired angle. It usually requires a sophisticated controller. Servo motors are classified in 2 variants, which are DC Servo motors and AC servo motors. In this sun-tracking mechanism, the type of motors used is the DC Servo motor as massive torque is not required. The servo motors used have a range of 0-180°. For the cleaning mechanism, a linear servo motor has been used, which has a precise linear movement that can be calculated and precisely controlled using the microcontroller. Adapting the hardware components, the prototype model has been built to track the solar radiations to extract the maximum power.

In the prototype, a dual-axis sun tracker is being made. For the dual-axis tracker, 4 LDRs have been incorporated. Each LDR is placed on four different corners of the panel. One set of LDRs used is to give pitch movement signals, and the other is for the yaw movement signal. As two sets of sensors are present, two different function blocks have been used to compare the intensity of light falling on LDRs. In the morning, the set of LDRs measuring the intensity of sunlight for the East-west arc gets low resistance from the LDR facing east. The LDR facing the west shows a high value of resistance. The signal compared is in the function block, and it depicts that solar radiation is more on the east.

IOP Conf. Series: Materials Science and Engineering **906** (2020) 012030 doi:10.1088/1757-899X/906/1/012030

Hence, the panel inclines itself towards the east using the servo motors. Here, the tracking is done in fixed intervals of 1 hour only. Instantaneous tracking using the actuators would lead to more power consumed. Hence, including instantaneous tracking would lead to reduction in net efficiency of the system. Doing this will reduce the power usage for the Servo motor.

Table 1 consists of the technical specifications of the components used and their cost has been shown below to give an overall idea of the cost of the prototype. The hardware prototype is shown below in the Fig 5.

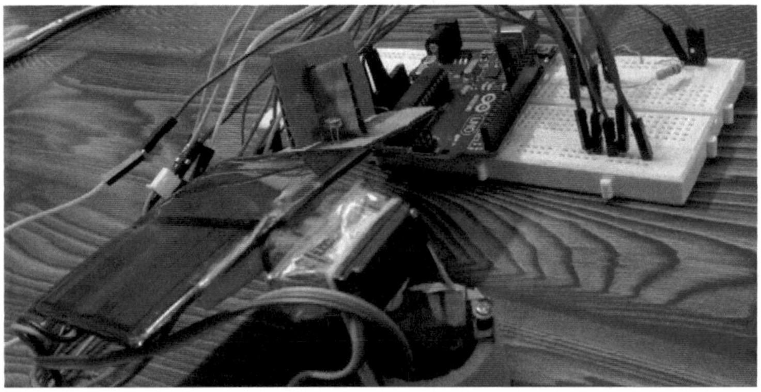

Figure 5. Sun tracking and wiping prototype model

Table 1: Technical Specification with cost description

Components	Specifications	Quantity	Total Cost
Solar Panel (70x70x3mm)	(6V/100mA)	1	160 INR
Light Dependent Resistor (LDR)	N/A	4	40 INR
Servo Motor SG90	Operating Vol:3.0V - 7.2V Stall Torque: 1.2 kg-cm	2	200 INR
Arduino UNO	ATmega328P – 8-bit AVR family Operating Voltage: 5V	1	350 INR
Linear Servo Motor	Voltage: DC 3-6 V Current: 100 mA-120mA	1	250 INR

4. Results and Discussion

The implementation of the Sun tracking solar panel was successfully carried out. The initial position of the solar tracker is depicted in Figure 3. For demonstration purpose it is considered that the servo motors to be along two axis which are horizontal and vertical. The horizontal servo would give the panel an east-west direction motion and on the other hand the vertical servo would give it a north-south direction motion as and when required. The vertical axis comes into picture in the higher latitude regions.

4.1. Case 1:

On increasing the intensity falling on the top right and bottom right LDRs the sunrise situation has been simulated. The horizontal and vertical servo motors are at the initial position. Considering that the Sun's position is on the right side of the panel; the average value of the right LDRs is greater than the average values of the Left LDRs. The average top, average bottom, average left and average right is calculated. There is no incident light falling on the Top left and bottom left LDRs. The difference between average top and average bottom is very small and hence there is an anti-clockwise drift in the vertical servo, whereas the average right value is greater than average left. Hence, the angle of the servo motor is incremented until the difference between average right and average left doesn't become zero. The direction of rotation of servo motor is also shown with the help of arrows.

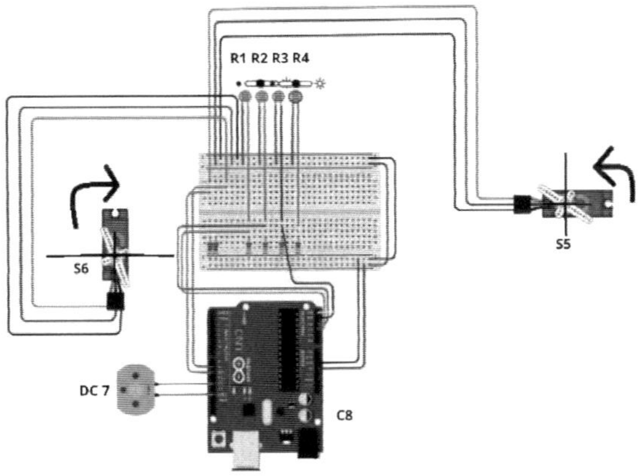

Figure 6. Case 1

(Top right and bottom right)

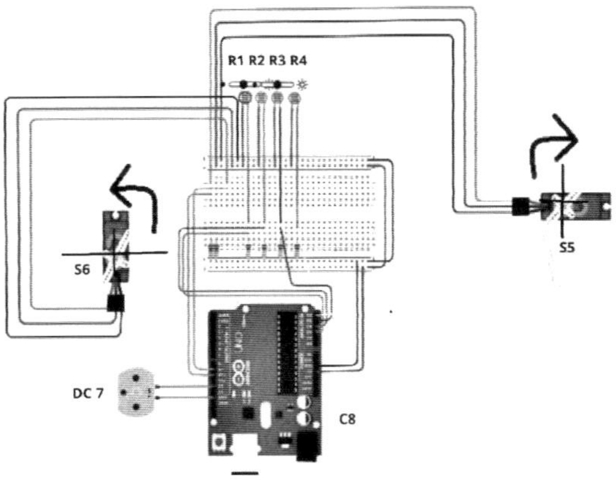

Figure 7. Case 2

(Top left and bottom left)

4.2. Case 2:

On increasing the intensity falling on the top left and bottom left LDRs the sunset situation has been simulated. The horizontal and vertical servo motors are at the initial position. Considering that the sun's position is on the left side of the panel; the average value of the left LDRs is greater than the average values of the right LDRs. The Average top, average bottom, average left and average right is calculated. There is no incident falling on the Top right and bottom right LDRs. The difference between average top and average bottom is very small and hence a clockwise drift is seen in the vertical servo, whereas the average left value is greater than average right. Hence, the angle of the servo motor is decremented until the difference between average right and average left doesn't become zero.

4.3. Case 3:

On increasing the intensity falling on the top right, top left, and bottom right LDRs 'Sun at the North-east' situation has been simulated. The horizontal and vertical servo motors are at initial positions; considering that the sun's position is on the top-right side of the panel; the average value of the right LDRs is greater than the average values of the Left LDRs and the average value of the top LDRs is greater than the average values of the bottom LDRs. The Average top, average bottom, average left and average right variables are calculated. There is very little light falling on the bottom LDRs and the left LDRs. Hence, average top is greater than average bottom, and the average right value is greater than average left. Hence, the angle of the horizontal servo motor is incremented until the difference

IOP Conf. Series: Materials Science and Engineering **906** (2020) 012030 doi:10.1088/1757-899X/906/1/012030

between average right and average left doesn't become zero. And, the angle of vertical servo motor is incremented until the difference between them doesn't become zero.

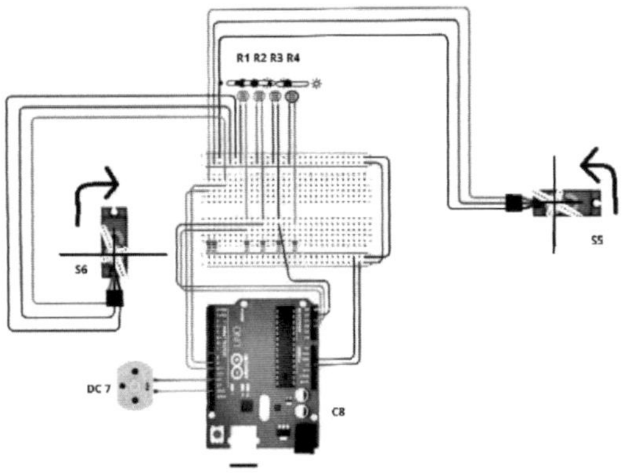

Figure 8. Case 3

(Top right, bottom right and top left)

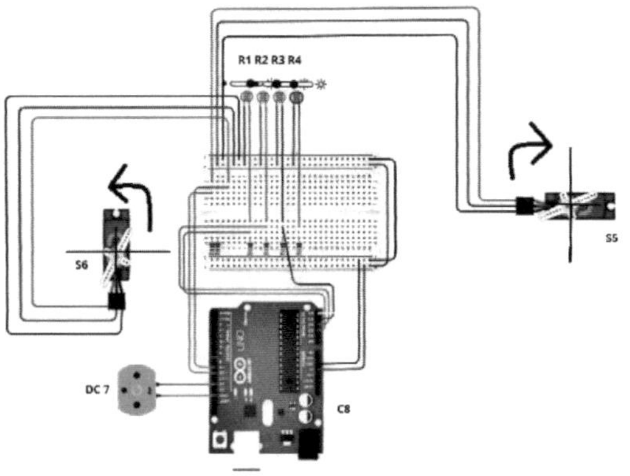

Figure 9. Case 4

(Bottom right, bottom left and top left)

4.4. Case 4:

On increasing the intensity falling on the bottom right, Bottom Left, and Top Left LDRs 'Sun at the South-west' situation has been simulated. The horizontal and vertical servo motors are at initial positions; considering that the sun's position is on the bottom-left side of the panel; the average value of the left LDRs is greater than the average values of the right LDRs and the average value of the bottom LDRs is greater than the average values of the top LDRs. The Average top, average bottom, average left and average right variables are calculated. There is very little light falling on the top LDRs and the right LDRs. Hence, Average bottom is greater than average top, and the average left value is greater than average right. Hence, the angle of the horizontal servo motor is decremented until the difference between average right and average left doesn't become zero. And, the angle of vertical servo motor is decremented until the difference between them doesn't become zero.

5. Tabulation

Table 2: Output comparison between proposed and conventional systems

Time	Uncleaned w/o tracking (Volts)	Cleaned w/o tracking (Volts)	Uncleaned with tracking (Volts)	Cleaned with tracking (Volts)
6:00	1.29	2.39	1.78	2.71
7:00	1.41	2.51	2.00	3.03
8:00	1.66	2.63	2.07	3.10
9:00	1.83	2.78	2.17	3.22
10:00	2.10	3.18	2.34	3.37
11:00	2.22	3.25	2.54	3.47
12:00	2.40	3.50	2.59	3.54
13:00	2.24	3.32	2.49	3.49
14:00	2.17	3.10	2.32	3.42
15:00	1.90	2.83	2.05	3.25
16:00	1.63	2.59	2.02	3.17
17:00	1.46	2.54	1.85	3.05

18:00	1.24	2.41	1.73	2.78

The hardware prototype which was built according to the concept mentioned was used to take the readings of the voltages given above in Table 2. Readings are taken for a cleaned panel with and without tracking and also for an unclean panel with and without tracking. The results were noted down for an entire day from 6AM to 6PM. By observing the output values, it can be seen that there is a notable difference when sun tracking and cleaning mechanism is applied as shown in Fig 10. There is a 64.3% increase in voltage compared to a dusty and non-solar tracking system. Also, it was observed that the sun tracking system does not show a significant difference for <10% of sun's movement. The increase of voltage is very small and thus for rotating the servo motor for small angles, there is a loss of energy in running them. Hence, the motor is considered to operate after 1 hour, to get significant change in movement and voltage values.

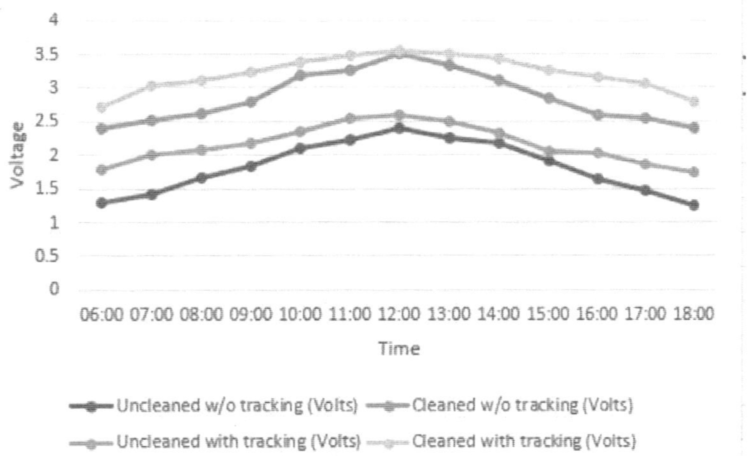

Figure 10. Output comparison between proposed and conventional systems

6. Conclusion

Solar energy is one of the most promising energy sources for the future. The challenge now is to maximize the utilization of this solar energy by absorbing it. Utilizing the Sun tracking solar panel, the results have shown that the efficiency of the system can be increased considerably. The system with the design and algorithm has been presented in the paper. The implementation was successfully carried out using Tinkercad software and the servo motor angles were also presented in the tabulation column. The motion of the servo motor was successfully visualized and controlled using the intensity values on the LDR. The automatic cleaning mechanism was visualized using the DC motor in the Tinkercad software.

7. Future Scope

Optimised by using Fuzzy Logic algorithm to increase the precision of the actuator movement. The proposed model in this paper uses external power to actuate the circuit, hence using its own power for the servo movement can make it a more reliable and increase monitoring and control. The cleaning mechanism can be improved by comparing the solar PV output with the predicted values to increase the efficiency.

8. References

[1] International Energy Anergy Agency (IEA). Technology Roadmap: Solar Photovolaic Energy. 2010

[2] M. Catelani, L. Ciani, L. Cristaldi, M. Faifer, M. Lazzaroni, and M. Rossi *Characterization of photovoltaic panels: the effects of dust* in Energy Conference and Exhibition (ENERGYCON), 2012 IEEE International, 2012, pp. 45-50.

[3] I. Nayshevsky, Q. Xu, G. Barahman and A. Lyons *Anti-reflective and anti-soiling properties of a KleanBoost, a superhydrophobic nano-textured coating for solar glass* 2017 IEEE 44th Photovoltaic Specialist Conference (PVSC), Washington, DC, 2017, pp. 2285-2290, doi: 10.1109/PVSC.2017.8366777

[4] Mo Yibo, Yang Liang, Huang Liuyan, Xu Qiongying, Chen Haifeng *Review of various solar power generation technologies* Dongfang Electric Review, 2018, 32(125):78- **82.**

[5] Mayank Kumar Lokhande (2014) *Automatic Solar Tracking System* Journal of Core Engineering & Management, Volume 1

[6] N. Bose, Arindam & Sarkar, Sounak& Das, Sayan. (2012). Helianthus – ***A*** *Low Cost High Efficient Solar Tracking System Using AVR Microcontroller* International Journal of Scientific and Engineering Research. 3..

[7] Maghami, Mohammad Reza &Hizam, Hashim & Gomes, Chandima&MohdRadzi, MohdAmran&Rezadad, Sina&Hajighorbani, Shahrooz. (2016) *Power Loss Due to Soiling on Solar Panel: A review. Renewable & Sustainable Energy Reviews* 59. 1307-1316. 10.1016/j.rser.2016.01.044.

[8] A. Kimber, L. Mitchell, S. Nogradi and H. Wenger *The Effect of Soiling on Large Grid-Connected Photovoltaic Systems in California and the Southwest Region of the United States* 2006 IEEE 4th World Conference on Photovoltaic Energy Conference, Waikoloa, HI, 2006, pp. 2391-2395, doi: 10.1109/WCPEC.2006.279690.

[9] B. K. S. Vastav, S. Nema, P.Swarnkar, and D. Rajesh *Automatic Solar Tracking System using DELTA PLC* Conference 2016 International Conference on Electrical Power and Energy Systems (ICEPES), 2016.

[10] B. Abhilash and A. K. Panchal *Self-cleaning and tracking solar photovoltaic panel for improving efficiency* 2016 2nd International Conference on Advances in Electrical, Electronics, Information, Communication and Bio-Informatics (AEEICB), Chennai, 2016, pp. 1-4, doi: 10.1109/AEEICB.2016.7538291.

[11] Ma Huihui, Bai Lianping *Dual-axis tracking of solar energy panels for maximum power* Journal of Beijing Information Science and Technology University, 2011, 26(5):93-**96.**

[12] V.R. Sushma, V. M. Sneha *PLC Based Solar Tracking Panel Assembly* International Journal of Engineering Trends and Technology (IJETT) – Volume 18 Number 5, pp 230-234, 2014.

[13] N. Barsoum and P. Vasant *Simplified solar tracking prototype* Global Journal of Technology and Optimization GJTO, vol. 1, pp. 38-45, 2010

[14] P. Rani, O. Singh and S. Pandey *An Analysis on Arduino based Single Axis Solar Tracker* 2018 5th IEEE Uttar Pradesh Section International Conference on Electrical, Electronics and Computer Engineering (UPCON), Gorakhpur, 2018, pp. 1-5, doi: 10.1109/UPCON.2018.8596874

[15] H. Kawamoto and M. Kato *Electrostatic Cleaning Equipment for Dust Removal from Solar Panels of Mega Solar Power Generation Plants* 2018 IEEE 7th World Conference on Photovoltaic Energy Conversion (WCPEC) (A Joint Conference of 45th IEEE PVSC, 28th PVSEC & 34th EU PVSEC), Waikoloa Village, HI, 2018, pp. 3648-3652, doi: 10.1109/PVSC.2018.8547468.

IOP Conf. Series: Materials Science and Engineering **906** (2020) 012031 doi:10.1088/1757-899X/906/1/012031

4G Signal Booster

P. Kavipriya[1], M.R. Ebenezar Jebarani[2], G. Jegan[3], P. Chitra[4] S. Lakshmi[5]

[1,2,3,4,5] Department of ECE, Sathyabama Institute of Science and Technology, Chennai -119

ecekavipriya@gmail.com

Abstract. A system which is capable of or adapted for turning easily from one to another of various tasks, fields of endeavor have been evolved since 1980's as time changes everything is of higher execution, higher limit and simplicity of getting to from anyplace on the world. There is a fast progression in remote correspondence innovation giving the system benefits anyplace and at whatever time 4G correspondence frameworks are being created to take care of the different issues the present correspondence frameworks (3G, 2.5G) are facing. In this paper we give a simple booster which works on the present network 4G.We talk about the innovations that exist in market and can be utilized to upgrade the system data transmission and execution of the right now living innovation in the market. Some of regions were ineffectively secured by wire line gadgets because of significant expense. 4G will be a canny innovation that will decrease the quantity of various advances to a solitary worldwide standard. This would lead to more efficient 4G mobile networks. As previously used Specifically, An artificial intelligent technique involving the use of Bat Algorithm (BA) and Artificial Neural Network (ANN) have been proposed. Nowadays everyone is preferred wireless then wired connection and everyone need wireless connectivity. It is trusted that this will fill in as a one-stop look for specialists and engineers in the significant field of remote sign promoters and extenders, who wish to realize what is accessible and existing difficulties.

1. Introduction

The time of remote framework had progressed from 1980's. Directly in the present day we can see more rapidly creating framework which is thought to comprehend the data correspondence with a speedier data rate and having far reaching framework scope around the globe. The season of remote framework started with unique (1G) towards 2G, 3G and 4G. A present survey in framework development advancement says that there will be enormous improvement in framework action by the year 2020. It is surveyed that number of supporters will create by 10 times and consistently there will be 100 times higher development in framework utilize [1]. There are more than 5 billion devices being utilized on the planet and it is well ordered extending well ordered.. Without range, no remote media transmission or remote web administrations would be conceivable. Radio invention is not psychological, it likewise incorporates a progressive change in how the range is controlled. Intellectual Radio and 4G are two reciprocal advancements that will reframe the universe of remote correspondences. 4G systems utilizing subjective radios are an answer that alters the media transmission industry, essentially changing the way we plan our remote frameworks and administrations. Modern 4G cell phones operate over a wide variety of bands, from 850 MHz to 2.5 GHz, in 4G standards .This places a lot of challenges on the cell phone antenna since it needs to be matched and with relatively optimum gain over a very wide frequency range. A matched antenna also results in higher total radiated power since the power amplifier operates in a more efficient manner when loaded. This is especially true nowadays with the linearization techniques applied in modern-day cell-phone power amplifiers. It has made possible services such as online banking, social networking, online markets, internet job markets and etc. All of these have contributed to making the Telecommunication Industry a major contributor to the gross domestic product (GDP) of many countries both the developed and the developing. In general, wireless communications has cut down the cost of living and doing business in several ways [2], [3].

2. Literature Summary

This section represents the survey for classification from previous references. There are many planning and previous references that are given in order to develop a different set of parameters and different types of algorithms are used to perform different types of techniques are followed in order to get the required output in different ways.

2.1 **Kevin Ming-Jiang Ho** et al presented the impedance matching networks. It is seen that a tunable matching network can improve the cell phone system noise figure and total integrated sensitivity (TIS) by at - 1 dB at 700 MHz - 3 GHz. The reason is that a mismatched low-noise amplifier has a higher noise figure and a lower gain than a noise- matched LNA, and therefore, a tunable matching network is important for noise matching in receiver applications. The project calculations based on realistic cell phone assumptions showing that a tunable impedance network capable of matching an antenna VSWR of 10 to a VSWR of 2.6 will result in 1dB improvement in the system NF and total integrated sensitivity. This work was supported by the UCSD Center for Wireless Communications.

2.2 **Elizabeth N. Onwuka** This paper presented a survey of cellular signal boosters. It made a detailed over view of cellular boosters, highlighting its operation, and pointed out issues for consideration in antenna selection. It discussed the design and operational working principles, cost determinants, good and poor booster technologies. It specifically discussed boosters fit for local environments like Nigerian, and mentioned some basic booster specifications. Research efforts in booster design, deployment and analysis across the generations of mobile communications were outlined and briefly discussed. All of these culminated to a look at future work where a signal booster design for given local environment is considered appropriate

2.3 **Divvela Santhoshn**the Tri-Band Cellular Repeater consists of Bidirectional amplifier, receiving and transmitting antennas. This paper discusses our assembling process, beginning with component selection and our difficulty in obtaining the required gain according to the user requirement in the process of testing. This cellular repeater can be operated in 3 different operating frequency bands namely, GSM 900, DCS and 3G. Hence, practically also proved the importance of repeater in enhancing the signal. Finally, we conclude by saying that repeaters can improve the signal strength in low signal coverage areas to such an extent that everyone can receive signals without having any disturbances.

2.4 **Salihu Oladimeji** The design, implementation, and testing of a GSM signal booster was carried out. The signal booster works in both the 900 and 1800 MHz bands. It was successfully tested on the four mobile networks available in Nigeria. The GSM signal booster achieved a maximum percentage boost of 73.59% on 2G network for which it was designed In situation where the signal is in a deep fade such that no signal is received by the outdoor antenna; the booster will not boost. However, as long as

IOP Conf. Series: Materials Science and Engineering **906** (2020) 012031 doi:10.1088/1757-899X/906/1/012031

there is available signal, not less than the sensitivity of receiving antenna, the booster will pick it up and boost it for mobile stations' reception

3. Existing System:

Telecommunication is increasing day by day by increasing the functions in telecommunications the disadvantage are mainly there are two types of disadvantages they are radiation and signal breakdown. The radiation is big problem in telecommunication sector. We want to fix the problem of radiation. The signal strength has to be decreased. If the signal strength is decreased the message signal strength did not have more capacity to travel longer distance apart of this problem.

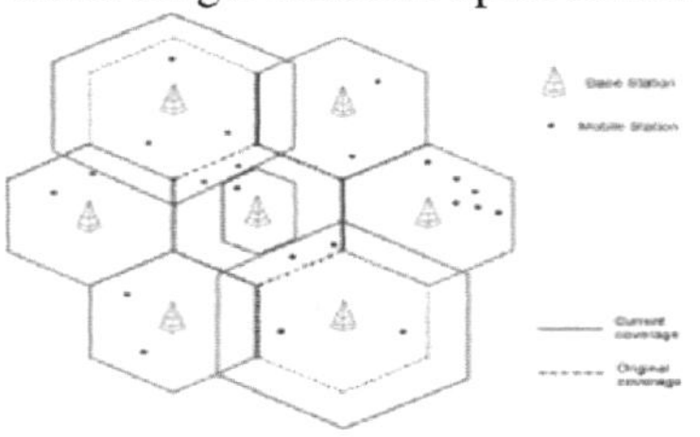

Fig 3.1 Existing System

There is another problem which telecom operator proposed some solution to fix this one [5]. The problem is signal strength decreases when it travels to longer distance so what telecom operator did is they increase the cell size so that the area present in the cell have the chance to get high capacity of signal. Even through if the number of users are high the signals automatically distracted so that high capacity of antenna are used to receiving end to get signal. The existing system which was done by telecom operator are increasing cell size

4. Methodology:

The signal is coming from the towers are high intensity so that by using Radio Frequency circuit it should be converted to the low intensity signal next it will send to be Band pass filter so that 4G bandwidth range signal are transferred to the power amplifier now power amplifier is used to amplify the signal so that the data present in the signal is recreated from that antenna transmits the signal at the destination path.

5. Proposed System

In communication by improving the telecommunication department by improving in the communication field there are many disadvantages mainly the radiation. Now from the earlier stage the telecommunication department faces a lot of problems especially the message signal transfer for longer distances. In India the telecommunication sector is trying to reduce this problem by increasing the frequency where there is a chance for the signal to travel long distances. By increasing the frequency means stringing the signal which leads to radiation. So the telecom sector decides the bandwidth range in the form of generations. There are 2G,3G,4G, LTE mode [3], [7]. At present we were in the 4G generation. In India, there are mainly 5 telecom operator namely, Airtel, BSNL, IDEA, JIO, Vodafone. Now coming to the point these 5 telecom operators are going to give the bandwidth. So Telecom Regularity of India has decided the 4g ranges and gives the tender, bids to these operators. As the signal decreases due to longer distances or a more number of users So this point of the era is raised in the operators. So they increased the cell size.

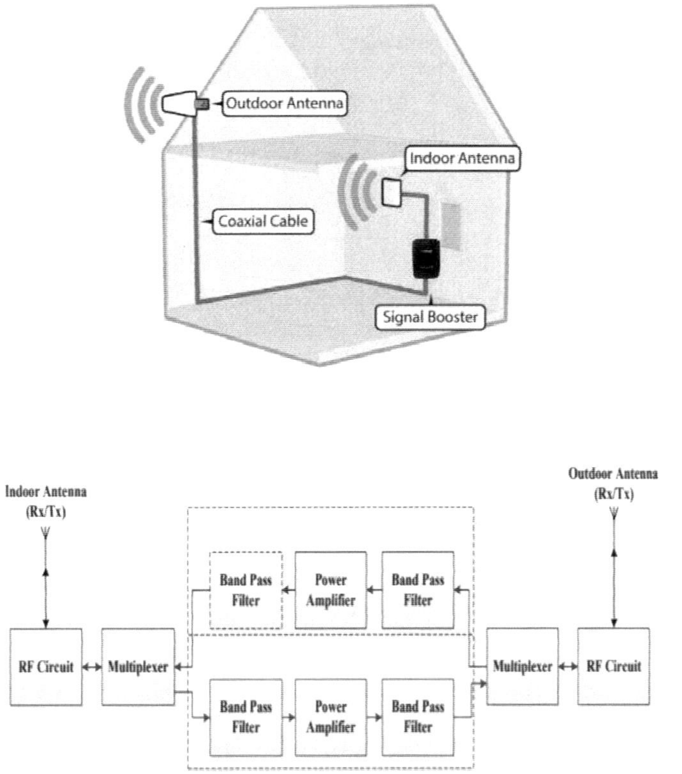

Fig 5.1 Proposed System

For a large range of areas a strong signal is providing by the operators. Even though there is a problem the thickness of wall increases leads to weakening the signal. To this a large definition of an antenna is used. It is difficult to implement. So at that time there is where the implementation of the booster arrived [6], [8].

These are less costly, less radiation, easy to implement, no need to study how to use this. As the name indicates it's boosting the message signal. The message signal from the operator is come to our receiving point due to the wall density there is a disturbance in the call or damage the message signal or weaken the signal. So an antenna is equipped in the terrace which is mainly used to collect the signal from the antenna [10]. Now the collected signal is weakened so that it undergoes some steps to back the original signal. Here the antenna which we used in the boosters is patch panel antenna. The main advantage of this antenna is Omni directional which means it collects the signal from any direction that's the main advantage of the antenna. It is less cost and lightweight. The collected signal from the antenna sends to the RF circuit. The RF circuit is used due to the high energy message signal is converted to a suitable powered signal [5], [9]. At last the recovered signal is again converted to the high energy message signal by using another RF circuit. The Bandpass filter is used to detect the range of 4G. Here there are so many signals so that we were designing the 4G signal booster so that it only collects the 4G range bandwidth signal. At last, the power amplifier is used to amplify the weak signal so that it back to the original signal now this message signal makes to original strong signal [11].

IOP Conf. Series: Materials Science and Engineering **906** (2020) 012031 doi:10.1088/1757-899X/906/1/012031

6. Results and Discussion:

The circuit which is tested in Two locations. The Location 1 (shollinganallur) signal strength status before and after boosting of different telecom operator.

Network Provider	Signal Before Boosting (-dBm)	Signal After Boosting (-dBm)
AIRTEL	113	105
IDEA	112	109
BSNL	83	79

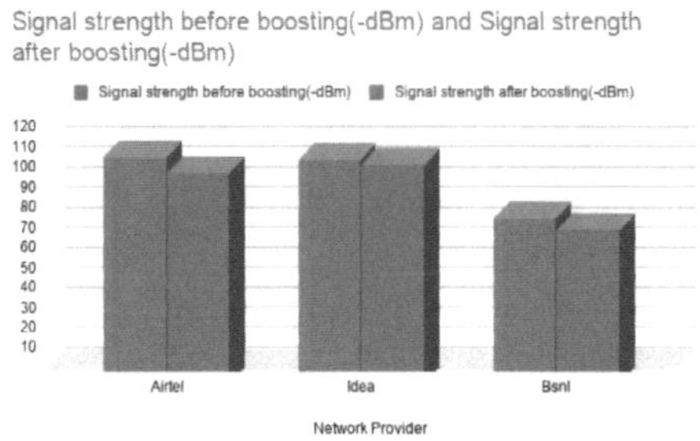

Fig 6.1 Comparison Chart before and after boosting at first location

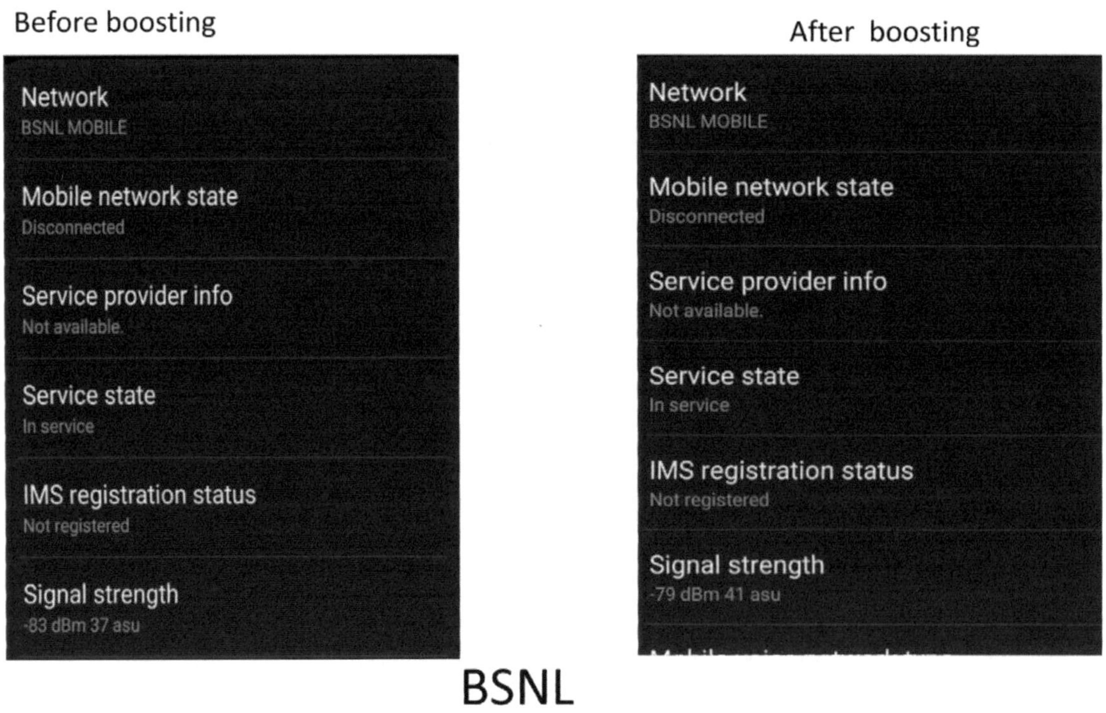

Fig: 6.2 BSNL signal strength status of the BSNL telecom operator at Shollinganallur location.

Analysis of signal strength status so from the above bar graph we concluded that the after boosting the signal strength is slightly varied.

The Location 2 signal strength status before and after boosting of different telecom operator.

Network Provider	Signal Before Boosting (-dBm)	Signal After Boosting (-dBm)
AIRTEL	98	89
IDEA	97	82
BSNL	96	85

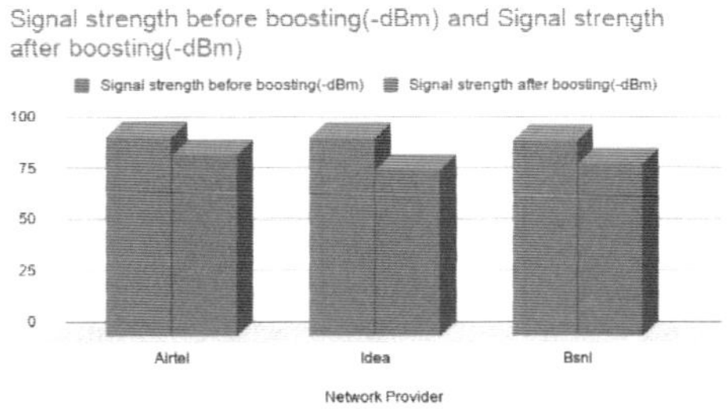

Fig 6.3 Comparison Chart before and after boosting at second location

Analysis of Signal Strength Status So from the above bar graph we concluded that the after boosting the signal strength is varied.

Before boosting

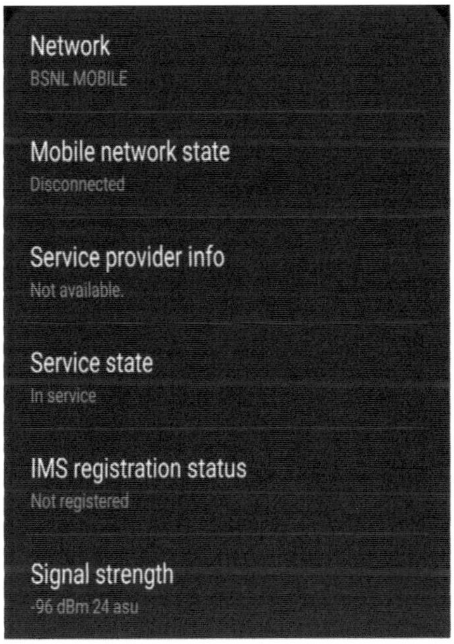

After boosting

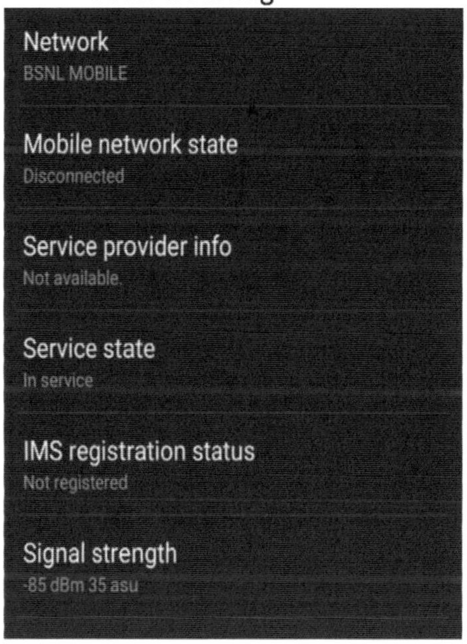

BSNL

Fig: 6.4: BSNL signal strength status of the BSNL telecom operator at Chemmenchery location.

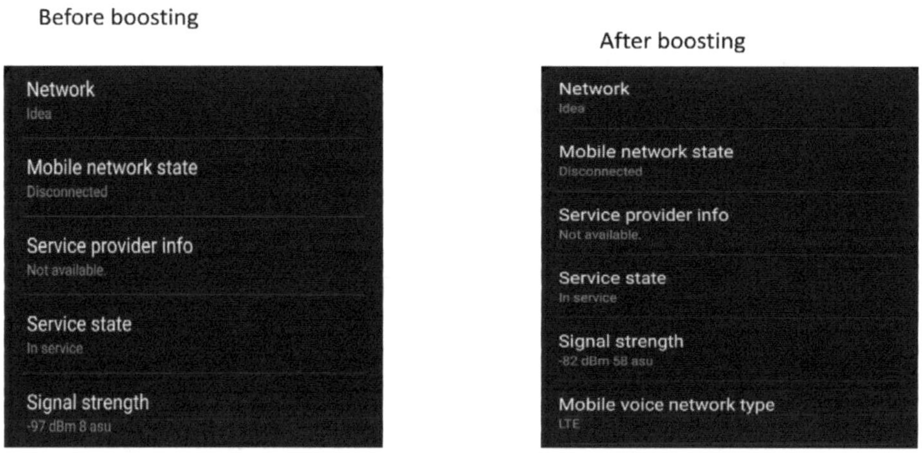

IDEA

Fig: 6.5: Idea signal strength status of the IDEA telecom operator at Chemmenchery location.

6.1 Meter Range values of BSNL:

Network Provider	Signal Before Boosting (-dBm)	Signal After Boosting (-dBm)
1 meter	89	71
5 meter	83	74
10 meter	82	75

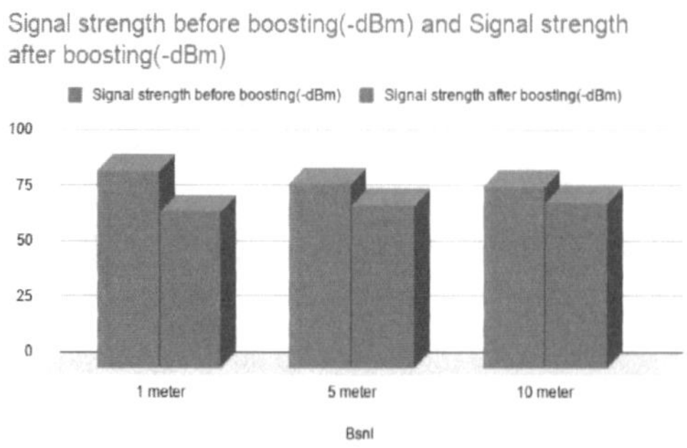

Fig 6.6 Comparison Chart before and after boosting for BSNL

Analysis of meter range values so from the above bar graph we concluded that the after boosting the signal strength is varied.

*6.2 **Meter Range values of IDEA:***

Network Provider	Signal Before Boosting (-dBm)	Signal After Boosting (-dBm)
1 meter	86	72
5 meter	87	76
10 meter	89	78

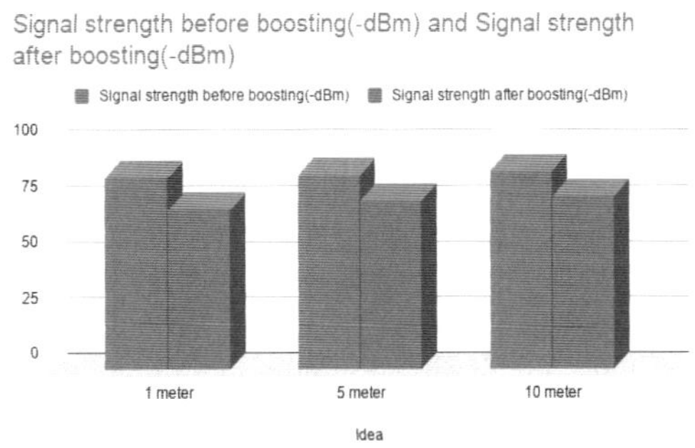

Fig 6.7 Comparison Chart before and after boosting for IDEA

Analysis of meter range values so from the above bar graph we concluded that the after boosting the signal strength is varied

7. Conclusion:

By the above experiment, we have to conclude that the booster that works for all type of Telecom operators present in India. Mainly they worked for all 4G bands. Here the output can be obtained by your mobile phone signal status in that mobile strength has to change. Actually mobile strength is from -60dbm to -140dbm. -60dbm means the best signal,-140dbm means poor signal. From the experiment we concluded that the mobile strength has been decreased in value that means increased in the signal strength. If someone wants to continue the project the output has to be taken from CRO so that the exact increase of the signal strength has been observed. This experiment was done in the Chennai location. As part of the paper we concluded that the 4G signal booster has increased signal strength of the different Telecom operators in 4G range

References

[1] Adegoke, A. S., Babalola, I. T., & Balogun, W. A. Performance evaluation of GSM mobile system in Nigeria. Pacific Journal of Science and Technology, 9(2), 436–441(2008)

[2] A. S. Morris and V. Steel, "Integrated Tunable Systems for Scalable 4G Radios," IEEE MTT-S Int. Microw. Symp. Dig., Jul. 2013.

IOP Conf. Series: Materials Science and Engineering **906** (2020) 012031 doi:10.1088/1757-899X/906/1/012031

[3] P.A. Tomatta and R. Gaddi, "Aperture Tuned Antennas for 4G Applications Using MEMS Digital Variable Capacitor," IEEE MTT-S Int. Microw. Symp. Dig., Jul. 2013.

[4] B. Baxter, T. Ranta, M. Facchini, D. Jung and D. Kelly, "The State-of-the-Art in Silicon-On-Sapphire Components for Antenna Tuning," IEEE MTT-S Int. Microw. Symp. Dig., Jul. 2013.

[5] Costa, M. Carroll, D. Kerr, C. Iversen, P. Mason and E. Spears, "RFCMOS SOl Technology for 4G Reconfigurable RF Solutions," IEEE MTT-S Int. Microw. Symp. Dig., Jul. 2013

[6] F. Ali, E. Lourandakis, R. Gloeckler, K. Abt, G. Fischer and R. Weigel, "Tunable Multiband Power Amplifier using Thin-Film BST Varactors for 4G Handheld Applications," IEEE Radio Wireless Symp. Dig., Jan. 2010

[7] K. V. S. Rao, P. V. Nikitin and S. Lam, "Antenna design for UHF RFID tags: A review and a practical application," IEEE Transactions on Antennas and Propagation, vol. 53, no. 12, pp. 38703876, Dec. 2005

[8] K.P. Ray and Y. Ranga, "CPW-fed modified rectangular printed monopole antenna with slot," Microwave and Optoelectronics Conference, 2007 IMOC 2007 SBMO/IEEE MTT-S International, pp.79-81, Oct. 29 2007-Nov. 1 2007

[9] C. Balanis, Antenna Theory, Analysis and Design, 3rd edition, New York: Wiley, 2000
[10] R.C. Baraniuk, V. Cevher, and M. B. Wakin, "Low-dimensional models for dimensionality reduction and signal recovery, a geometrical perspective," Proc. IEEE, vol. 98, no. 6, pp. 959–971,Jun. 2010
[11] O. M. Bucci, G. D'Elia, and M. D. Migliore, "A new strategy to reduce the truncation error in near-field far-field transformations," Radio Sci.,vol. 35, no. 1, pp. 3–17, Jan.–Feb. 2000.

AUTHOR INDEX